Springer
*Berlin
Heidelberg
New York
Hong Kong
London
Milan
Paris
Tokyo*

Christian Koeberl
Francisca C. Martínez-Ruiz (Eds.)

Impact Markers in the Stratigraphic Record

with 115 Figures and 23 Tables

Springer

Professor Christian Koeberl
Department of Geological Sciences
University of Vienna
Althanstrasse 14
1090 Vienna
Austria
Email: christian.koeberl@univie.ac.at

Dr. Francisca C. Martínez-Ruiz
Instituto Andaluz de Ciencias de la Tierra (CSIC-UGR)
Facultad de Ciencias
Campus Fuentenueva
18002 Granada
Spain
Email: fmruiz@ugr.es

ISBN 3-540-00630-3 Springer-Verlag Berlin Heidelberg New York

Cataloging-in-Publication Data applied for

Bibliographic information published by Die Deutsche Bibliothek
Die Deutsche Bibliothek lists this publication in the Deutsche Nationalbibliografie;
detailed bibliographic data is available in the Internet at <http://dnb.ddb.de>.

Springer-Verlag Berlin Heidelberg New York
a member of BertelsmannSpringer Science+Business Media GmbH

http://www.springer.de

Camera ready by editors
Cover design: E. Kirchner, Heidelberg

Printed on acid-free paper   32/3130/as 5 4 3 2 1 0

Preface

The present volume is an outcome of the scientific programme "Response of the Earth System to Impact Processes" (IMPACT) by the European Science Foundation (ESF). The ESF is an association of 67 national member organizations devoted to scientific research in 24 European countries. The IMPACT programme is aimed at understanding meteorite impact processes and their effects on the Earth System. Launched in 1998 for duration of 5 years, 15 ESF member organizations now participate in this programme, which will officially end in late 2003, although the momentum gained for European (and worldwide) impact research will be carried on in other programs and organizations. The programme deals with all aspects of meteorite impact research and operates through workshops, exchange programs, publications, and short courses. This particular book is the third in an informal series on "Impact Studies", which is published by Springer and intended to go beyond the ESF IMPACT programme by providing a venue for high quality (and peer-reviewed) monographs and conference and workshop proceedings on general topics connected to impact cratering and related research.

The 6th ESF-Impact workshop "Impact makers in the stratigraphic record" was held in Granada (Spain) on May 2001, with about sixty scientists from Europe, Taiwan, and North America attending the workshop. During the workshop 30 oral, 32 poster, and 3 keynote contributions were presented. The workshop also included, as a special session, a round table discussion on the Azuara structure in Spain, which has been listed as an impact structure for some time now, but recent research has precipitated a new debate on this topic. A post-workshop field trip took the participants to the Cretaceous-Tertiary (K/T) boundary sections of Agost and Caravaca, located in the SE of Spain. These are two of the most complete K/T boundary sections, where the distal ejecta from Chicxulub crater is well preserved. Scientifically, the workshop focused on the stratigraphic record of impacts events throughout geological time, covering the following themes:

- Identification of impact ejecta
- Impact effect on climate and biosphere
- Ejecta-crater correlation
- Impact signatures and geochemical record
- Impact glasses, tektites, and microkrystites
- Impact cratering throughout geological time

This volume contains fourteen papers resulting from the workshop. All the manuscripts were reviewed by two referees and were selected on the basis of originality and scientific value. The general theme of the volume deals with the record of impact events on Earth, in particular, the stratigraphic record. Contributions to this book have been arranged by placing overview and general chapters first, followed by descriptions of particular impact events and boundaries in order of increasing age, starting with the Late Eocene Popigai impact structure and ending with an enigmatic, possibly impact-related, breccia of Middle Ordovician age.

The volume begins with an introductory chapter by the editors, providing a brief overview of the stratigraphic record of impact events, followed by a proposed revision of the stratigraphic nomenclature of impact-related materials by King and Petruny. These authors review the history of studies on the nomenclature of impact-derived and impact-related materials and discuss how extant stratigraphic nomenclature of the North American Commission on Stratigraphic Nomenclature and the International Subcommission on Stratigraphic Classification may apply to these materials, making some recommendations on future usage of stratigraphic nomenclature.

Next, Muñoz-Espadas and colleagues review the main geochemical signatures related to meteorite impacts in terrestrial rocks. Such signatures are used to identify impact structures and for inferring the impactor type. The most common geochemical criteria used are the detection of a positive siderophile element anomaly, the Re-Os isotopic system, and the Mn-Cr isotopic system.

Jones and co-workers, in a stimulating and somewhat controversial contribution, examine the potential for decompression melting beneath a large terrestrial impact crater as a mechanism for generating melt to auto-obliterate the crater. They also model massive melting events that may trigger long-lived mantle up-welling or an impact plume, which could potentially resemble a mantle hotspot. This model provides a potential explanation for the formation of komatiites and other high degree partial melts and for most primitive geochemical signatures currently attributed to plumes as originating from the deep mantle or outer core.

Shuvalov gives a model of impact cratering to describe the excavation flow and expansion of ejecta through the atmosphere, considering condensed ejecta as an aggregate of discrete particles. The motion of particles and the exchange of heat and momentum with air and vapor are described by equations of multi-phase hydrodynamics. This model is applied to the study of the vertical impact of a spherical asteroid of 1 km radius on a granite target and the results demonstrate a sorting of condensed ejecta particles by size, which considerably influences the final distribution of the deposited ejecta.

Masaitis describes the lithology and distribution of proximal ejecta facies that formed during the late stages of formation of the 35 Ma Popigai impact crater, Siberia. Three types of ejecta facies occur in the upper parts of the crater fill: lithic microbreccias and suevites (A type), suevites with minor tagamites (B type), and suevites (C type). The preliminary interpretation of the mode of origin of these ejecta facies are air fall (A), pyroclastic-like flow (B), and base surge (C) deposits.

Two papers are aimed at the study and analysis of the Boltysh impact crater in the Ukraine, which initially was thought to be 88 million years old. Valter and Plotnikova analyze the ejecta of the Boltysh impact crater from outcrops at distances of 25 – 30 km NW from the center of the crater. These authors report two types of breccia: breccia composed of disintegrated basement granites (lower unit), and breccia with different rock fragments cemented by sand-like and more fine-grained material (upper unit). From their observations, the time range of the crater formation, in comparison with recent chronobiostratigraphic data, is determined to be 66.8 - 65 Ma.

Gurov et al. provide a detailed study of the ejecta blanket that is preserved over an area of ~6500 km^2. Fall-back ejecta from the Boltysh impact event include a suevite breccia layer, which now occurs only within the crater rim, where it overlies an impact melt sheet and the top of the central uplift in the crater. The Boltysh ejecta were deposited on top of Precambrian crystalline basement of the Ukrainian Shield over almost the entire extent of the ejecta blanket. New data reported here (and also data in a companion paper by Kelley and Gurov, in "Meteoritics and Planetary Science", vol. 37, 2002), derived by laser stepped heating and spot $^{40}Ar/^{39}Ar$ dating of impact melt rocks, yields an age of 65.17 ± 0.64 Ma. This age agrees also with the biostratigraphic studies of Valter and Plotnikova (previous chapter), indicating that the Boltysh crater was formed simultaneously with, or within a few hundred thousand years of, the Cretaceous-Tertiary boundary age Chicxulub impact structure. This poses an interesting question: was the Chicxulub-forming impact event just the largest of several coeval impact events? Only detailed future impact-stratigraphic studies will hopefully provide an answer.

The K/T boundary ejecta themselves are analyzed by King and Petruny, in the second contribution of these authors to the present volume. Here they provide new data on the proximal K/T deposits at Albion Island in northern Belize. These deposits consist of a basal impactoclastic clay layer (~ 1 to 2 m thick) and an upper carbonate-rich, coarse impactoclastic breccia layer (up to 15 m thick). The paper focuses on the stratigraphy and sedimentology of the coarse impactoclastic breccia of the upper layer, suggesting that its mode of emplacement during the impact aftermath was similar to that of a very large volcanic debris avalanche. These authors speculate that each sedimentation unit at Albion may represent a separate emplacement event during the process of ejecta-curtain collapse, perhaps owing to variations in atmospheric interaction with the debris.

Griscom and colleagues discuss data from electron-spin-resonance (ESR) studies of calcites of K/T boundary sediments from Belize, Northern Mexico and distal ejecta deposits (Sopelana, Caravaca and Blake Nose), in which they analyzed the geochemical behavior of the Mn^{2+} and SO_3^- ions. The ESR method is rarely applied in impact-related studies; thus this contribution also goes into some methodical details. At distal ejecta sites, anomalies in SO_3^- and/or Mn^{2+} intensities were noted by these authors at the K/T boundary relative to the corresponding background levels in the rocks above and below. Absolute ESR quantitative analyses of proximal impact deposits from Belize and southern Mexico group naturally into three distinct fields in a two dimensional $[SO_3^-]$ versus $[Mn^{2+}]$ scatter plot. These fields contain (I) limestone ejecta clasts, (II) accretionary lapilli, and (III) a variety of SO_3^- enriched impact deposits. These authors propose that (a) field I represents calcites from the Yucatán Platform, and that the Mn^{2+} depleted signature can be used as an indicator of primary Chicxulub ejecta in deep marine environments and (b) field II represents calcites that include a component formed in the vapor plume, either from condensation in the presence of CO_2/SO_3^- -rich vapors, or reactions between CaO and CO_2/SO_3 rich vapors, and that this SO_3^--enhanced signature can be used as an indicator of impact vapor plume deposits.

Reimold and Koeberl analyze the petrography and geochemistry of a deep drill core from the Edge of the Morokweng impact structure in South Africa. This impact structure is important because its age has previously been determined to coincide with the age of the Jurassic-Cretaceous boundary, about 145 million years ago. Since the discovery of this structure, about 6 years ago, there has been a debate about its size, and values ranging from about 70 to 340 km have been debated. The determination of the exact size is not easy, as the structure is eroded and almost completely buried, thus necessitating to rely on the interpretation of geophysical data and drill core information. In the present contribution, these authors discuss data from a deep drill core about 40 km from the center of the Morokweng structure. U-Pb SHRIMP dating of representative samples of felsic granophyre established that these rocks are Archean in age and are unrelated to the nearby 145 Ma impact melt rock. The general lack of deformation in the drill core rocks strongly suggests that this borehole was sunk outside or, at best, close to the edge, of the Morokweng impact structure – providing a strong case for a maximum diameter of about 70 to 80 km for this structure.

The Permian-Triassic boundary is examined by Schwindt and co-workers. This boundary (~253 Ma) is associated with the most severe mass extinction of marine species and terrestrial vertebrates and plants. These authors have studied the stratigraphy, paleomagnetism, and palynology of the Carlton Heights section in the southern Karoo Basin, South Africa and propose that the extinction of mammal-like reptiles at the end of the Permian may have preceded the fungal event and land-plant extinction within an interval of less than ~100,000 years, and possibly less than ~25,000 years.

The Alamo Breccia in Nevada is an extensive deposit (covering an area of about 10^4 km^2) of carbonate-dominated breccia and lapilli of Late Devonian age, containing some shocked quartz, thus confirming the impact origin of this deposit. Koeberl and colleagues describe their attempt to geochemically detect an extraterrestrial component in samples from this breccia. They used multiparameter coincidence spectrometry after neutron irradiation of the samples to determine the contents of the element iridium, which is often used as a proxy for the presence of a meteoritic component in impact-related rocks. Even though the method has detection limits in the parts-per-trillion range, the results are ambiguous and do not clearly indicate the presence of an extraterrestrial component in this Breccia, thus leaving the question regarding the projectile type of the impactor unanswered.

Finally, Suuroja and colleagues describe their analysis of the so-called Osmussaar Breccia, which occurs in beds of the ~475 Ma basal Middle Ordovician (Arenig and Llanvirn series) siliciclastic-carbonate rocks of northwestern Estonia. This breccia consists of fragmented and slightly displaced (sandy) limestones, which are penetrated by veins and bodies of strongly cemented, breccia-like, lime-rich sandstone injections. The Osmussaar Breccia covers an area of more than 5000 km^2 and is distributed in an east-west oriented elliptical half-circle that is centered approximately at Osmussaar Island. Although several hypotheses have been proposed to explain the origin of the Osmussaar event, the authors conclude that this breccia does not correspond to any known impact structure of this age in Baltoscandia, and suggest that a devastating ~475 Ma earthquake, with an epicenter close to Osmussaar, caused its formation.

It is the hope of the editors that this volume will remain a useful source of information on some interesting aspects of impact-related research for many years to come. If it stimulates further research, we have succeeded in our task.

Acknowledgements

The editors would like to thank the ESF IMPACT programme and the all institutions that have provided financial support for organizing the workshop and the research described in this book. We thank all those who contributed to this volume by submitting their manuscripts, and we would especially like to extend our gratitude to the reviewers, who devoted their time and effort to help improve the papers. We are also grateful to M.J. Román Alpiste (University of Granada) for help with preparing the camera-ready manuscript of this book.

Christian Koeberl   Francisca C. Martínez-Ruiz
Institute of Geochemistry   Instituto Andaluz de Ciencias de la Tierra
University of Vienna   CSIC-University of Granada
Austria   Spain
(christian.koeberl@univie.ac.at)   (fmruiz@ugr.es)

January 2003

Contents

List of Contributors

Ainsaar, L.
Institute of Geology, University of Tartu
Vanemuise 46, 51014 Tartu, Estonia
(arps@ut.ee)

Beltrán-López, V.
ICN, Universidad Nacional Autónoma de México
04510 México D.F., México

Clegg, R.
Dynamics House
Hurst Road, Horsham, W Sussex, RH12 2DT. United Kingdom
(all@centdyn.demon.co.uk)

DeCarli, P.S.
Department of Geological Sciences, University College London
Gower Street, London, WC1E 6BT, United Kingdom
(paul.decarli@sri.com)

Eshet, Y.
Geological Survey of Israel
Jerusalem, 95501, Israel

Griscom, D.L.
Laboratoire de Minéralogie et Cristallographie de Paris, Université de Paris
6, 4 place Jussieu, 75252 Paris, France
(dlgriscom@netscape.net)

Gurov, E.P.
Institute of Geological Sciences, National Academy of Sciences of Ukraine
55 b O. Gonchar Street, 01054 Kiev, Ukraine
(ep_gurov@ukr.net)

Huber, H.
Institute of Geochemistry, University of Vienna
Althanstrasse 14, A-1090 Vienna, Austria
(heinzhuber@ati.ac.at)

Jones, A.P.
Department of Geological Sciences, University College London
Gower Street, London, WC1E 6BT, United Kingdom
(adrian.jones@ucl.ac.uk)

Kelley, S.P.
Department of Earth Sciences, Open University
Milton Keynes MK7 6AA, United Kingdom
(S.P.Kelley@open.ac.uk)

King Jr, D.T.
Department of Geology, Auburn University
Auburn, Alabama 36849-5305, USA
(kingdat@auburn.edu)

Kirsimäe, K.
Institute of Geography, University of Tartu
Vanemuise 46, 51014 Tartu, Estonia
(arps@ut.ee)

Koeberl, C.
Institute of Geochemistry, University of Vienna
Althanstrasse 14, A-1090 Vienna, Austria
(christian.koeberl@univie.ac.at)

Kohv, M.
Institute of Geology, University of Tartu
Vanemuise 46, 51014 Tartu, Estonia
(arps@ut.ee)

Lunar, R.
Departamento de Cristalografía y Mineralogía, Facultad de Ciencias Geológicas,
Universidad Complutense de Madrid
Avenida Complutense s/n, 28040 Madrid, Spain
(lunar@geo.ucm.es)

Mahaney, W.C.
Geomorphology and Pedology Lab, York University
4700 Keele St., North York, Ontario, Canada, M3J IP3

Martínez-Frías, J.
Centro de Astrobiología, CSIC-INTA
Carretera de Torrejón a Ajalvir, 28850 Torrejón de Ardoz, Madrid, Spain
(martinezfrias@mncn.csic.es)

Masaitis, V.L.
Karpinsky All-Russia Geological Research Institute (VSEGEI)
Sredny Prospect 74, 199106 St. Petersburg, Russia
(vicmas@vsegei.sp.ru)

Morgan, M.
Colorado School of Mines
Golden, Colorado 80401-1887, USA
(mmorgan@mines.edu)

Muñoz-Espadas, M.J.
Departamento de Geología, Museo Nacional de Ciencias Naturales, CSIC
José Gutiérrez Abascal 2, 28006 Madrid, Spain
(majem@mncn.csic.es)

Ocampo, A.C.
Jet Propulsion Laboratory
Pasadena, CA 91109, USA
(Adriana.Ocampo@rssd.esa.int)

Petruny, L.W.
Astra-Terra Research
Auburn, Alabama 36831-3323, USA
(lpetruny@att.net)

Plotnikova, L.
Institute of Geological Sciences of National Academy of Science
55B Gonchara str., Kiev 54, 01601, Ukraine
(ignnanu@geolog.freenet.kiev.ua)

Pope, K.O.
Geo Eco Arc Research, Inc.
16305 St. Mary's Church Rd., Aquasco, MD 20608, USA
(kpope@starband.net)

Price, D.G.
Department of Geological Sciences, University College London
Gower Street, London, WC1E 6BT, United Kingdom
(d.price@ucl.ac.uk)

Price, N.
Department of Geological Sciences, University College London
Gower Street, London, WC1E 6BT, United Kingdom
(n.price@ucl.ac.uk)

Rampino, M.R.
Earth and Environmental Science Program, New York University,
100 Washington Square East, New York, NY 10003, USA
(mrr1@nyu.edu)

Reimold, W.U.
Impact Cratering Research Group, School of Geosciences,
University of the Witwatersrand
Private Bag 3, P.O. Wits 2050, Johannesburg, South Africa
(reimoldw@geosciences.wits.ac.za)

Shuvalov, V.V.
Institute for Dynamics of Geospheres
Leninsky Prospect 38-6, 117979 Moscow, Russia
(shuvalov@idg.chph.ras.ru)

Schwindt, D.M.
Earth and Environmental Science Program, New York University
100 Washington Square East, New York, NY 10003, USA

Steiner, M.B.
Department of Geology and Geophysics, University of Wyoming Laramie
WY 82071, USA

Suuroja, K.
Geological Survey of Estonia
12618 Tallinn, Estonia

Suuroja, S.
Geological Survey of Estonia
12618 Tallinn, Estonia

Valter, A.
Institute of Applied Physics of National Academy of Science
Department N 50, Nauki Avenue 46, Kiev 39, 03650, Ukraine
(avalter@iop.kiev.ua)

Warme, J. E.
Morgan, M.
Colorado School of Mines
Golden, Colorado 80401-1887, USA
(jwarme@mines.edu)

The Stratigraphic Record of Impact Events:
A Short Overview

Christian Koeberl[1] and Francisca Martinez-Ruiz[2]

[1]Institute of Geochemistry, University of Vienna, Althanstrasse 14, A-1090 Vienna, Austria.
(christian.koeberl@univie.ac.at)
[2]Instituto Andaluz de Ciencias de la Tierra (CSIC-UGR), Facultad de Ciencias,
Campus Fuentenueva, 18002 Granada, Spain. (fmruiz@ugr.es)

Abstract. In contrast to many other planets (and moons) in the solar system, the recognition of impact craters on the Earth is difficult, because active geological and atmospheric processes on our planet can obscure or erase the impact record in geologically short times. Impact craters are recognized from the study of actual rocks – remote sensing can only provide supporting information. Petrographic studies of rocks at impact craters can lead to the discovery of impact-characteristic shock metamorphic effects, and geochemical studies may yield information on the presence of meteoritic components in these rocks. Apart from studying meteorite impact craters per se, large amounts of information can also be gained from the study of impact ejecta. Such ejecta are found within the normal stratigraphic record, where they can provide excellent time markers, and allow to relate an impact event directly to possible biological effects. Impact ejecta are commonly divided into two groups - proximal ejecta (those that are deposited closer than 5 crater radii from the crater rim), and distal ejecta. In some cases, impact events have been identified solely from the discovery and study of regionally extensive or globally distributed impact ejecta. A well known case in point is the Cretaceous-Tertiary boundary, where the discovery of an extraterrestrial signature, together with the presence of shocked minerals, led not only to the identification of an impact event as the cause of the end-Cretaceous mass extinction, but also to the discovery of a large buried impact structure about 200 km in diameter, the Chicxulub structure. Tektites are another form of distal impact ejecta, the source craters of which have long remained elusive. To date only three of the four known Cenozoic tektite strewn fields have been connected to source craters. Distal impact ejecta allow to gain information about impact processes and the connection to biological events.

1
Introduction

All planets, moons, asteroids, etc., in the solar system that have solid surfaces are covered by craters. This observation is well documented for our Moon from centuries of telescopic observation, and, since the early 1960s, for the other planets and satellites (and our own Moon) also by extensive spacecraft photography. Studies over most of the 20th century have documented that almost all these craters are the result of high-energy impact events, i.e., the Moon and the other bodies are covered by impact craters. This makes impact cratering is either the most important, or one of the most important processes that affects the shaping of their surfaces. On Earth, one of the four inner planets along with Mercury, Venus, and Mars, the situation seems to be different. Craters of any sort an not obvious and common landform. As it is unlikely (if not impossible) that the Earth is somehow shielded from the cosmic bombardment that affects all other planetary bodies around it, other causes must be invoked. On Earth, endogenic forces cover or obliterate the traces of all impact events within geologically short times.

About 170 impact structures are currently (at the turn of 2002 to 2003) known on Earth (updates are available on the internet, see the "Earth Impact Database", which was originally assembled by the Canadian Geological Survey, but has now been transferred to the Planetary and Space Science Centre at the University of New Brunswick Department of Geology, at: (http://www.unb.ca/passc/Impact Database/index.html). A few of these structures were only recognized because layers of impact ejecta were found and eventually traced back to their source crater (e.g., Chicxulub, from studies of the K/T boundary impact layer). Considering that some impact events demonstrably affected the geological and biological evolution on Earth, and that even small impacts can disrupt the biosphere and lead to local and regional devastation (Chapman and Morrison 1994), the understanding of impact structures and the processes by which they form should be of interest not only to earth scientists, but also to society in general. This chapter is intended as a short introduction into the recognition of impact structures, and also gives an overview of the stratigraphic record of impact events, especially those of far-reaching consequences (such as the end-Cretaceous impact event), in the geological record. Most of the material here is summarized and updated from the book by Montanari and Koeberl (2000) and recent reviews by Koeberl (2001, 2002).

2
Impact Structures

We do not seem to have any historic record that documents the observation of a large impact event by humans over the last several thousand years (which is, of course, not a geologically long period of time). Thus, impact experiments and the detailed investigations of impact craters on Earth are indispensable for our understanding of these features. It is useful to find criteria to help distinguish volcanic structures from meteorite impact craters. Meteorite impact craters are surficial features without deep roots, whereas in volcanic craters or calderas the disturbances continue to (or, rather, emerge from) great depth. Impact craters are almost always circular. On the Moon, with its surface covered by impact craters, only a small number of non-circular features are known.

When discussing morphological aspects, we need to mention the distinction between an impact *crater*, i.e., the feature that results from the impact, and an impact *structure*, which is what we observe today, i.e., long after formation and modification of the crater. Thus, unless a feature is fairly fresh and unaltered by erosion, it should rather be called an "impact structure" than an "impact crater". Impact craters (before post-impact modification by erosion and other processes) occur on Earth in two distinctly different morphological forms. These are known as *simple* craters (small bowl-shaped craters) with diameters of up to ≤2 to 4 km (see Fig. 1), and *complex* craters, which are larger and have diameters of ≥2 to 4 km. The exact change-over diameter between simple and complex crater depends on the composition of the target and the gravity of the body on which the impact happens – on the Moon the changeover between simple and complex craters is around 15-20 km.

Fig. 1. An example of a typical simple impact crater on Earth, the Roter Kamm crater in Namibia (2.5 km in diameter, ca. 3.7 million years old).

Complex craters are characterized by a central uplift (see Fig. 2). Craters of both types have an outer rim and are filled by a mixture of fallback ejecta and material slumped in from the walls and crater rim during the early phases of formation. Such crater infill may include brecciated and/or fractured rocks, and impact melt rocks. Fresh simple craters have an apparent depth (crater rim to present-day crater floor) that is about one third of the crater diameter. For complex craters, this value is closer to one fifth or one sixth. The central structural uplift in complex craters consists of a central peak or of one or more peak ring(s) and exposes rocks that are usually uplifted from considerable depth. On average, the actual stratigraphic uplift amounts to about 0.1 of the crater diameter (e.g., Melosh 1989).

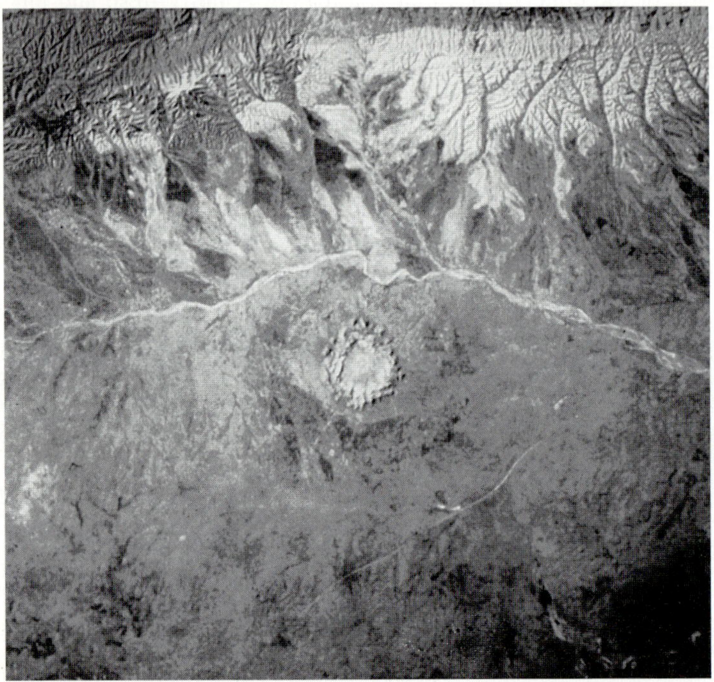

Fig. 2. A typical eroded complex impact structure on Earth, the a. 143 million year old Gosses Bluff impact structure in central Australia. The crater originally had a diameter of ca. 25 km; this is now only visible as a discoloration due to the drainage pattern. The visible rim is in fact the eroded remnant of the central uplift, ca. 6 km in diameter (Space Shuttle photograph).

The relationship between number (areal density) of impact structures and diameter on the Earth is quite different from that on (e.g.,) the Moon, because on Earth basically all of the small craters are relatively young. This size relationship is biased and results from typical terrestrial processes. The erosional processes that obliterate small (0.5–10 km diameter) craters after a few million years result in a severe deficit of such small craters, compared to the crater count that would be expected from the (observed) number of larger craters and from astronomical considerations. This effect also explains why most small craters known on Earth are young. Older craters of larger initial diameter also suffer erosional degradation leading to the destruction of the original topographical expression, or to burial of the structures under post-impact sediments. For details on crater morphology, see, e.g., Melosh (1989).

On, for example, the Moon, Mercury, or Mars, structures that are larger than the simple and complex craters have been identified. These are the so-called *multiring basins*, which have diameters ranging from a few hundred to at least 2000 km. It is not clear if the largest impact structures known on Earth (e.g., Chicxulub with ca. 200 km diameter, or Vredefort with an initial diameter of possibly 300 km) may represent terrestrial examples of multiring basins or not.

3
Criteria for Recognition of Impact Structures

On the Moon and other planetary bodies that lack an appreciable atmosphere, it is usually easy to recognize impact craters on the basis of morphological characteristics. On the Earth, complications arise as a consequence of the obliteration, deformation, or burial of impact craters. Thus, it is ironical that despite the fact that impact craters on Earth can be studied directly in the field, they may be much more difficult to recognize than on other planets. This dilemma necessitated the development of diagnostic criteria for the identification and confirmation of impact structures on Earth (see also French 1998; Montanari and Koeberl 2000; Koeberl 2002). The most important of these characteristics are: a) crater morphology, b) geophysical anomalies, c) evidence for shock metamorphism, and d) the presence of meteorites or geochemical evidence for traces of the meteoritic projectile.

Of these characteristics, only the presence of diagnostic shock metamorphic effects and, in some cases, the discovery of meteorites, or traces thereof, are generally accepted to provide unambiguous evidence for an impact origin. Shock deformation can be expressed in macroscopic form (shatter cones) or in microscopic form. The same two criteria apply to distal impact ejecta layers and allow to confirm that material found in such layers originated in an impact event at a possibly still unknown location. So far (2002), the presence of such evidence led to the confirmation of about 170 impact structures (some exposed on the surface, others are subsurface structures) on Earth.

Remote sensing, including morphological observations, provides important initial data regarding the recognition of a potential impact structure, but cannot provide confirming evidence – this requires the study of actual rock samples. Geological structures with a circular outline that are located in places with no other obvious mechanism for producing near-circular features may be of impact origin and at least deserve further attention. Geophysical methods are also useful in identifying candidate sites for further studies, especially for subsurface features. In complex craters the central uplift usually consists of dense basement rocks and usually contain severely shocked material. This uplift is often more resistant to erosion than the rest of the crater, and, thus, in old eroded structures the central uplift may be the only remnant of the crater that can be identified.

Table 1. Criteria for the Recognition of Terrestrial Impact Structures

Criterion	Confirming Evidence?
Morphology (Remote sensing)	
Circular diameter	no
Rim structure	no
Central structure	no
(breccia lens; central rise)	
Geophysics	
Gravity anomaly	no
Magnetic anomaly	no
Seismic studies	
Petrography and Geochemistry	
Occurrence of polymict breccia	no
Occurrence of melt rocks	no
Shock metamorphic features	
in minerals and rock clasts	yes
Extraterrestrial component in	
breccia or melt rocks	yes

Geophysical characteristics of impact craters include gravity, magnetic properties, reflection and refraction seismics, electrical resistivity, and others (see, e.g., Grieve and Pilkington 1996, for a review). In general, simple craters have negative gravity anomalies due to the lower density of the brecciated rocks compared to the unbrecciated target rocks, whereas complex craters often have a positive gravity anomaly associated with the central uplift that is surrounded by an annular negative anomaly. Magnetic anomalies can be more varied than gravity anomalies. Seismic data often show the loss of seismic coherence due to structural disturbance, slumping, and brecciation. Such geophysical surveys are important for the recognition of anomalous subsurface structural features, which may be deeply eroded craters or impact structures covered by post-impact sediments.

Nevertheless, only the petrographic and geochemical study of actual rocks from the potential impact structure will bring final clarification. In case of a structure that is not exposed on the surface, drill core samples are essential. Good materials for the recognition of evidence for an impact origin are various types of breccia (see below) and melt rocks. These rocks often carry unambiguous evidence for the impact origin of a structure in the form of shocked minerals or an extraterrestrial contamination (see below for details).

4
Impactite Nomenclature (Breccias, Melt Rocks)

Before discussing the petrographic and geochemical details of confirming an impact structure, some definitions are necessary. For the nomenclature of impactites there are some well-established and widely accepted (although not internationally standardized) classification criteria (e.g., Stöffler and Grieve 1994a, b; French 1998). Some authors prefer a terminology that is specific to the particular crater they are studying, but such local terms make it very difficult to compare rock units from different impact structures. A possible shortcoming of the general impactite terminology may be the lack of genetic information, for example, with regard to the possibility that more than one type of suevitic or fragmental impact breccia may have been deposited, perhaps by different processes, such as atmospheric fall-out, or base surge deposition, within the crater. For another view on the classification of impactites, see King and Petruny (this volume).

The definitions of Stöffler and Grieve (1994b) and French (1998) are fairly easy to understand and describe the most important impact formations. An *impactite* is a collective term for all rocks affected by (an) impact(s) resulting from collision(s) of planetary bodies. The classification scheme for impactites uses criteria that combine a) lithological components, texture, and degree of shock metamorphism, and b) mode of occurrence (in- or outside the crater). In terms of location, we distinguish between parautochthonous rocks beneath and allochthonous (or allogenic) rocks that fill the crater (crater-fill units, e.g., breccias and melt rocks) and also occur as ejecta around the crater. French (1998) distinguished four locations in and around an impact structure, which are, with their associated rock units: a) sub-crater (parautochthonous rocks, cross-cutting allogenic units, pseudotachylite); b) crater interior (allogenic crater-fill deposits: lithic [fragmental] breccias, suevitic breccias, impact melt breccias); c) crater rim region (proximal ejecta deposits), and d) distal ejecta.

Parautochthonous rocks may include target rocks that were subjected to shock metamorphism but remained in place, as well as impactites (e.g., monomict fragmental impact breccia that remained in situ, but was internally brecciated, and where breccia clasts were subjected to small-scale movements or rotation). Another breccia type, *pseudotachylite*, a friction melt, has been reported from sub-crater basement, in the form of small veins and dikes hardly ever exceeding a few centimeters width. Only in very large impact structures (e.g., Sudbury, Canada, or Vredefort, South Africa) have large amounts of such breccia been observed. Pseudotachylite needs to be distinguished from other dike breccias in crater floor locations (including impact melt breccia, fragmental, and suevitic breccia injections, or clastic breccias, such as cataclasites, or even pre- or post-impact tectonically produced pseudotachylites; see Reimold 1995, 1998).

The crater fill contains a variety of breccia types. *Fragmental impact breccia* is a "monomict or polymict impact breccia with clastic matrix containing shocked and unshocked mineral and lithic clasts, but lacking cogenetic impact melt particles" (Stöffler and Grieve 1994b). These rocks have also been termed *lithic breccia* (French 1998). *Impact melt breccia* has been defined by Stöffler and Grieve (1994b) as an "impact melt rock containing lithic and mineral clasts displaying variable degrees of shock metamorphism in a crystalline, semihyaline or hyaline matrix (crystalline or glassy impact melt breccias)" (with an *impact melt rock* being a "crystalline, semihyaline or hyaline rock solidified from impact melt"). *Suevite* (or suevitic breccia) is defined as a "polymict breccia with clastic matrix containing lithic and mineral clasts in various stages of shock metamorphism including cogenetic impact melt particles which are in a glassy or crystallized state". Figure 3 shows a typical suevite from the Bosumtwi crater in Ghana. The distribution of the rock types is a function of their formation and the order in which they formed. For example, lithic breccias can occur not only inside, but also outside a crater. For the identification of meteorite impact structures, suevites and impact melt breccias (or impact melt rocks) are the most commonly studied units. It is easy to distinguish between the two impact formations, as suevites are polymict breccias that contain inclusions of melt rock (or impact glass), i.e., they are clast-dominated ("melt fragment breccias"), and impact melt breccias have a melt matrix with a variable amount of (often shocked) rock fragment inclusions (they are matrix-dominated breccias that also have been termed "melt-matrix breccias"). Whether these various breccia types are indeed present and/or preserved in a crater depends on factors including the size of the crater, the composition, and the porosity of the target area, and the level of erosion.

The rocks in the crater rim zone are usually only subjected to relatively low shock pressures (commonly <2 GPa), leading mostly to fracturing and brecciation, and often do not show shock-characteristic deformation. Even at craters of several kilometers in diameter crater rim rocks that are *in situ* rarely show evidence for shock deformation. In well-preserved impact structures the area immediately outside the crater rim is covered by a sequence of different impactite deposits (see, e.g., French, 1998), which often allow the identification of these structures as being of impact origin. Distal ejecta can only be recognized as such if they include either shocked minerals or rock fragments, and/or meteoritic components (see, e.g., Koeberl 2001, for a review). Tektites and microtektites are natural glasses that an important group of distal ejecta (see below).

Fig. 3. Suevitic breccia from the Bosumtwi impact crater in Ghana.

5
Proximal vs. Distal Ejecta

At this point it might be useful to explain some basic concepts. From studies of ejecta on the Earth, Moon, and other planets, as well as from impact experiments, researchers have arrived (many decades ago) at a fairly clear definition of proximal and distal ejecta. Proximal ejecta are those that are found in the immediate vicinity of an impact crater, within <5 crater radii from the rim. In contrast, distal ejecta are those ejecta that occur at considerable distances from the source crater (>5 crater radii from the crater rim; see, e.g., Melosh 1989). This definition has been derived from lunar studies and it may be debated if the 5-crater-radii limit should not be different for ejecta distribution on the Earth, which has a higher gravitational attraction than the Moon, and also where the distribution of ejecta is hampered by the presence of an atmosphere. It could probably be argued that on the Earth this number should be lower – maybe 3 or 4 crater radii, but a detailed discussion of scaling is beyond the scope of this short review.

Proximal ejecta often consist of coherent ejecta blankets that also preserve the (inverted) target stratigraphy. These can include a variety of breccias and melt rocks (fallout and fallback material). Distal ejecta can either consist of (usually fine-grained) rock and mineral fragments or be of glassy consistency. It is often not immediately possible to recognize that they are directly connected to a specific impact structure. However, distal ejecta can act as a guide to major impact events and even lead to the discovery of large impact structures. During crater formation, at the end of the excavation stage, about 90% of all material ejected (excavated) from the crater is deposited as proximal ejecta. This conclusion has been obtained from impact experiments, simulation calculations, and comparison with nuclear and chemical explosions. The mass of the ejecta decreases outwards: about half of the ejected material can be found within about one crater radius from the rim. At greater distances the ejecta blankets become increasingly thinner and discontinuous. Most material is ejected from the crater in ballistic trajectories to a distance (range) R_b, which, for short distances, follows the simple equation:

$$R_b = (v^2_e/g)\sin(2\Phi)$$

where v_e is the ejection velocity, g is the surface acceleration of gravity, and Φ is the ejection angle. Again, this equation applies to bodies without an atmosphere. The presence of an atmosphere significantly influences the ejection and distribution mechanisms. On an airless planet, ejecta will be emplaced ballistically. On Earth (or another planet with an atmosphere), ejecta near the crater can also be deposited in a base surge, which is a gravity-driven density current (composed of air and dust entrained by the fireball) that flows down (and outwards) from the expanding mushroom-shaped cloud.

Oberbeck (1975) concluded that material launched early in the crater formation will still be in flight after the crater is completely excavated and after the more massive local ejecta have been deposited to form the continuous deposits around the crater. Material that is ejected late during the crater formation (when the shock wave has already decayed somewhat) is ejected at lower velocities and will be deposited close to the crater rim, whereas material that was ejected early and at higher velocities is deposited at greater distances from the crater. The low velocity ejecta deposited at the crater rim often preserve the initial stratigraphic relationship, which leads to the inverted stratigraphy (the overturned flap) at the crater rim (because material from greater depth at the target is deposited last on the rim). This relationship also indicates that material that from greater stratigraphic depth at a crater location is, in general, deposited close to the crater, whereas distal ejecta more commonly consist of the uppermost stratigraphic layers of the target. The higher energies available early in the crater formation sequence make it also more likely that the earliest ejecta are molten or at least strongly shocked, while the later ejecta may only be slightly shocked or not at all. On Earth the interaction with the atmosphere complicates matters somewhat. Early ejecta may also be entrained in the rapidly expanding fireball and, if the event is large enough, they will be ejected outside of the atmosphere, leading to a global distribution. Size sorting will result in larger particles settling out first and finer dust being deposited later.

Over the past several decades, mainly as a result of lunar studies, several researchers have empirically derived formulas that describe the thickness of ejecta blankets with distance from the crater (e.g., McGetchin et al. 1973; Stöffler et al. 1975; see Melosh 1989, p 90). One of the most commonly used ones, after

McGetchin et al. (1973), gives a power law for the ejecta blanket thickness t as a function of distance r from the crater center:

$$t = 0.14R^{0.74}(r/R)^{-3.0} \ \ \text{for } r \geq R$$

where R is the radius of the transient cavity and dimensions are in meters. The exact values for the function of R are debated, and Melosh (1989) proposed a more general version in which the term $0.14R^{0.74}$ is replaced by f(R), a poorly defined function that depends on a variety of parameters and may be different for different planets.

6
Shock Metamorphism

As mentioned above, the discovery of shock metamorphic effects constitutes confirming evidence for impact processes. In nature, shock metamorphic effects are uniquely characteristic of shock levels associated with hypervelocity impact. And the only natural process known to date that leads to shock metamorphic effects is the hypervelocity impact of extraterrestrial bodies. The response of materials to shock has been the subject of study over much of the second half of the 20th century, in part stimulated by military research. Controlled shock wave experiments, which allow the collection of shocked samples for further studies, using various techniques, have led to a good understanding of the conditions for the formation of shock metamorphic products and a pressure-temperature calibration of the effects of shock pressures up to about 100 GPa (see, e.g., French and Short 1968; Stöffler 1972; Stöffler and Langenhorst 1994; Huffman and Reimold 1996; Koeberl 2002; and references therein). Shock metamorphic effects are best studied in the various breccia types that are found within and around the crater structure (see above). During impact, shock pressures of ≥ 100 GPa and temperatures $\geq 3000°C$ are produced in large volumes of target rock. These conditions are significantly different from conditions for endogenic metamorphism of crustal rocks, with maximum temperatures of 1200°C and pressures of usually <2 GPa. Shock compression is not a thermodynamically reversible process, and most of the structural and phase changes in minerals and rocks are uniquely characteristic of the high pressures (5–>50 GPa) and extreme strain rates (10^6–10^8 s^{-1}) associated with impact. The products of static compression, as well as those of volcanic or tectonic processes, differ from those of shock metamorphism, because of lower peak pressures and strain rates that are different by many orders of magnitude. The descriptions in this chapter follow in part the review by Montanari and Koeberl (2000) and by Koeberl (2002).

A wide variety of shock metamorphic effects has been identified, with the most common ones listed in Table 1. The best diagnostic indicators for shock metamorphism are features that can be studied easily by using the polarizing microscope. They include planar microdeformation features, optical mosaicism, changes in refractive index, birefringence, and optical axis angle, isotropization (e.g., formation of diaplectic glasses), and phase changes (high pressure phases; melting). Kink bands (mainly in micas) have also been described as a result of shock metamorphism, but can also be the result of normal tectonic deformation.

Shatter cones may be the only good macroscopic indicator of impact-generated deformation, and a variety of structures were proposed to be of impact origin on the basis of shatter cone occurrences. Shatter cones form in a variety of target rocks, including crystalline igneous and metamorphic rocks, sandstones, shales, and carbonates, but they are best developed in fine-grained rocks, such as limestone (see, e.g., French 1998, pp 38–40, for details). However, conclusive criteria for the recognition of "true" shatter cones have not yet been defined, and it is easy to confuse them with concussion features, pressure-solution features (cone-in-cone structure), or abraded or otherwise striated features with shatter cones.

Mosaicism is a microscopic shock-characteristic feature that has been observed in a number of rock-forming minerals and appears as an irregular mottled optical extinction pattern, which is distinctly different from undulatory extinction that, for example, occurs in tectonically deformed quartz. Mosaicism can be semiquantitatively defined by X–ray diffraction study of the asterism of single crystal grains, where it shows up as a characteristic increase (with increasing shock pressure) of the width of individual lattice diffraction spots in diffraction patterns. Highly shocked quartz crystals show a diffraction pattern that becomes similar to a powder pattern, because of shock-induced polycrystallinity. Many shocked quartz grains that show planar microstructures also show mosaicism.

Planar microstructures_are the most characteristic expressions of shock metamorphism and occur as planar fractures (PFs) and planar deformation features (PDFs). The presence of PDFs in rock-forming minerals (e.g., quartz, feldspar, or olivine) provides diagnostic evidence for shock deformation, and, thus, for the impact origin of a geological structure or ejecta layer (see, e.g., French and Short 1968; Stöffler 1972; Stöffler and Langenhorst 1994; Huffman and Reimold 1996; Grieve et al. 1996; French 1998; Montanari and Koeberl 2000). PFs, in contrast to irregular, non-planar fractures, are thin fissures, spaced about 20 µm or more apart, which are parallel to rational crystallographic planes with low Miller indices. To the inexperienced observer, it is not always easy to distinguish "true" PDFs from other lamellar features (fractures, fluid inclusion trails).

The most important characteristics of PDFs are: they are extremely narrow, closely and regularly spaced, completely straight, parallel, extend through the whole grain, and usually show more than one set per grain. This way they can be distinguished from features that are produced at lower strain rates, such as the tectonically formed Böhm lamellae, which are not completely straight, occur only in one set, usually consist of bands that are >10 µm wide, and are spaced at distances of >10 µm. It was demonstrated from Transmission Electron Microscopy (TEM) studies (see, e.g., Goltrant et al. 1991) that PDFs consist of amorphous silica, i.e., they are planes of amorphous quartz that extend throughout the quartz crystal. This allows them to be preferentially etched by, e.g., hydrofluoric acid, emphasizing the planar deformation features (see, e.g., Montanari and Koeberl 2000). PDFs occur in planes that correspond to specific rational crystallographic orientations. In quartz, the (0001) or c (basal), {$10\bar{1}3$} or

Table 2. Characteristics of Shock Deformation Features in Rocks and Minerals

Pressure (GPa)	Features	Target Characteristics	Feature Characteristics
2–45	Shatter cones	Best developed in homogeneous fine-grained, massive rocks.	Conical fracture surfaces with subordinate striations radiating from a focal point.
5–45	Planar fractures and Planar deformation features (PDFs)	Highest abundance in crystalline rocks; found in many rock-forming minerals; e.g., quartz, feldspar, olivine and zircon.	PDFs: Sets of extremely straight, sharply defined parallel lamellae; may occur in multiple sets with specific crystallographic orientations.
30—40	Diaplectic glass	Most important in quartz and feldspar (e.g., maskelynite from plagioclase)	Isotropization through solid-state transformation under preservation of crystal habit as well as primary defects and sometimes planar features. Index of refraction lower than in corresponding crystal but higher than in fusion glass.
15–50	High-pressure polymorphs	Quartz polymorphs most common: coesite, stishovite; but also ringwoodite from olivine, and others,	Recognizable by crystal parameters, confirmed usually with XRD or NMR; abundance influenced by post-shock temperature and shock duration; stishovite is temperature-labile.
>15	Impact diamonds	From carbon (graphite) present in target rocks; rare	Cubic (hexagonal?) form; usually very small but occasionally up to mm-size; inherits graphite crystal shape.
45–>70	Mineral melts	Rock-forming minerals (e.g., lechatelierite from quartz)	Impact melts are either glassy (fusion glasses) or crystalline; of macroscopically homogeneous, but microscopically often heterogeneous composition.

XRD = X-ray diffraction; NMR= nuclear magnetic resonance; PDF = planar deformation features.
Table after Montanari and Koeberl (2000) and Koeberl (2002).

ω, and $\{10\bar{1}2\}$ or π orientations are the most common ones (for details, see, e.g., Stöffler and Langenhorst 1994; Grieve et al. 1996; French et al. 1998). With increasing shock pressure, the distances between the planes decrease, and the PDFs become more closely spaced and more homogeneously distributed over the grain, until at about ≥35 GPa the grains show complete isotropization. Depending on the peak pressure, PDFs are observed in about 2 to 10 orientations per grain.

Figure 4 shows an example of multiple sets of PDFs in quartz from an impact structure. To confirm the presence of PDFs, it is necessary to measure their crystallographic orientations by using either a universal stage or a spindle stage, or to characterize them by TEM (see, e.g., Goltrant et al. 1991). Because PDFs are well developed in quartz (Stöffler and Langenhorst 1994), and because their crystallographic orientations are easy to measure in this mineral, most studies report only shock features in quartz. However, other rock-forming minerals, as well as accessory minerals (e.g., Leroux et al. 1999), develop PDFs as well (see also Stöffler 1972).

Fig. 4. Shocked quartz grain from the K-T boundary at DSDP 596, Western Pacific Ocean (46-47 cm layer). The grain has been etched by hydrofluoric acid to more clearly show the PDFs, which consist of amorphous quartz along crystallographically oriented planes. The planes typically occur in multiple orientations (SEM photograph courtesy B.F. Bohor).

A decrease of the density of shocked quartz with increasing shock pressure was noted (e.g., Stöffler and Langenhorst 1994). Optical properties, such as the birefringence of quartz and the refractive index, show also an inverse relationship with shock pressure in the 25 to 35 GPa range. At pressures of about 35 GPa, diaplectic glass forms. This isotropic phase preserves the crystal habit, original crystal defects, and, in some cases, planar features, and forms at shock pressures in excess of about 35 GPa (Table 2) without melting by solid-state transformation. Diaplectic glass has a refractive index that is slightly lower, and a density that is slightly higher, than that of synthetic quartz glass. At pressures that exceed about 50 GPa, lechatelierite, a "normal" mineral melt, forms by fusion of quartz.

At higher pressures, phase transitions to high-pressure polymorphs occur. These are a result of a solid state transformation process. The high-pressure polymorphs of quartz are coesite and stishovite, and both have been found at impact craters. Stishovite forms at lower shock pressures than coesite, probably because stishovite forms directly during shock compression, whereas coesite crystallizes during pressure release. The first time that coesite and stishovite were found in nature was in impactites, and stishovite has so far not been found in any other natural rocks. There are a few rare occurrences of coesite in metamorphic rocks of ultra-high pressure origin or in kimberlites, but it is easy to distinguish these coesites from those in impactites because they occur in significantly different mineral assemblages (see also Grieve et al. 1996; Glass and Wu, 1993). Another interesting high-pressure phase are diamonds that form from carbon in graphite- or coal-bearing target rocks (e.g., Gilmour 1998).

As the passage of shock waves through rocks generate temperatures that are far beyond those reached even in volcanic eruptions, and at pressures exceeding ca. 60 GPa, rocks undergo complete (bulk) melting. The high temperatures are demonstrated by the presence of inclusions of high-temperature minerals, such as lechatelierite, which is the monomineralic quartz melt and forms from pure quartz at temperatures >1700°C, or baddeleyite, which is the thermal decomposition product of zircon, forming at a temperature of about 1900°C. Lechatelierite is not found in any other natural rock, except in fulgurites, which form by fusion of soil or sand when lightning hits the ground. Lechatelierite does not occur in any volcanic igneous rocks. Depending on the initial temperature, the location within the crater, the composition of the melt, and the speed of cooling, impact melts result in either impact glasses (if they cool fast), or in fine-grained impact melt rocks (if they cool slow). As mentioned above, suevitic breccias contain inclusions of glass fragments or melt clasts, whereas impact melt rocks contain clasts of shocked minerals or lithic clasts. Recently, carbonate melts have been identified in impact structures.

Because glass slowly devitrifies, impact glasses are usually found at young impact craters rather than at old ones. Very fine-grained recrystallization textures are often characteristic for devitrified impact glasses. Impact glasses have chemical and isotopic compositions that are very similar to those of individual target rocks or mixtures of several rock types. For example, it is possible to use the rare earth element (REE) distribution patterns, or the isotopic composition, which are identical to those of the (often sedimentary or metasedimentary) target rocks, to distinguish the impact melt rocks from intrusive or volcanic rocks. Impact glasses have also lower water contents (about 0.001–0.05 wt%) than volcanic or other natural glasses (e.g., Beran and Koeberl 1997).

Impact melt rocks are true igneous rocks that have formed by cooling and crystallization of high-temperature silicate melts. Even though they often have textures and mineral compositions that are similar to those of volcanic igneous rocks, evidence for an impact origin can be obtained from evidence for shock metamorphism (e.g., PDFs in rock-forming minerals; lechatelierite). Geochemical studies may also provide evidence for an impact origin of a melt rock. For example, the isotopic composition is different for volcanic rocks and locally melted crustal rocks, or the presence of a meteoritic component in such rocks can be established by geochemical analyses (see below). Impact melts and glasses are often are the most suitable material for the dating of an impact structure (see, e.g., the review by Deutsch and Schärer 1994).

7
Meteoritic Components in Impactites

The detection of small amounts of meteoritic matter in breccias and melt rocks can also provide confirming evidence of an impact event, but is extremely difficult. Only elements that have high abundances in meteorites, but low ones in terrestrial crustal rocks are useful – for example, the siderophile platinum-group elements (PGEs; Ru, Rh, Pd, Os, Ir, and Pt) and other siderophile elements (e.g., Co, Ni) in the case of the Cretaceous-Tertiary (K-T) boundary layer. Elevated siderophile element contents in impact melts, compared to target rock abundances, can be indicative of the presence of either a chondritic or an iron meteoritic component. Achondritic projectiles (stony meteorites that underwent magmatic differentiation) are much more difficult to discern, because they have significantly lower abundances of the key siderophile elements. It is also necessary to sample all possible target rocks to determine the so-called indigenous component (i.e., the contribution to the siderophile element content of the impact melt rocks from the target). So far, meteoritic components have been identified for just over 40 impact structures out of the more than 170 impact structures currently identified on Earth. This number reflects mostly the extent to which these structures have been studied in detail, as only a few of these impact structures were first identified by finding a meteoritic component (the majority has been confirmed by the identification of shock metamorphic effects). Iridium is most often determined as a proxy for all PGEs, because it can be measured with the best detection limit of all PGEs by neutron activation analysis (which was, for a long time, the only more or less routine method for Ir measurements at sub-ppb abundance levels in small samples). Studies of impact glasses from some small craters for which the meteorite has been partly preserved (e.g., Meteor Crater, Wabar, Wolf Creek, Henbury) indicated that the siderophile elements are significantly and variably fractionated in their interelement ratios compared to the initial ratios in the impacting meteorite. Other types of fractionation have also been observed, for example impact ejecta from the Acraman structure in Australia show deviations from chondritic PGE patterns due to low-temperature hydrothermal alteration (e.g., Gostin et al. 1989; cf. also Colodner et al. 1992). Since the late 1970s, several studies tried to determine the type or class of meteorite for the impactor from analyses of impact melt rock or glass, but these attempts were not always successful, as it is difficult to distinguish among different chondrite types. Other problems may arise if the target rocks have high abundances of siderophile elements or if the siderophile element concentrations in the impactites are very low. In such cases, the use of the osmium (e.g., Koeberl and Shirey 1997) and chromium (Shukolyukov and Lugmair 1998; Shukolyukov et al. 2000) isotopic systems can help to establish the presence of a meteoritic component in impact melt rocks and breccias. More information on the topic of meteoritic components in impactites is given by Koeberl (1998) and Muñoz-Espadas (this volume).

8
Tektites and Microtektites

Tektites are chemically homogeneous, often spherically symmetric natural glasses, with most being a few centimeters in size. Mainly due to chemical studies, it is now commonly accepted that tektites are the product of melting and quenching of terrestrial rocks during hypervelocity impact on the Earth. The chemistry of tektites is in many respects identical to the composition of upper crustal material. Tektites are currently known to occur in four strewn fields of Cenozoic age on the surface of the Earth. Strewn fields can be defined as geographically extended areas over which tektite material is found. The four strewn fields are: the North American, Central European (moldavite), Ivory Coast, and Australasian strewn fields. Tektites found within each strewn field have the same age and similar petrological, physical, and chemical properties. Relatively reliable links between craters and tektite strewn fields have been established between the Bosumtwi (Ghana), the Ries (Germany), and the Chesapeake Bay (USA) craters and the Ivory Coast, Central European, and North American fields, respectively. The source crater of the Australasian strewn field has not yet been identified. Tektites have been the subject of much study, but their discussion is beyond the scope of the present review. For details on tektites see the reviews by, e.g., Koeberl (1986, 1994), Montanari and Koeberl (2000), and Koeberl (2001). Here we just provide a few basic observations and inferences.

In addition to the "classical" tektites on land, microtektites (<1 mm in diameter) from three of the four strewn fields have been found in deep-sea cores (see, e.g., Glass 1967, 1972). Microtektites have been very important for defining the extent of the strewn fields, as well as for constraining the stratigraphic age of tektites, and to provide evidence regarding the location of possible source craters. Microtektites have been found together with melt fragments, high-pressure phases, and shocked minerals (e.g., Glass 1989; Glass and Wu 1993) and, therefore, provide confirming evidence for the association of tektites with an impact event. The variation of the microtektite concentrations in deep-sea sediments with location increases towards the assumed or known impact location (Glass and Pizzuto 1994).

There has been some discussion about how to define a tektite, but the following characteristics should probably be included (see Koeberl 1994; Montanari and Koeberl 2000): 1) they are glassy (amorphous); 2) they are fairly homogeneous rock (not mineral) melts; 3) they contain abundant lechatelierite; 4) they occur in geographically extended strewn fields (not just at one or two closely related locations); 5) they are distal ejecta and do not occur directly in or around a source crater, or within typical impact lithologies (e.g., suevitic breccias, impact melt breccias); 6) they generally have low water contents and a very small extraterrestrial component; and 7) they seem to have formed from the uppermost layer of the target surface (see below). Thus, it is recommended to use the term "tektite" only for glasses that fulfill (most) of the above points, and, if in doubt, use the (probably much better) general term "impact glass".

Fig. 5. Muong Nong-type (or layered) tektite from Vietnam. The layered structure and the large vesicles are obvious.

An interesting group of tektites are the Muong Nong-type tektites (Fig. 5), which, compared to "normal" (or splash-form) tektites are larger, more heterogeneous in composition, of irregular shape, have a layered structure, and show a much more restricted geographical distribution (for details on these tektites, see Koeberl 1992a). They are also important because they contain relict mineral grains that indicate the nature of the parent material and contain shock-produced phases that indicate the conditions of formation (e.g., Glass and Barlow 1979). The occurrence of relict minerals in some tektites points to sedimentary source rocks. Muong Nong-type tektites contain unmelted relict inclusions, including zircon, chromite, quartz, rutile, and monazite, all showing evidence of various degrees of shock metamorphism. Coesite, stishovite, and shocked minerals were found in the North American and Australasian microtektite layers (Glass 1989; Glass and Wu 1993).

Despite knowing the source craters of three of the four tektite strewn fields, we still do not know exactly when and how during the impact process tektites form. Detailed reviews of the parameters that have to be considered were provided by, for example, Koeberl (1994) and Montanari and Koeberl (2000). These characteristics include, for example, the following observations: a) vapor fractionation played no major role in tektite formation; b) tektites are very poor in water with contents ranging from about 0.002 to 0.02 wt%; c) bubbles in tektites contain residues of the terrestrial atmosphere at low pressures; d) meteoritic components in tektites are very low or below the detection limit; e) the ^{10}Be content of Australasian tektites cannot have originated from direct irradiation with cosmic rays in space or on Earth, but can only have been introduced from sediments that have absorbed ^{10}Be that was produced in the terrestrial atmosphere. Tektites might be produced in the earliest stages of impact, which are poorly understood. It is clear, however, that tektites formed from the uppermost layers of terrestrial target material (otherwise they would not contain any ^{10}Be) However, the question which process was responsible for tektite production and distribution remains the subject of further research.

9
Case Studies for Distal Ejecta

Relatively few distal ejecta layers are known in the stratigraphic record; most of them have been linked to specific source craters. In some cases the ejecta led to the source craters, and in some cases the reverse is true. In a few cases, no source crater is know, or he confirmation of some layers as impactoclastic is still outstanding. In the following paragraphs we attempt a short summary of some important distal ejecta layers and their possible source craters, beginning with the most important (and best studied) one – the Cretaceous-Tertiary boundary. Other impact layers are then discussed in order of increasing age. The topic of distal ejecta layers is discussed more extensively in Montanari and Koeberl (2000) and Koeberl (2001).

9.1
The Cretaceous-Tertiary Boundary (65 Ma)

The impact ejecta at the Cretaceous-Tertiary (K-T) boundary were not the first distal ejecta layer to be described, as microtektite-bearing layers were already discovered in the late 1960s. However, the study of the K-T boundary ejecta provided the most influence for the discussion about the importance of impact events. It is easy to detect the K-T boundary layer in the field, where is appears as a distinct break in lithology, with a thin (usually up to 1 or 2 cm thick) layer commonly composed of clay or claystone at distal marine sections (see below for details; Fig. 6). Good descriptions of field relations are given, for example, by

Smit (1999), and references therein. Information on the paleontology of the K-T boundary can be found, for example, in the "Snowbird" series of conference proceedings (Silver and Schultz 1982; Sharpton and Ward 1990; Ryder et al. 1996; Koeberl and MacLeod 2002).

The first physical evidence pointing to a contribution of extraterrestrial material that was discovered was the presence of anomalously high PGE abundances in K-T boundary clay in Italy (Alvarez et al. 1980) and other locations around the world (e.g., Smit and Hertogen 1980). The contents of Ir and other PGEs were found to be enriched in these K-T boundary clay layers by up to four orders of magnitude compared to average terrestrial crustal abundances. Also, the interelement ratios of the PGEs in K-T boundary clay samples are very similar to the values observed in chondritic meteorites. Osmium- and Cr-isotopic studies of the K-T boundary provided further evidence of an extraterrestrial component. Shukolyukov and Lugmair (1998) presenting Cr-isotope data for a meteoritic component that is in better agreement with a carbonaceous chondritic composition of the impactor, rather than an ordinary chondritic composition.

Many K-T boundary locations around the world show evidence for global wildfires in the form of a charcoal and soot layer that coincides with the Ir-rich layer (Wolbach et al. 1985). The insoluble carbon fraction after acid dissolution is dominated by kerogen and elemental carbon, which also show a marked change in their isotopic composition across the K-T boundary (Gilmour et al. 1990). The total amount of soot in the atmosphere due to the global wildfires at the end of the Cretaceous has been estimated at $7 \cdot 10^{16}$ g, which must have had a large influence on the environment.

Most important, however, was the discovery of clear evidence of shock metamorphism at the K/T boundary (Bohor et al. 1984, 1987). The shocked quartz grains show multiple intersecting sets of planar deformation features (PDFs) with shock-characteristic crystallographic orientations. Shocked zircons with planar features were discovered by Bohor et al. (1993). Impact glass was found at the K-T boundary (in Haiti) as well (e.g., Sigurdsson et al. 1991; Koeberl and Sigurdsson 1992). Blum et al. (1993) showed that the Haitian glasses are mixtures of silicate rocks of upper crustal composition with a high CaO-endmember (e.g., limestone). Koeberl (1992b) measured the water content in glasses from Haiti and found a range of 0.013 to 0.021 wt% H_2O, which agrees with an origin by impact, as impact glasses are extremely dry. Using Os isotope analyses, Koeberl et al. (1994) found a small meteoritic component in the Haitian glasses. Precise age determinations on the Haitian glasses have shown that the materials have an age indistinguishable from that of the K/T boundary, at 65 Ma (e.g., Swisher et al. 1992). Impact-related nano- and micro-diamonds were found at some K-T boundary locations (e.g., Hough et al. 1997; Gilmour 1998).

Spinel (magnesioferrite) of various compositions were first reported by Montanari et al. (1983) from magnetic spherules at the Petriccio, Italy, K-T boundary section. Spinel at the K-T boundary can be used as an event marker, as it shows an abundance peak similar to that observed for the PGEs.

Despite all the evidence supporting a major impact event at the end of the Cretaceous, no large impact structure with a corresponding age was known until the early 1990s, when finally the ca. 200-km-diameter Chicxulub impact structure was proposed and confirmed as the elusive K-T boundary crater (e.g., Hildebrand et al. 1991, Sharpton et al. 1992, 1993). A detailed discussion of the Chicxulub impact structure is beyond the scope of this short overview (for details, see Koeberl 1996 and Montanari and Koeberl 2000). However, all geochemical and geochronological data confirm that Chicxulub is indeed the long-sought K-T impact structure. The detailed study of a distal impact ejecta layer had led to the discovery of one of the largest impact structures on Earth.

For the topic of the present volume, it is instructive to review the characteristics of the K-T boundary layer in some more detail. The characteristics of the K-T boundary deposits depend mainly on the distance to the Chicxulub crater. The deposition of the material ballistically ejected from the crater resulted in a layer which clearly separates Cretaceous from Tertiary sediments. The distinctive dispersed ejecta produced by this impact (e.g., Alvarez et al. 1995; Pierazzo and Melosh 1999) were mostly derived from the turbulent front of the melted target rocks, and the vertically expanding hot vapor plume of vaporized bolide with entrained melted target rocks and were deposited in the proximity of the crater site and globally, respectively. Thus, greater contributions of the ejecta blanket derived from the target rocks are reported at locations proximal to Chicxulub, whereas distal sections contain a relatively higher contribution of extraterrestrial material.

Sites that are relatively close to the Chicxulub structure (e.g., Yucatan, Belize, S-Mexico, NE-Mexico, Central Mexico, Guatemala, Southern USA, Haiti, DSDP/ODP sites) comprise 1- to 11-m-thick complex clastic deposits transported by gravity flows, landslides, and tsunami waves. Nearby ejecta can be divided into: 1) a continuous ejecta blanket extending radially up to some 600 km on the Yucatan peninsula consisting of polymictic mega- and microbreccias composed of moderately shocked clasts of extremely variable size (this would be the proximal ejecta unit), 2) an area extending south of the crater to about 1000 km characterized by carbonate megabreccias, and 3) another subarea occurring around the Gulf of Mexico and in the Caribbean characterized by thick tsunami deposits overlain by ballistically transported debris and melt spherules (e.g., Pope et al. 1996; Claeys et al. 1998; Smit 1999). At these locations K/T boundary deposits comprise materials of quite variable composition, ranging from pure carbonate to silicate melt and mixtures of these two end-members. Such materials are commonly altered to clay minerals. Thus, in the Guatemala sections (El Caribe, El Ceibo) the ejecta layer is composed of glass spherules altered to well-crystallized Ca-Mg-(Na)-rich smectite (Debrabant et al., 1999). At the Belize site (Albion Island section), smectite is present in the breccia matrix and clay-forming spheroids as results of the alteration of impact-generated material (Pope et al. 1999). In the Haiti section, the ejecta layer consists of a basal layer very rich in smectite and spherules, overlain by a unit consisting of a mixture of smectite spherules in a carbonate/smectite matrix. Spherical to ellipsoidal smectite bodies derive from impact-glass alteration (Izett et al. 1991; Kring and Boynton 1991; Koeberl and Sigurdsson 1992). In the El Tecolote section (NE Mexico), spherules are usually altered to chlorite (Soria et al. 2001; Mata et al. 2001).

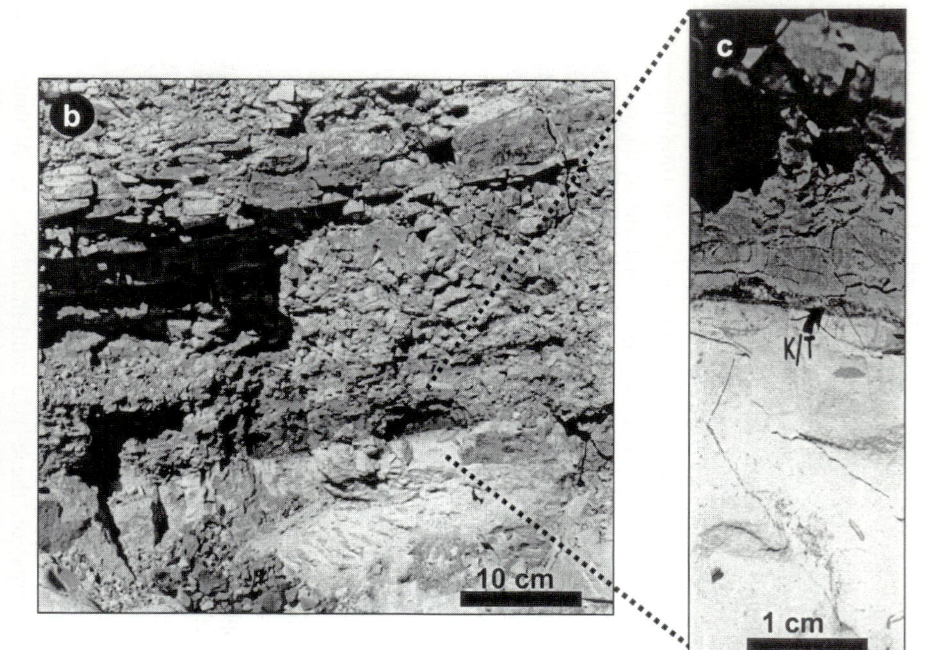
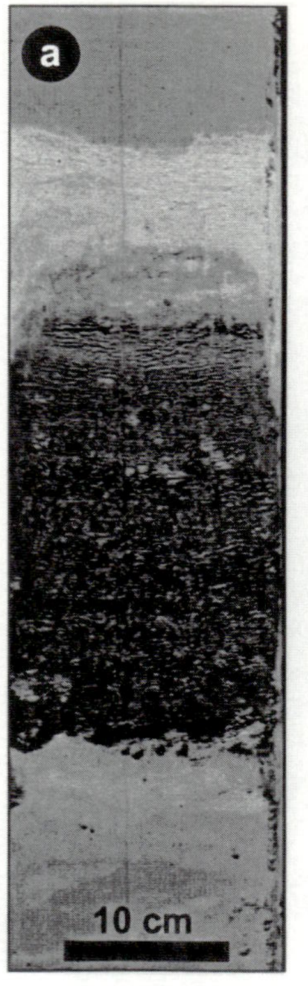

Fig. 6. Macroscopic appearance of the K-T boundary layer. a) Core photograph of the spherule bed that marks the K-T boundary drilled at Hole 1049A (ODP Leg 171B) in Blake Nose Plateau, b) Field photograph showing the K/T boundary interval at the Agost section (SE Spain), and c) detail of the ejecta layer at Agost section.

About 2000 km to the Northeast on the North American margin some excellent K-T boundary record has been reported at Blake Nose (ODP Leg 171B Hole 1049), Bass River (ODP Leg 174AX), and DSDP Leg Hole 603B. At Blake Nose, the K-T boundary is marked by a 9- to 17-cm-thick ejecta layer (Fig. 6), mainly consisting of spherical and oval-shaped spherules, but also containing some lithic fragments, Cretaceous foraminifera, and clasts of Cretaceous material, which indicates reworking of the spherule bed material. This observation is further supported by the variable thickness of the spherule bed in the three holes drilled at Site 1049 in the Blake Nose Plateau (Klaus et al. 2000). Spherules are altered here to smectite, and different smectite compositions resulted from the alteration of dark-green and pale-yellow spherules derived from different precursor glass types (Fig. 7), which are richer in Si and Ca, respectively (Martínez-Ruiz et al. 2001, 2002). The mineralogy and morphologies of the Blake Nose spherules are similar to those reported by Klaver et al. (1987) from DSDP Hole 603B and by Olsson et al. (1997) from Bass River sections. All of these spherules represent the same diagenetically altered impact ejecta from the Chicxulub crater.

At distal locations 2500-4000 km from the Chicxulub crater in the Western Interior of North America (e.g., Madrid, Starkville, Sugarite, Raton), the K/T boundary interval consists of a 3-cm-thick, two-layered, clay-rich unit. The dual nature of this K/T boundary sequence supports different ejection and dispersal mechanisms of the ejecta material (e.g., Pollastro and Pillmore 1987; Pollastro and Bohor 1993). The lower claystone layer (ejecta layer) is mainly kaolin minerals, derived from silicic glass formed from melted target rocks. The upper laminated layer (fireball layer) mostly consists of altered vitric dust and abundant shocked minerals. The fireball layer is mostly altered to smectite from a mafic glass condensed from the vaporized chondritic bolide, along with some kaolinite formed from blebs of melted silicic target material entrained in the vapor plume cloud during ejection (Pollastro and Bohor 1993).

The most distal sites (> 4000 km), such as those from the Mediterranean area (e.g., Agost and Caravaca in Spain, El Kef and Elles II in Tunisia, and Petriccio and Gubbio in Italy) and NE Atlantic regions (Zumaya, Monte Urko, Sopelana and Biarritz in the Basque-Cantabrian Basin, and Stevns Klint in Denmark) are characterized by a 2- to 3-mm-thick layer. In some Mediterranean sections, such as Agost, Caravaca (e.g., Martínez-Ruiz et al. 1997; Smit 1999), and El Kef, the K/T boundary layer is better preserved (Lindinger 1988; Adatte et al. 2002) than in the Basque-Cantabrian basin (Ortega-Huertas et al. 1998, 2002). At Stevns Klint (Denmark), the K-T boundary is marked by a red-rust basal layer overlain by a black marl layer (e.g., Schmitz 1985; Elliott 1993). In all these distal sections, this red layer is equivalent to the uppermost layer of the two-layered clay unit described for the Western Interior of North America sections (Pollastro and Bohor 1993). In the Agost, Caravaca, Petriccio, El Kef and Elles II sections in the Mediterranean Domain, the fireball layer consists of almost pure smectite, derived from the alteration of distal ejecta material, and abundant spherules (microkrystites, Fig. 8) (e.g., Martínez-Ruiz et al. 1997; Smit 1999). In the Gubbio section, the boundary-layer clays contain less expandable minerals and have a

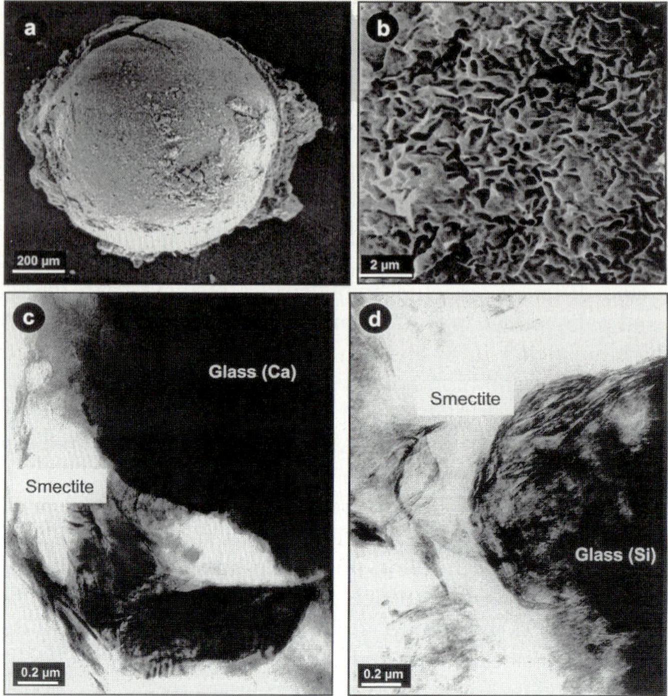

Fig. 7. Spherules from the K-T boundary at Blake Nose. a) SEM image showing an example of smectite spherules from Blake Nose (ODP Leg 171B, Hole 1049A), b) detail on the surface of spherules from Blake Nose showing smectite morphologies, c, d) TEM images showing the alteration of the impact-generated glass (Ca-rich and Si-rich) into smectite.

composition apparently dominated by detrital illite and kaolinite (Rampino and Reynolds 1983). In the Basque-Cantabrian Basin, smectite is not detected as a consequence of the extensive diagenetic evolution undergone by these sequences (Ortega-Huertas et al. 1998). At Stevens Klint (Denmark) the fireball layer is composed exclusively of pure magnesian smectite that formed by alteration of glass spherules at a low water/rock ratio (Kastner et al. 1984). At this section, the existence of abundant 10-20 nm-diameter iron oxides with 10% Ni and minor Zn intergrown with smectite suggests the presence of altered meteorite fragments (Bauluz et al. 2000).

In marine sections, the sudden decrease in ocean productivity resulted in the deposition of the lowermost Danian clay layer above the impact-generated materials, which also characterizes the K-T boundary in most of these sections. The clay boundary layer represents a reduced sedimentation deposit and records the faunal crisis resulting from the impact event. It may also contain some extraterrestrial contamination, but its composition mainly reflects the environmental conditions prevailing after the impact.

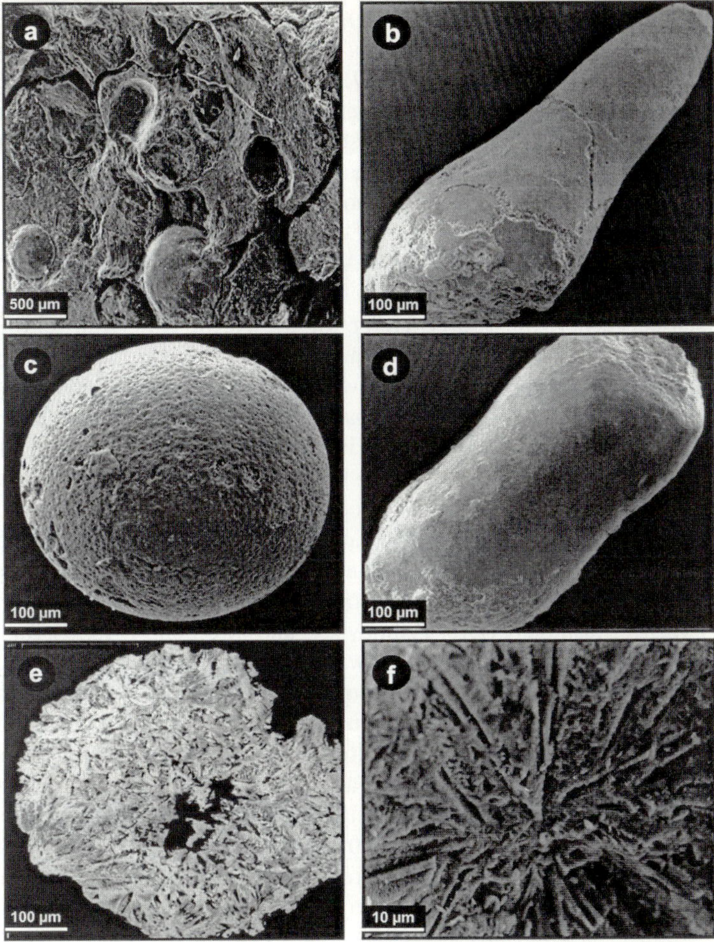

Fig. 8. Electron microphotographs of spherules from various K-T boundary locations at distal sites. a) Secondary electron microscope (SEM) images showing a general view of the K-T ejecta layer from Agost section showing smectite containing abundant microkrystites (a), the morphologies of microkrystites (b, c, d) and their internal textures: e) back-scattered electron (BSE) image showing fibroradial textures from K-feldspar spherules from Caravaca section; f) SEM image showing the fibroradial texture of Fe-oxide spherules from Agost section.

9.2
Late Eocene Impactoclastic Layers (35 Ma)

Late Eocene marine sediments around the world contain evidence for at least two closely spaced impactoclastic layers. One layer known from the eastern U.S. coast, the Caribbean, and the Gulf of Mexico is correlated with the North

American tektite strewn field (see above). This layer contains microtektites (i.e., glassy - not recrystallized - spherules), shocked minerals, and high-pressure phases (e.g., coesite) (e.g., Glass 1989), but no marked siderophile element anomaly. The presence of crystalline spherules composed mostly of clinopyroxene (cpx) was detected in the same deep sea sediments and initially it was considered that these spherules also belong to the North American tektite strewn field; however, the cpx spherules were found not only in the Caribbean and the Gulf of Mexico, but also in the Pacific Ocean. Despite suggestions for more layers, the presence of these two layers (the North American microtektite layer and the cpx spherule layer) is now accepted (e.g., Wei 1995).

As discussed earlier, the source crater for the North American tektite strewn field has now been identified with a certain degree of confidence as being the 35 Ma Chesapeake Bay impact structure, which has a diameter of about 90 km (Koeberl et al. 1996; Poag 1997). An impact event that created a crater of this size would be capable of globally distributing its distal ejecta (e.g., Langenhorst 1996). There is a second large crater with an age that is indistinguishable from that of the Chesapeake Bay structure and the two ejecta layers, namely the 100-km-diameter Popigai impact structure in Siberia, which has been dated by Bottomley et al. (1997) at 35.7±0.8 Ma. The Popigai structure is exposed in Archean crystalline rocks of the Anabar Shield, with overlying Proterozoic to Mesozoic sedimentary sequences (e.g., Masaitis 1994; Vishnevsky and Montanari 1999), and is the largest Cenozoic crater on Earth. It is now commonly assumed that the global late Eocene microkrystite layer originated from the Popigai impact event, but this link has yet to be confirmed, probably by using isotope geochemical methods, as radiometric age determinations do not allow to resolve an age difference of 10 or 20 k.y. It is also interesting to note that Farley et al. (1998) found much enhanced levels of 3He coinciding with the two late Eocene impactoclastic layers. This isotope is a proxy for the influx of extraterrestrial dust, and as interpreted as indicating that during the late Eocene there was a time of enhanced comet activity in the inner solar system, probably resulting in a higher impact rate than usual.

9.3
Manson Impact Structure and Ejecta Layer (74 Ma)

The Manson impact structure is well-preserved complex impact structure with a diameter of about 35 to 37.8 km in north-central Iowa. Early dating attempts of rocks from the Manson structure by Ar-Ar dating by Kunk et al. (1989) indicated that the structure may have formed at 65.7 Ma, an age indistinguishable from that of the K-T boundary, leading to a drilling program in 1991-1992. One of the first products of the investigation of the core materials recovered was a new and more accurate age for the formation of the Manson structure. $^{40}Ar/^{39}Ar$ analyses of sanidine feldspar from impact melt rocks by Izett et al. (1993) demonstrated that the Manson structure was formed at 74.1±0.1 Ma – long before the K-T boundary [It is interesting to note, though, that another impact structure – Boltysh – which was initially dated at 88 Ma is now confirmed to have an age more or less identical to that of the K-T boundary – see Kelley and Gurov (2002), Valter and Plotnikova, this volume, and Gurov et al., this volume. Izett et al. (1993, 1998) and Witzke et al. (1996) also described the discovery of a distal ejecta layer related to the Manson impact structure in the Crow Creek Member of the Cretaceous Pierre Shale in South Dakota and Nebraska.

9.4
Morokweng and the Jurassic-Cretaceous Boundary (145 Ma)

The existence of a large subsurface impact structure in the area around Morokweng, South Africa (centered at 23°32' E and 26°31' S), was inferred on the basis of gravity and magnetic investigations, and confirmed by petrographic and geochemical investigations of drill core samples. There was a debate about its diameter, with values ranging from about 70 to 340 km being discussed, but new data seem to show that a diameter of 70 to 80 km is probably the correct answer (Henkel et al. 2002; Reimold and Koeberl, this volume). The discovery of the Morokweng structure caused some interest because dating of zircons extracted from the impact melt rocks yielded an age of 146.2±1.5 Ma (Kocbcrl ct al. 1997), which is indistinguishable from the currently accepted age for the Jurassic-Cretaceous (J-K) boundary, which is placed at 145 Ma at the base of the Berriasian. The J-K boundary situation is somewhat different from that of the K-T boundary, as the J-K boundary is poorly defined. There is a report of platinum group element enrichment at a J-K boundary site in Siberia, but these data are not yet confirmed, and even the identity of this layer with the J-K boundary is uncertain (Rampino and Haggerty 1996). Also, there seems to be a difference between the placement of the J-K boundary in the Boreal and Tethyan provinces (e.g., Rampino and Haggerty 1996) The top of the Volgian (just below the boreal J-K boundary) offshore Norway is marked by impact ejecta from the nearby 40-km-diameter Mjølnir impact structure (Dypvik et al. 1996). In addition, the Gosses Bluff impact structure in central Australia (ca. 25 km in diameter) also has an age of about 143 Ma. However, it is difficult to use biostratigraphical correlations to determine if the J-K boundary represents a global event, because of limited faunal exchange between different paleobiogeographic provinces. Also, at the present time no well-defined distal impact ejecta layer has been identified at the J-K boundary.

9.5
Triassic-Jurassic Boundary (200 Ma)

The Triassic-Jurassic (Tr-J) boundary is marked by yet another major mass extinction - one the big five mentioned above. Bice et al. (1992) reported on the discovery of shocked quartz grains from a Tr-J boundary location in northern Italy, although the identification of the PDFs has been questioned. A search for shocked quartz at Tr-J boundary locations in Nova Scotia (Canada) was negative.

The 100-km-diameter Manicouagan impact structure in Quebec has an age that is comparable to that of the Tr-J boundary, but currently available dates indicate that with an age of 214 Ma (Hodych and Dunning 1992) it slightly predates the Tr-J boundary. Spray et al. (1998) discussed new evidence for a multiple impact event in the late Triassic, probably slightly predating the Tr-J boundary. However, as past experience has shown, it is very difficult to correlate radiometric ages obtained from impact melt rocks with biostratigraphic ages obtained from the sedimentary record, as a correlation between the two records implies the use of the same time scale, which is basically never the case. Thus, confirmation of an impact signature from other Tr-J boundaries might be the next step in the investigation of the end-Triassic event.

Recent analysis of tetrapod footprints and skeletal material from over 70 localities in eastern North America shows that large theropod dinosaurs appeared less than 10 thousand years after the Triassic-Jurassic boundary and less than 30 thousand years after the last Triassic taxa, synchronous with a terrestrial mass extinction (Olsen et al. 2002). These authors found also that this extraordinary turnover is associated with an Ir anomaly (up to 0.28 ppb, average 0.14 ppb) and a fern spore spike, suggesting that a bolide impact was the cause. Eastern North American dinosaurian diversity reached a stable maximum less than 100 thousand years after the boundary, marking the establishment of dinosaur-dominated communities that prevailed for the next 135 million years. Thus, it could be that impact events also had other influences on the biosphere.

In addition, recent work indicated the possible presence of an ejecta layer in England (Walkden et al. 2002), which may be related to the 214 Ma (Hodych and Dunning 1992) Manicouagan impact event.

9.6
Permian-Triassic Boundary (253 Ma)

The Permian-Triassic (P-Tr) boundary is associated with the largest mass extinction known in Earth history. Following the association of the K-T boundary mass extinction with a large impact event, speculations bloomed that other major mass extinctions might also be related to impact events. However, so far the evidence in favor of such a proposal is controversial. Siderophile element anomalies (e.g., enhanced Ir contents) were found at some P-Tr boundary locations (e.g., Holser et al. 1989), but their source is not clear and confirmation of their extraterrestrial nature is still pending. Recent research, however, succeeded in demonstrating the P-Tr boundary event was a much shorter event than thought before, and that the severe environmental changes that resulted in the mass extinction were brought on within less than a few hundred thousand years (Bowring et al. 1998). These authors also found that at Meishan, China, a negative excursion in the carbon isotopic composition had a duration of less than about 160,000 years and suggested that it could be the result of the impact of an icy, carbon-rich comet (but also scenarios involving volcanic processes are conceivable). Retallack et al. (1998) reported on the possible discovery of shocked quartz grains from P-Tr boundary locations in Australia and Antarctica, but the exact association of the quartz-bearing layers with layers that have enhanced Ir contents and with the P-Tr boundary is still unclear.

Kaiho et al. (2001) reported sulfur isotope and chemical data for samples from the Meishan (China) Permo-Triassic (P-Tr) boundary section. They interpreted S-isotope data, as well as the occurrence of Fe- and Ni-rich particles, as evidence for a large-scale impact event that penetrated the Earth's mantle and formed a crater ~1000 km in diameter. Koeberl et al. (2002) gave a detailed discussion why the hypothesis of Kaiho et al. (2001) is a complete failure. The lack of shocked quartz implies an oceanic impact event is misleading. Shock metamorphic effects are not restricted to quartz, but occur in all rock-forming (and accessory) minerals, which are abundant in ocean floor rocks. Impact-induced volcanism or excavation of mantle material in impact events have been postulated before, but such effects are physically implausible, and that no known impact on Earth has ever had such consequences. Kaiho et al. (2001) made a fundamental mistake in that they assume complete vaporization of target material inside the crater cavity. From their calculated degassed sulfur volume the authors arrive at a crater diameter of 600 to 1200 km. However, in reality this is an estimate of the zone of vaporization of the crater so that, in fact, the actual size of the crater should be much larger. To produce such a crater a projectile with a diameter of 750 – 1500 km would be necessary – which is implausible as the largest main belt asteroid has a diameter of 1000 km and that the largest crater formed on the terrestrial planets in the last 500 Myr is Mead Crater on Venus with a diameter of ~280 km. This would seem to be an upper limit of a crater size we should assume for possible catastrophic impacts during the Phanerozoic on Earth.

Koeberl et al. (2002) noted that none of the points raised by Kaiho et al. (2001) provide conclusive evidence – or even vague suggestions – of an impact event at the P-Tr boundary. Attempts to utilize the questionable interpretations by Kaiho et al. in an attempt to support the equally controversial (cf. Farley and Mukhopadhyay 2001) claims for the presence of extraterrestrial [3]He in fullerenes at the P-Tr boundary represent circular logic. Thus, it seems as if the jury is still out on the cause of the P-Tr boundary mass extinction event.

9.7
Late Devonian Impact Layer and Alamo Breccia (367 Ma)

The Frasnian-Famennian (F-F) boundary in the Late Devonian is associated with one of the five largest mass extinctions in the geological record. There is some indication for impact evidence around this time, although the evidence does not seem to support the presence of a uniform coeval global event. Glassy spherules, probably similar to microtektites, have been discovered in Late Devonian sections in south China slightly above the F-F boundary (e.g., Wang 1992) and also in Belgium (e.g., Claeys et al. 1992; Claeys and Casier 1994). The spherules have all the characteristics of having formed by impact (Claeys et al. 1992; Wang 1992). In China and Australia the spherule layer seems to be correlated with a minor Ir anomaly, which is absent from the Belgian sections (Claeys et al. 1996). It is possible that the spherule layers in China and in Belgium do not belong to precisely the same layer, as conodont stratigraphy indicates a slight time difference, but this has not yet been confirmed. No source crater for the distal impact layer(s) has yet been identified, except for the 54-km-diameter Siljan impact structure in Sweden, which is of Late Devonian age, but is probably too small to cause any mass extinctions. Also, the relation between the microtektite layer(s) and the F-F mass extinction has not been explored in any detail.

There is evidence for another large impact event in the Late Devonian. A large-scale impact event, dated from conodont stratigraphy at about 367 Ma, occurred in a nearshore marine setting, and resulted in the deposition of the wide-spread Alamo Breccia in Nevada (e.g., Warme and Sandberg 1996; Warme and Kuehner 1998). This megabreccia, which contains shocked quartz with multiple sets of PDFs, altered spherules, and possibly an Ir anomaly, is spread discontinuously over a semi-circular zone of about 200 km diameter and has a total thickness of more than 100 m in some locations. The breccia show a variation in lithology and thickness as a function of increasing distance from the inferred center, but no crater has yet been found, and any such crater may well have been eroded or tectonized since then. The age of the Alamo event does not seem to coincide with the ages inferred for the Belgian and Chinese microtektite horizons. Some more details of the Alamo Breccia are discussed by Koeberl et al. (this volume).

9.8
Acraman Impact Structure and Ejecta Layer (590 Ma)

An impactoclastic layer was found within late Precambrian shales of the 590 Ma Bunyeroo Formation in the Adelaide geosyncline, South Australia (Gostin et al. 1986). The ejecta occur in outcrops and drill cores over a distance of several hundred kilometers. At the same time, Williams (1986, 1994) identified the Acraman structure in South Australia as an impact structure, and confirmed it to be the source crater of the Bunyeroo impact ejecta layer. Gostin et al. (1989) and Wallace et al. (1990) detected enrichments of the PGEs in the ejecta layer; however, post-formational redistribution had altered the PGE patterns. The diameter of the Acraman structure is at least 90 km, with some outer arcuate features at 150 km diameter (Williams 1994). Impact ejecta have been found at distances of up to 450 km from the Acraman structure (i.e., about 10 crater radii), making this a true distal ejecta layer.

9.9.
South African and Australian Archean Spherule Layers (2.6 – 3.4 Ga)

Spherule layers in the ~3.4 Ga Barberton Greenstone Belt, South Africa, have been interpreted (e.g., Lowe and Byerly 1986) as the result of large asteroid or comet impacts onto the early Earth. These spherule layers show extreme enrichments in the PGEs, unlike modern ejecta deposits, which caused Koeberl and Reimold (1995) to question the impact interpretation. In the meantime, though, Shukolyukov et al. (2000) found Cr isotopic anomalies in samples from these layers that seem to support the presence of an extraterrestrial component in these layers. Other occurrences of unusual spherule layers were reported by Simonson (1992) from the Hamersley Basin in Western Australia. On the basis of similarities to microtektites and mikrokrystites, Simonson (1992) interpreted the spherules as having formed in an impact even and having been redeposited in a sediment gravity flow. Later, three additional spherule-bearing layers were found in the Hamersley Basin sequence, which were also interpreted to be of impact origin (e.g., Simonson et al. 1998). None of these spherules are associated to with shocked minerals, which Simonson et al. (1998) suggested to be the result of impact into an oceanic target, where quartz is not a major component. Simonson et al. (2000) also reported on the discovery of a similar spherule layer (ca. 2.6 Ma) in the Monteville Formation of the Transvaal Supergroup in South Africa, which might be correlated with one of the Australian layers.

However, all in all the identification of Precambrian impact deposits (especially distal ejecta) remains a largely unresolved problem. Unfortunately, so far no definitive criteria for the identification of Archean impact deposits are known. For none of the South African (Barberton and Monteville) or Australian spherule layers has a source crater been found; given the scarcity of the geological record it is likely that it will never be found. It is not clear why impact events in the Archean would predominantly produce large volumes of spherules, which are mostly absent from post-Archean impact deposits (i.e., those for which source craters are known). On the other hand, none of these spherule layers is associated with any shocked minerals, which are the hallmark for all confirmed impact structures and ejecta. Even rocks from the 2 Ga Vredefort impact structure contain abundant shocked minerals, so it is unlikely that Archean impacts would, for some reason, not produce shocked minerals. The question regarding how to identify Archean impact deposits remains open and will hopefully be addressed in future studies (but see Simonson and Harnik 2000, for some interesting thoughts on the subject). Nevertheless, the discovery of these various spherule layers provides interesting material for the discussion about the importance of impact events in the Earth's history.

10
Conclusions: Impact in the Stratigraphic Record

Distal ejecta ("impactoclastic layers") can be used as markers for impact events in the stratigraphic record. "Impact markers" are a variety of chemical, isotopic, and mineralogical species derived from the encounter of cosmic bodies (such as cometary nuclei or asteroids) with the Earth, as explained in more detail by Montanari and Koeberl (2000). Such markers are important for the to detection and study of accretionary events in the sedimentary record, to identify their origin, and to evaluate their possible role in global change and on the Earth's biotic and climatic evolution throughout geological time. Distal ejecta layers can be used to study a possible relationship between biotic changes and impact events, because it is possible to study such a relationship in the same outcrops, whereas correlation with radiometric ages of a distant impact structure is always associate with larger errors. Impactoclastic layers are composed of distal ejecta. In the past the discovery and detailed study of distal ejecta layers has led to the discovery of previously unknown large impact structures (for example, Chicxulub and Acraman).

Recent investigations (for example, the discovery of a possible ejecta layer in England, possibly derived from the Manicouagan impact event) indicate that there is wide-spread interest in the study of impact markers, allowing identification of smaller events and the study of their effects. Mader et al. (2002) reported on their (so far unsuccessful) search for ejecta in central Italy, about 600 km from the from the 24-km-diameter Ries impact structure (Southern Germany). Following the demonstration that the Boltysh impact structure (Ukraine) has an age that is within error of that of the Chicxulub impact structure, Gurov et al. (this volume) and Valter and Plotnikova (this volume) propose to study drill core samples from the Ukrainian Shield area to determine the possible relationship between the K-T boundary layer and ejecta from the Boltysh event to determine if they are really coeval. Thus, the search for (and study of) impact markers in the sedimentary record, and, more specifically, at various paleontological boundaries, is an important component of impact-related research. It may lead to the discovery of previously unknown impact events and structures. Detailed analyses of impact markers yields important information regarding the physical and chemical conditions of their formation, such as temperature, pressure, oxygen fugacity, composition of the atmosphere. New techniques and methods may be applied to the study of impact-derived minerals (e.g., Gucsik et al. 2002). We need more and better methods to help identify impact layers in the field and in the laboratory, given the importance of impact events for the geological and biological evolution of the Earth (as discussed in recent compilations by, e.g., Gilmour and Koeberl 2000; Buffetaut and Koeberl 2002; Koeberl and MacLeod 2002).

Acknowledgments

This work has been supported by the Austrian Fonds zur Förderung der wissenschaftlichen Forschung, project Y58-GEO. The support of the ESF IMPACT programme is appreciated for this paper and the whole book.

References

Adatte T, Keller G, Stinnesbeck W (2002) Late Cretaceous to early Paleocene climate and sea-level fluctuations: the Tunisian record. Palaeogeography Palaeoclimatology Palaeoecology 178: 165-196

Alvarez LW, Alvarez W, Asaro F, Michel HV (1980) Extraterrestrial cause for the Cretaceous-Tertiary extinction. Science 208: 1095-1108

Alvarez W, Claeys P, Kieffer S (1995) Emplacement of Cretaceous-Tertiary boundary shocked quartz from Chicxulub crater. Science 269: 930-935

Beran A, Koeberl C (1997) Water in tektites and impact glasses by FTIR spectrometry. Meteoritics and Planetary Science 32: 211-216

Bice DM, Newton CR, McCauley S, Reiners PW, McRoberts CA (1992) Shocked quartz at the Triassic-Jurassic boundary in Italy. Science 255: 443-446

Blum JD, Chamberlain CP, Hingston MP, Koeberl C, Marin LE, Schuraytz BC, Sharpton VL (1993) Isotopic comparison of K-T boundary impact glass with melt rock from the Chicxulub and Manson impact structures. Nature 364: 325-327

Bohor BF, Foord EE, Modreski PJ, Triplehorn DM (1984) Mineralogical evidence for an impact event at the Cretaceous/Tertiary boundary. Science 224: 867-869

Bohor BF, Modreski PJ, Foord EE (1987) Shocked quartz in the Cretaceous/Tertiary boundary clays: Evidence for global distribution. Science 236: 705-708

Bohor BF, Betterton WJ, Krogh TE (1993) Impact-shocked zircons: discovery of shock-induced textures reflecting increasing degrees of shock metamorphism. Earth and Planetary Science Letters 119: 419-424

Bottomley RJ, Grieve RAF, York D, Masaitis V (1997) The age of the Popigai impact event and its relations to events at the Eocene/Oligocene boundary. Nature 388: 365-368

Bowring SA, Erwin DH, Jin YG, Martin MW, Davidek K, Wang W (1998) U/Pb zircon geochronology and tempo of the end-Permian mass extinction. Science 280: 1039-1045

Buffetaut E, Koeberl C (eds) (2001) Geological and Biological Effects of Impact Events. Impact Studies, vol. 1. Springer Verlag, Heidelberg, 312 pp

Chapman CR, Morrison D (1994) Impacts on the earth by asteroids and comets: Assessing the hazard. Nature 367: 33-40

Claeys P, Casier JG (1994) Microtektite-like impact glass associated with the Frasnian-Famennia boundary mass extinction. Earth and Planetary Science Letters 122: 303-315

Claeys P, Casier JG, Margolis SV (1992) Microtektites and mass extinctions: Evidence for a late Devonian asteroid impact. Science 257: 1102-1104

Claeys P, Kyte FT, Herbosch A, Casier JG (1996) Geochemistry of the Frasnian-Famennian boundary in Belgium: Mass extinction, anoxic oceans and microtektite layer, but not much iridium? In: Ryder G, Fastovsky D, Gartner S (eds) New Developments Regarding the KT Event and Other Catastrophes in Earth History. Geological Society of America, Special Paper 307, pp 491-504

Claeys P, Smit J, Montanari A, Alvarez W (1998) The Chicxulub impact crater and the Cretaceous-Tertiary boundary in the Gulf of Mexico region. Bulletin de la Societé géologique France 169: 20-27

Colodner DC, Boyle EA, Edmond JM, Thomson J (1992) Post-depositional mobility of platinum, iridium and rhenium in marine sediments. Nature 358: 402-404

Deutsch A, Schärer U (1994) Dating terrestrial impact events. Meteoritics 29: 301-322

Debrabant P, Fourcade E, Chamley H, Rocchia R, Robin E, Bellier JP, Gardin S, Thiébault F (1999) Les argiles de la transition Crétacé-Tertiaire au Guatemala, témoins d'un impact d'astéroide. Bulletin de la Société géologique de France 170: 643-660

Dypvik H, Gudlaugson ST, Tsikalas F, Attrep M Jr, Ferrell RE Jr, Krinsley DH, Mork A, Faleide JI, Nagy J (1996) Mjolnir structure: An impact crater in the Barents Sea. Geology 24: 779-782

Elliott WC (1993) Origin of the Mg-smectite at the Cretaceous/Tertiary (K/T) boundary at Stevns Klint, Denmark. Clays and Clay Minerals 41: 442-452

Farley KA, Mukhopadhyay S (2001) An extraterrestrial impact at the Permian-Triassic boundary? Science 293: 2343a

Farley KA, Montanari A, Shoemaker EM, Shoemaker CS (1998) Geochemical evidence for a comet shower in the late Eocene. Science 280: 1250-1253

French BM (1998) Traces of catastrophe: A handbook of shock-metamorphic effects in terrestrial meteorite impact structures. LPI Contribution 954, Lunar and Planetary Institute, Houston, 120 pp

French BM, Short NM (eds) (1968) Shock metamorphism of natural materials. Mono Book Corp, Baltimore, 644 pp

Gilmour I (1998) Geochemistry of carbon in terrestrial impact processes. In: Grady MM, Hutchison R, McCall GJH, Rothery DA (eds) Meteorites: Flux with Time and Impact Effects. Geological Society of London, Special Publication 140, pp 205-216

Gilmour I, Koeberl C (eds) (2000) Impacts and the Early Earth. Lecture Notes in Earth Sciences, Vol. 91, Springer Verlag, Heidelberg, 445 pp

Gilmour I, Wolbach WS, Anders E (1990) Early environmental effects of the terminal Cretaceous impact. In: Sharpton VL, Ward PD (eds) Global Catastrophes in Earth History. Geological Society of America, Special Paper 247, pp 383-390

Glass BP (1967) Microtektites in deep sea sediments. Nature 214: 372-374

Glass BP (1972) Bottle green microtektites. Journal of Geophysical Research 77: 7057-7064

Glass BP (1989) North American tektite debris and impact ejecta from DSDP Site 612. Meteoritics 24: 209-218

Glass BP, Barlow RA (1979) Mineral inclusions in Muong Nong-type indochinites: Implications concerning parent material and process of formation. Meteoritics 14: 55-67

Glass BP, Pizzuto JE (1994) Geographic variation in Australasian microtektite concentrations: Implications concerning the location and size of the source crater. Journal of Geophysical Research 99: 19075-19081

Glass BP, Wu J (1993) Coesite and shocked quartz discovered in the Australasian and North American microtektite layers. Geology 21: 435-438

Goltrant O, Cordier P, Doukhan JC (1991) Planar deformation features in shocked quartz: A transmission electron microscopy investigation. Earth and Planetary Science Letters 106: 103-115

Gostin VA, Haines PW, Jenkins RJE, Compston W, Williams IS (1986) Impact ejecta horizon within late Precambrian shales, Adelaide Geosyncline, south Australia. Science 233: 198-200

Gostin VA, Keays RR, Wallace MW (1989) Iridium anomaly from the Acraman impact ejecta horizon: Impacts can produce sedimentary iridium peaks. Nature 340: 542-544

Grieve RAF, Pilkington M (1996) The signature of terrestrial impacts. AGSO Journal Australian Geology and Geophysics 16: 399-420

Grieve RAF, Langenhorst F, Stöffler D (1996) Shock metamorphism in nature and experiment: II. Significance in geoscience. Meteoritics and Planetary Science 31: 6-35

Gucsik A, Koeberl C, Brandstätter F, Reimold WU, Libowitzky E (2002) Cathodoluminescence, electron microscopy, and Raman spectroscopy of experimentally shock-metamorphosed zircon. Earth and Planetary Science Letters 202: 495-509

Henkel H, Reimold WU, Koeberl C (2002) Magnetic and gravity model of the Morokweng impact structure, South Africa. Journal of Applied Geophysics 49: 129-147

Hildebrand AR, Penfield GT, Kring DA, Pilkington M, Carmargo ZA, Jacobsen SB, Boynton WV (1991) Chicxulub crater: A possible Cretaeceous-Tertiary boundary impact crater on the Yucatan Peninsula, Mexico. Geology 19: 867-871

Hodych JP, Dunning GR (1992) Did the Manicouagan impact trigger end-of-Triassic mass extinction? Geology 20: 51-54

Holser WT, Schönlaub HP, Attrep M Jr, Boeckelmann K, Klein P, Magaritz M, Orth CJ, Fenninger A, Jenny C, Kralik M, Mauritsch H, Pak E, Schramm JM, Stattegger K, Schmöller R (1989) A unique geochemical record at the Permian-Triassic boundary. Nature 337: 39-44

Hough RM, Gilmour I, Pillinger CT, Langenhorst F, Montanari A (1997) Diamonds from the iridium-rich K-T boundary layer at Arroyo el Mimbral, Tamaulipas, Mexico. Geology 25: 1019-1022

Huffman AR, Reimold WU (1996) Experimental constraints on shock-induced microstructures in naturally deformed silicates. Tectonophysics 256: 165-217

Izett GA, Dalrymple GB, Snee LW (1991) ^{40}Ar / ^{39}Ar Age of Cretaceous-Tertiary boundary Tektites from Haiti. Science 252: 1539-1542

Izett GA, Cobban WA, Obradovich JD, Kunk MJ (1993) The Manson impact structure: ^{40}Ar-^{39}Ar age and its distal impact ejecta in the Pierre shale in southeastern South Dakota. Science 262: 729-732

Izett GA, Cobban WA, Dalrymple GB, Obradovich JD (1998) ^{40}Ar/^{39}Ar age of the Manson impact structure, Iowa, and correlative impact ejecta in the Crow Creek Member of the Pierre Shale (Upper Cretaceous), South Dakota and Nebraska. Geological Society of America Bulletin 110: 361-376

Kaiho K, Kajiwara Y, Nakano T, Miura Y, Kawahata H, Taziki K, Ueshima M, Chen Z, Shi GR (2001) End-Permian catastrophe by bolide impact: Evidence of a gigantic release of sulfur from the mantle. Geology 29: 815-818

Kastner M, Asaro F, Michel HV, Alvarez W, Alvarez LW (1984) The precursor of the Cretaceous-Tertiary boundary clays at Stevns Klint and DSDP Hole 465A. Science 226: 137-143

Kelley SB, Gurov E (2002) Boltysh, another end-Cretaceous impact. Meteoritics and Planetary Science 37: 1031-1043

Klaus A, Norris RD, Kroon D, Smit J (2000) Impact-induced mass wasting at the K-T boundary: Blake Nose, western North Atlantic. Geology 28: 319-322

Klaver GT, van Kempen TMG, Bianchi FR, van der Gaast S (1987) Green spherules as indicators of the Cretaceous/Tertiary boundary in Deep Sea Drilling Project Hole 603B. Initial Reports of the Deep Sea Drilling Project 93: 1039-1056

Koeberl C (1986) Geochemistry of tektites and impact glasses. Annual Reviews of Earth and Planetary Science 14: 323-350

Koeberl C (1992a) Geochemistry and origin of Muong Nong-type tektites. Geochimica et Cosmochimica Acta 56: 1033-1064

Koeberl C (1992b) Water content of glasses from the K/T boundary, Haiti: Indicative of impact origin. Geochimica et Cosmochimica Acta 56: 4329-4332

Koeberl C (1994) Tektite origin by hypervelocity asteroidal or cometary impact: Target rocks, source craters, and mechanisms. In: Dressler BO, Grieve RAF, Sharpton VL (eds) Large meteorite impacts and planetary evolution. Geological Society of America, Special Paper 293, pp 133-152

Koeberl C (1996) Chicxulub - The K-T boundary impact crater: A review of the evidence, and an introduction to impact crater studies. Abhandlungen der Geologischen Bundesanstalt (Wien) 53: 23-50

Koeberl C (1998) Identification of meteoritical components in impactites. In: Grady MM, Hutchison R, McCall GJH, Rothery DA (eds) Meteorites: Flux with Time and Impact Effects. Geological Society of London, Special Publication 140, pp 133-152

Koeberl C (2001) The sedimentary record of impact events. In: Peucker-Ehrenbrink B, Schmitz B (eds) Accretion of Extraterrestrial Matter throughout Earth's History. Kluwer Academic/Plenum Publishers, pp 333-378

Koeberl C (2002) Mineralogical and geochemical aspects of impact craters. Mineralogical Magazine 66: 745-768

Koeberl C, MacLeod K (eds) (2002) Catastrophic Events and Mass Extinctions: Impacts and Beyond. Geological Society of America, Special Paper 356, 746 pp

Koeberl C, Reimold WU (1995) Early Archaean spherule beds in the Barberton Mountain Land, South Africa: no evidence for impact origin. Precambrian Research 74: 1-33

Koeberl C, Shirey SB (1997) Re-Os isotope systematics as a diagnostic tool for the study of impact craters and ejecta. Palaeogeography, Palaeoclimatology, Palaeoecology 132: 25-46

Koeberl C, Sigurdsson H (1992) Geochemistry of impact glasses from the K/T boundary in Haiti: Relation to smectites, and a new type of glass. Geochimica et Cosmochimica Acta 56: 2113-2129

Koeberl C, Sharpton VL, Schuraytz BC, Shirey SB, Blum JD, Marin LE (1994) Evidence for a meteoritic component in impact melt rock from the Chicxulub structure. Geochimica et Cosmochimica Acta 58: 1679-1684

Koeberl C, Poag CW, Reimold WU, Brandt D (1996) Impact origin of Chesapeake Bay structure and the source of North American tektites. Science 271: 1263-1266

Koeberl C, Armstrong RA, Reimold WU (1997) Morokweng, South Africa: A large impact structure of Jurassic-Cretaceous boundary age. Geology 25: 731-734

Koeberl C, Gilmour I, Reimold WU, Claeys P, Ivanov BA (2002) Comment on "End-Permian catastrophe by bolide impact: Evidence of a gigantic release of sulfur from the mantle" by Kaiho et al. Geology 30: 855-856

Kring DA, Boynton WV (1991) Altered spherules of impact melt and associated relic glass from the K/T boundary sediments in Haiti. Geochimica et Cosmochimica Acta 55: 1737-1742

Kunk MJ, Izett GA, Haugerud RA, Sutter JF (1989) [40]Ar-[39]Ar dating of the Manson impact structure: A Cretaceous-Tertiary boundary crater candidate. Science 244: 1565-1568

Langenhorst F (1996) Characteristics of shocked quartz in late Eocene impact ejecta from Massignano (Ancona, Italy): clues to shock conditions and source crater. Geology 24: 487-490

Leroux H, Reimold WU, Koeberl C, Hornemann U, Doukhan JC (1999) Experimental shock deformation in zircon: A transmission electron microscopic study. Earth and Planetary Science Letters 169: 291-301

Lindinger M (1988) The Cretaceous-Tertiary boundaries of El Kef and Caravaca: sedimentological, geochemical and clay mineralogical aspects. PhD. thesis, Swiss Federal Institute of Technology (ETH), Zürich, Switzerland 253 pp

Lowe DR, Byerly GR (1986) Early Archean silicate spherules of probable impact origin, South Africa and Western Australia. Geology 14: 83-86

Mader D, Montanari A, Gattacceca J, Koeberl C, Handler R, Coccioni R (2001) ^{40}Ar/^{39}Ar dating of a Langhian biotite-rich clay layer in the pelagic sequence of the Cònero Riviera, Ancona, Italy. Earth and Planetary Science Letters 194: 111-126

Martínez-Ruiz F, Ortega-Huertas M, Palomo I, Acquafredda P (1997) Quench textures in altered spherules from the Cretaceous-Tertiary boundary layer at Agost and Caravaca, SE Spain. Sedimentary Geology 113: 137-147

Martínez-Ruiz F, Ortega-Huertas M, Palomo I, Smit J (2001) K-T boundary spherules from Blake Nose (ODP Leg 171B) as a record of the Chicxulub ejecta deposits. Geological Society of London Special Publications 183: 149-161

Martínez-Ruiz F, Ortega-Huertas M, Palomo I, Smit J (2002) Cretaceous-Tertiary boundary at Blake Nose (Ocean Drilling Program Leg 171B): A record of the Chicxulub impact ejecta. In: Koeberl C, MacLeod KG (eds) Catatrophic Events and Mass Extinctions: Impacts and Beyond. Geological Society of America Special Paper 356: 189-199

Masaitis VL (1994) Impactites from Popigai crater. In: Dressler BO, Grieve RAF, Sharpton VL (eds) Large Meteorite impacts and planetary evolution. Geological Society of America, Special Paper 293, pp 153-162

Mata P, Peacor DR, Soria AR, Liesa C, Meléndez A (2001) The spherule facies at the Cretaceous-Tertiary (K/T) boundary in the El Tecolote (Northeastern Mexico): A TEM study [abs.]. In: Martinez-Ruiz F, Ortega-Huertas M, Palomo I (eds) Abstracts, ESF Workshop on Impact Markers in the Stratigraphic Record. Universidad de Granada, pp 77-78

McGetchin TR, Settle M, Head JW (1973) Radial thickness variation in impact crater ejecta: Implications for lunar basin deposits. Earth and Planetary Science Letters 20: 226-236

Melosh HJ (1989) Impact cratering - A geologic process. Oxford University Press, New York, 245 pp

Montanari A, Koeberl C (2000) Impact Stratigraphy - The Italian Record. Springer Verlag, Heidelberg, 364 pp

Montanari A, Hay RL, Alvarez W, Asaro F, Michel HV, Alvarez LW (1983) Spheroids at the Cretaceous-Tertiary boundary are altered droplets of basaltic composition. Geology 11: 668-671

Oberbeck VR (1975) The role of ballistic erosion and sedimentation in lunar stratigraphy. Reviews of Geophysics and Space Physics 13: 337-362

Olsen PE, Kent DV, Sues H-D, Koeberl C, Huber H, Montanari A, Rainforth EC, Fowell SJ, Szajna MJ, Hartline BW (2002) Ascent of dinosaurs linked to Ir anomaly at Triassic-Jurassic boundary. Science 296: 1305-1307

Olsson RK, Miller KG, Browning JV, Habib D, Sugarman PJ (1997) Ejecta layer at the Cretaceous-Tertiary boundary, Bass River, New Jersey (Ocean Drilling Program Leg 174AX). Geology 25: 759-762

Ortega-Huertas M, Palomo I, Martínez-Ruiz F, González I (1998) Geological factors controlling clay mineral patterns across the Cretaceous-Tertiary boundary in Mediterranean and Atlantic sections. Clay Minerals 33: 483-500

Ortega-Huertas M, Martínez-Ruiz F, Palomo I, Chamley H (2002) Review of the mineralogy of the Cretaceous-Tertiary boundary clay: evidence supporting a major extraterrestrial catastrophic event. Clay Minerals 37: 395-411

Pierazzo E, Melosh JH (1999) Hydrocode modeling of Chicxulub as an oblique impact event. Earth and Planetary Science Letters 165: 163-176

Poag CW (1997) The Chesapeake Bay bolide impact: A convulsive event in Atlantic Coastal Plain evolution. Sedimentary Geology 108: 45-90

Pollastro RM, Pillmore CL (1987) Mineralogy and petrology of the Cretaceous-Tertiary boundary clay bed and adjacent clay-rich rocks, Raton Basin, New Mexico and Colorado. Journal of Sedimentary Petrology 57: 456-466

Pollastro RM, Bohor BF (1993) Origin and clay-mineral genesis of the Cretaceous/Tertiary boundary unit, western interior of North America. Clays and Clay Minerals 41: 7-25

Pope KO, Ocampo AC, Kinsland GL, Smith R (1996) Surface expression of the Chicxulub crater. Geology 24: 527-530

Pope KO, Ocampo AC, Fischer AG, Alvarez W, Fouke BW, Webster CL, Vega FJ. Smit J. Fritsche AE, Claeys P (1999) Chicxulub impact ejecta from Albion Island, Belize. Earth and Planetary Science Letters 170: 351-364

Rampino MR, Haggety BM (1996) Impact crises and mass extinctions: A working hypothesis. In: Ryder G, Fastovsky D, Gartner S (eds) New Developments Regarding the KT Event and Other Catastrophes in Earth History. Geological Society of America, Special Paper 307, pp 11-30

Rampino MR, Reynolds RC (1983) Clay mineralogy of the Cretaceous-Tertiary Boundary Clay. Science 219: 495-498

Reimold WU (1995) Pseudotachylite in impact structures - generation by friction melting and shock brecciation?: A review and discussion. Earth-Science Reviews 39: 247-265

Reimold WU (1998) Exogenic and endogenic breccias: a discussion of major problematics. Earth-Science Reviews 43: 25-47

Retallack GJ, Seyedolali A, Krull ES, Holser WT, Ambers CP, Kyte FT (1998) Search for evidence of impact at the Permian-Triassic boundary in Antarctica and Australia. Geology 26: 979-982

Ryder G, Fastovsky D, Gartner S (eds) (1996) The Cretaceous-Tertiary Event and other Catastrophes in Earth History. Geological Society of America, Special Paper 307, 576 pp

Schmitz B (1985) Metal precipitation in the Cretaceous/Tertiary boundary clay at Stevns Klint, Denmark. Geochimica et Cosmochimica Acta 49: 2361-2370

Sharpton VL, Ward PD (eds) (1990) Global Catastrophes in Earth History. Geological Society of America, Special Paper 247, 631 pp

Sharpton VL, Dalrymple GB, Marin LE, Ryder G, Schuraytz BC, Urrutia-Fucugauchi J (1992) (1992) New links between the Chicxulub impact structure and the Cretaceous/Tertiary boundary. Nature 359: 819-821

Sharpton VL, Burke K, Camargo-Zanoguera A, Hall SA, Lee S, Marín LE, Suárez-Reynoso G, Quezada-Muñeton JM, Spudis PD, Urrutia-Fucugauchi J (1993) Chicxulub multiring impact basin: Size and other characteristics derived from gravity analysis. Science 261: 1564-1567

Shukolyukov A, Lugmair GW (1998) Isotopic evidence for the Cretaceous-Tertiary impactor and its type. Science 282: 927-929

Shukolyukov A, Kyte FT, Lugmair GW, Lowe DR, Byerly GR (2000) The oldest impact deposits on earth – first confirmation of an extraterrestrial component. In: Gilmour I, Koeberl C (eds) Impacts and the Early Earth. Lecture Notes in Earth Sciences 91. Springer, Heidelberg-Berlin, pp

Sigurdsson H, D'Hondt S, Arthur MA, Bralower TJ, Zachos JC, van Fossen M, Channell ET (1991) Glass from the Cretaceous/Tertiary boundary in Haiti. Nature 349: 482-487

Silver LT, Schultz PH (eds) (1982) Geological Implications of Impacts of Large Asteroids and Comets on the Earth. Geological Society of America, Special Paper 190, 528 pp

Simonson BM (1992) Geological evidence for a strewn field of impact spherules in the early Precambrian Hamersley Basin of Western Australia. Geological Society of America Bulletin 104: 829-839

Simonson BM, Harnik P (2000) Have distal impact ejecta changed through geologic time? Geology 28: 975-978

Simonson BM, Davies D, Wallace M, Reeves S, Hassler SW (1998) Iridium anomaly but no shocked quartz from Late Archean microkrystite layer: Oceanic impact ejecta? Geology 26: 195-198

Simonson BM, Koeberl C, McDonald I, Reimold WU (2000) Geochemical evidence for an impact origin for a late Archean spherule layer, Transvaal Supergroup, South Africa. Geology 28: 1103-1106

Smit J (1999) The global stratigraphy of the Cretaceous-Tertiary boundary impact ejecta. Annual Reviews of Earth and Planetary Science 27: 75-113

Smit J, Hertogen J (1980) An extraterrestrial event at the Cretaceous-Tertiary boundary. Nature 285: 198-200

Soria AR, Liesa C, Mata MP, Arz JA, Alegret L, Arenillas I, Meléndez A (2001) Slumping and a sandbar deposit at the Cretaceous-Tertiary boundary in the El Tecolote section (Northeastern Mexico): An impact-induced sediment gravity flow. Geology 29: 231-234

Spray JG, Kelley SP, Rowley DB (1998) Evidence for a late Triassic multiple impact event on Earth. Nature 392: 171-173

Stöffler D (1972) Deformation and transformation of rock-forming minerals by natural and experimental shock processes: 1. Behaviour of minerals under shock compression. Fortschritte der Mineralogie 49: 50-113

Stöffler D, Grieve RAF (1994a) Classification and nomenclature of impact metamorphic rocks: A proposal to the IUGS subcommission on the systematics of metamorphic rocks [abs.]. Lunar and Planetary Science 25: 1347-1348

Stöffler D, Grieve RAF (1994b) Classification and nomenclature of impact metamorphic rocks: A proposal to the IUGS subcommission on the systematics of metamorphic rocks. In: Montanari A, Smit J (eds) Post-Östersund Newsletter, European Science Foundation (ESF) Scientific Network on Impact Cratering and Evolution of Planet Earth, Strasbourg, pp 9-15

Stöffler D, Langenhorst F (1994) Shock metamorphism of quartz in nature and experiment: I. Basic observations and theory. Meteoritics 29: 155-181

Stöffler D, Gault DE, Wedekind J, Polkowski G (1975) Experimental hypervelocity impact into quartz sand: Distribution and shock metamorphism of ejecta. Journal of Geophysical Research 80: 4062-4077

Swisher CC, Grajales-Nishimura JM, Montanari A, Margolis SV, Claeys P, Alvarez W, Renne P, Cedillo-Pardo E, Maurrasse FJMR, Curtis GH, Smit J, McWilliams MO (1992) Coeval ^{40}Ar/^{39}Ar ages of 65.0 million years ago from Chicxulub crater melt rock and Cretaceous-Tertiary boundary tektites. Science 257: 954-958

Vishnevsky S, Montanari A (1999) Popigai impact structure (Arctic Siberia, Russia): Geology, petrology, geochemistry, and geochronology of glass-bearing impactites. In: Dressler BO, Sharpton VL (eds) Large meteorite impacts and planetary evolution II. Geological Society of America, Special Paper 339: 19-60

Walkden G, Parker J, Kelley S (2002) A late Triassic impact ejecta layer in southwestern Britain. Science 298: 2185-2188

Wallace MW, Gostin VA, Keays RR (1990) Acraman impact ejecta and host shales: Evidence for low-temperature mobilization of iridium and other platinoids. Geology 18: 132-135

Wang K (1992) Glassy microspherules (microtektites) from an Upper Devonian limestone. Science 256: 1547-1550

Warme JE, Kuehner H-C (1998) Anatomy of an anomaly: The Devonian catastrophic Alamo impact breccia of southern Nevada. International Geology Review 40: 189-216

Warme JE, Sandberg CA (1996) Alamo megabreccia: Record of a late Devonian impact in southern Nevada. GSA Today 6(1): 1-7

Wei W (1995) How many impact-generated microspherule layers in the upper Eocene? Palaeogeography Palaeoclimatology Palaeoecology 114: 101-110

Williams GE (1986) The Acraman impact structure: Source of ejecta in late Precambrian shales, South Australia. Science 233: 200-203

Williams GE (1994) Acraman, South Australia: Australia's largest meteorite impact structure. Proceedings of the Royal Society of Victoria 106: 105-127

Witzke BJ, Hammond RH, Anderson RR (1996) Deposition of the Crow Creek Member, Campanian, South Dakota and Nebraska. In: Koeberl C, Anderson RR (eds) The Manson Impact Structure, Iowa: Anatomy of an Impact Crater. Geological Society of America, Special Paper 302, pp 433-456

Wolbach WS, Lewis RS, Anders E (1985) Cretaceous extinctions: Evidence for wildfires and search for meteoritic material. Science 230: 167-170

Application of stratigraphic nomenclature to terrestrial impact-derived and impact-related materials

David T. King, Jr.[1] and Lucille W. Petruny[2]

[1]Department of Geology, Auburn University, Auburn, Alabama 36849-5305, USA. (kingdat@ auburn.edu)
[2]Astra-Terra Research, Auburn, AL 36831-3323, USA, and Department of Curriculum and Teaching, Auburn University, Auburn, Alabama 36849, USA. (lpetruny@ att.net)

Abstract. In this paper, we briefly recount the history of studies on the nomenclature of impact-derived and impact-related materials before reviewing and discussing in detail how extant stratigraphic nomenclature of the North American Commission on Stratigraphic Nomenclature and the International Subcommission on Stratigraphic Classification may apply to these materials. In the course of the review and discussion, we cite relevant examples and make some recommendations on future usage of stratigraphic nomenclature as applied to impact-derived and impact-related materials.

1
Introduction

The two most influential sources on general stratigraphic nomenclature, principles, and procedure are the respective treatises of the North American Commission on Stratigraphic Nomenclature (NACSN 1983) and the International Union of Geological Sciences' International Subcommission on Stratigraphic Classification (or IUGS-ISSC; Salvador 1994). Perhaps not surprisingly, both of these works fail to specifically address stratigraphic classification of impact-derived and impact-related Earth materials. In fact, it is clear from careful analysis of both treatises that impact-derived and impact-related Earth materials were not considered at all in preparing these classificatory schemes.

With slight broadening of some concepts, extant NACSN and IUGS-ISSC systematics could be expanded to potentially accommodate future needs of the impact-geology and impact-stratigraphy communities for stratigraphic nomenclature of impact-derived and impact-related materials. However, some new terminology and concepts may be required. In this paper, we examine how extant stratigraphic nomenclature of the NACSN and IUGS-ISSC could be used to classify and properly name various impact-derived and impact-related materials, depending upon the goals of the investigations. Further, we make some suggestions about changes in extant stratigraphic nomenclature, which could potentially improve scientific communication.

Our goal for this paper is to invigorate discussion on the topic of how stratigraphic nomenclature can accommodate impact-derived and impact-related Earth materials. As demonstrated by the long-term positive effects of wide acceptance of established NACSN and IUGS-ISSC stratigraphic nomenclature, scientific progress can be enhanced by the reduction in misunderstanding that accompanies usage of standardized terminology.

Even though the North American Stratigraphic Code (NACSN 1983) was written "by and for North American Earth scientists," it has broad applicability and is used globally as an aid in understanding stratigraphic classification. Much like the Code, the International Stratigraphic Guide, 2nd edition (Salvador 1994) was written under the auspices of an international working group whose goal was to promote basic stratigraphic principles and standard terminology and classificatory procedure "across national boundaries." In our work, we find that reference to both works, despite their slight differences, to be most useful in understanding current views on stratigraphic classification.

2
Review of Impact-Related Petrologic Classification

Raikhlin et al. (1980) briefly addressed "problems of classification in rocks of terrestrial impact craters" noting that such a classification at that time had "only been slightly touched upon in literature." Their abstract touches upon both stratigraphic and petrographic classification schemes. Regarding stratigraphic relations, they note "three main rock groups," including the broad categories called "brecciated rocks and breccias formed in-situ within a crater" (i.e., authigenous (*sic*) breccias), "lithic breccias filling a crater interior" (i.e., allogenic (*sic*) breccias), and "impactites within a crater (including impact melts) and impactites within ejecta blankets." In their paper, Raikhlin et al. (1980) proposed no specific stratigraphic classification scheme. Similarly, Laznicka (1988) provided a diagrammatic inventory of "coarse fragmentites in meteoric (*sic*) impact craters" within his lengthy treatise on breccias. His inventory of 16 categories addressed only petrologic nomenclature of coarse impact-generated deposits, not their stratigraphic aspects.

Deutsch and Langenhorst (1994) addressed the issue of "geological formations in and around impact structures" and recognized four essentially different "formations," including "crater basement," "allochthonous breccia deposits," "coherent impact melt layers," and "distant ejecta." Crater basement, in their view, included "monomict brecciated megablocks," "parauthochthonous and authochthonous (authigenic) breccias that are either polymict or monomict," and "fragmental, pseudotachylitic and impact-melt" dikes. In their view, allochthonous breccia deposits included crater-filling breccia lenses and proximal continuous ejecta. Coherent impact melt layers included all intra-crater melt rocks containing any significant amount of impact-melted material (i.e., clast-free melt rock to "polymict breccia with a melt matrix that contains shocked and unshocked rock and mineral fragments"). Finally, distant ejecta included impact-derived mineral and rock fragments, impact glass (including spherules, microtektites, tektites, and fragmental glass), microkrystites and exotic spinels, and sediment enriched in siderophile elements such as platinum-group elements and other exotic components like extraterrestrial fullerines and extraterrestrial amino acids. In their paper, Deutsch and Langenhorst (1994) proposed no specific stratigraphic classification scheme.

Reimold (1995, 1998) reviewed nomenclature of impact-related materials, particularly terminology related to breccias and pseudotachylite dikes. His discussion relates how easily some generalized terminology can cause confusion within and between workers in the tectonic and impact communities. He also emphasized how important it is to understand timing of emplacement of breccias, pseudotachylitic breccia dikes, and associated features when relating them to impact events and structures. Reimold's papers strongly argued that *objective criteria* should be applied to identification and mapping of impact-derived and impact-related materials. He proposed no new terminology and suggested no new classification schemes.

Currently accepted petrologic classification and nomenclature of impact-derived and impact-related materials has been established provisionally by a study group of the International Union of Geological Sciences' Subcommission on Systematics of Metamorphic Rocks group. Stöffler and Grieve (1994a; 1994b) have presented this classification, with a pertinent glossary, in two reports. The primary focus of these reports was on impact petrology. Thus, a holistic *stratigraphic* perspective was not taken into account. Recently, French (1999) and Montanari and Koeberl (2000) have reviewed current views on petrologic classification of impactites and have supported use of the terminology advanced by Stöffler and Grieve (1994a; 1994b).

3
Application of Current Schemes of Stratigraphic Nomenclature

3.1
Overview

The North American Stratigraphic Code (NACSN 1983) is a lengthy description and discussion of "recommended procedures for classifying and naming stratigraphic and related units." The North American Stratigraphic Code addresses sedimentary and volcanic strata, "the events recorded therein," and "all Earth materials, not solely strata." The emphasis of the Code is upon *formal* stratigraphic units, which are those units "named in accordance with an established scheme of classification" and whose "stability of nomenclature" deserves protection by recognized classification procedures (see Appendix). In contrast, *informal* stratigraphic units, "whose unit terms are ordinary nouns," are innovative concepts that generally arise for "economic or scientific reasons" (NACSN 1983).

The North American Stratigraphic Code's formal "material categories *based on content or physical limits*," which are applicable potentially in this instance, include (1) lithostratigraphic units, (2) lithodemic units, and (3) allostratigraphic units (NACSN 1983; Table 1). In addition, the Code's formal "material categories *used to define temporal spans*," which are applicable potentially in this instance, are limited to the category of chronostratigraphic units. The Code also addresses eight other types of material and non-material units, but these other types (e.g., biostratigraphic and pedostratigraphic units) have no obvious application to impact-derived and impact-related materials.

Informal lithostratigraphic units are not discussed with sufficient detail in the North American Stratigraphic Code to make many specific comments on how they may apply in this instance. However, it is worth noting that the Code specifically assigns *all types of facies* to the category "informal" (NACSN 1983). We think it obvious that a lithologic facies, as an informal unit of impact-derived and impact-related materials, would potentially be quite useful in some situations, as noted later on.

Both the Code and Guide have specific requirements for the definition and naming of formal stratigraphic units. We will review these in the Appendix at the end of this paper rather than present all details here.

3.2
Formal Lithostratigraphic Units, their Hierarchy, and Applicability

Lithostratigraphic units are defined as bodies of "sedimentary, extrusive igneous, metasedimentary, or metavolcanic strata, which (are) distinguished and delimited on the basis of lithic characteristics and stratigraphic position." Further, the Code notes, "a lithostratigraphic unit generally conforms to the Law of Superposition and commonly is stratified and tabular in form" (Table 1). Lithostratigraphic units are the "basic units of general geologic work and serve as the foun-dation for delineating strata, local and regional structure, economic resources, and geologic history in regions of stratified rocks" (NACSN 1983). Lithostratigraphic units are intended to be practical, objective units, as noted in Code remarks citing "independence from inferred geologic history" and "independence from time concepts." On the first point, the Code notes that "inferred geologic history … (has) no place in the definition of a lithostratigraphic unit." On the latter, the Code says "inferred time spans, however measured, play no part in differentiating or determining the boundaries of any lithostratigraphic unit."

Lithostratigraphic units have a familiar hierarchical list of rank terms (NACSN 1983; Table 2). From highest to lowest rank, these terms are: *supergroup, group, formation, member* (or *lens, lentil,* or *tongue*), and *bed* (or *flow*). The Code says that such units (and also other categories of units like lithodemic and allostratigraphic, both discussed below) will have a geographic term (referring to a locale where the unit is well developed) and an appropriate rank term. The formality of a unit is conveyed in writing by capitalization of the rank term (e.g., Exmore Breccia (Poag et al. 1994) or Crow Creek Member of the Pierre Shale (Izett et al. 1998)).

Formation is the fundamental lithostratigraphic rank term (NACSN 1983; Table 2). The formation is defined in the Code as "a body of rock identified by its lithic characteristics and stratigraphic position; it is prevailingly but not necessarily tabular and is mappable at the Earth's surface or traceable in the subsurface." In lieu of the rank term formation, a "simple lithic term" may substitute, if appropriate, for example breccia for formation, but non-common or compound lithic terms *cannot* be substituted (NACSN 1983).

Member is "the formal lithostratigraphic unit next in rank below a formation and is always a part of some formation." Members are recognized as "a named entity within a formation because it possesses characteristics distinguishing it from adjacent parts of the formation." A member may "extend from one formation to another." A *lens* (or, the equivalent, *lentil*) is an alternative name for a "geographically restricted member that terminates on all sides within a formation." A *tongue* is a "wedging member that extends outward beyond a formation or wedges ... out with another formation" (NACSN 1983). Members, lenses, lentils, and tongues may have names composed of a proper geographic term plus a lithic term plus a rank term (e.g., Alamo Breccia Member (Sandberg et al. 1997)) or may not have a lithic term (e.g., "Alamo Member").

Bed(s) is (are) the smallest formal lithostratigraphic unit(s) and their usage "should be limited to certain distinctive bed(s) whose recognition is (are) particularly useful." A *flow* is the smallest lithostratigraphic unit of volcanic rocks (NACSN 1983).

A *group* is the "lithostratigraphic unit next higher in rank to formation; a group may consist entirely of named formations, or alternatively, need not be composed entirely of named formations." Groups, in the view of the Code, are defined "to express natural relationships of associated formations (NACSN 1983). A *supergroup* is a "formal assemblage of related or superposed groups, or of groups and formations" (NACSN 1983).

The International Stratigraphic Guide, 2nd edition (Salvador 1994), which is a rather comprehensive guide to stratigraphic nomenclature much like the North American Stratigraphic Code, also recognizes the basic hierarchy of the well-established formal lithostratigraphic units described above. However, the Guide varies from the Code in some minor ways. For example, in the view of the Guide, lithostratigraphic units are "bodies of rocks, bedded and unbedded, that are defined and characterized on the basis of their observable lithologic properties." The Guide goes on to say that "lithostratigraphic units are the basic units of geological mapping ... and an important key to geological history even if no ages are available." The Guide's definition would include all classes of rock, including igneous and metamorphic, and, thus, is different from the Code's definition, wherein rocks to which the Law of Superposition does not apply are placed in a different category (i.e., lithodemic units; NACSN 1983). Lastly, the Guide does not require, as does the Code, that lithostratigraphic units be "prevailingly but not necessarily tabular."

Further, the Guide recognizes two different lithostratigraphic units, the "complex" and the "lithostratigraphic horizon" (or "lithohorizon;" Salvador 1994). According to the Guide, the complex is a "lithostratigraphic unit composed of diverse types of any class or classes or rock (sedimentary, igneous, metamorphic) and characterized by irregularly mixed lithology or by highly complicated structural relations to the extent that the original sequence of the component rocks

Table 1. Extant formal stratigraphic classification of the North American Stratigraphic Code (NACSN 1983) and the International Stratigraphic Guide, 2nd edition (Salvador 1994) pertinent to impact-derived and impact-related materials (see text).

I. _**Units to which the Law of Superposition applies and that are generally tabular**_

    A. <u>Units that lack synchronous boundaries and are not isochronous or units whose boundary characteristics respecting synchroneity cannot be determined or are not relevant</u>

          1. _Units that may or may not have distinctive bounding discontinuities_
            a. <u>Mappable units</u>
                • Lithostratigraphic units of the Code and Guide
           b. <u>Non-mappable units (too thin to map, but useful in correlation)</u>
                • Lithostratigraphic horizons (or lithohorizons) of the Guide
        2. _Units that have distinctive bounding discontinuities_
                • Allostratigraphic units of the Code

    B. <u>Units that have synchronous boundaries and are isochronous</u>

          1. _Mappable units_
                • Chronostratigraphic units of the Code and Guide
        2. _Non-mappable units (too thin to map, but useful in correlation)_
                • Chronostratigraphic horizon (or chronohorizon) of the Guide
                • Lithochronozones of the Code

II. _**Units to which the Law of Superposition does not apply and are not generally tabular**_

                • Lithodemic units of the Code
                • Lithostratigraphic "complexes" of the Guide

---

Table 2. Hierarchy of stratigraphic nomenclature among units pertinent to stratigraphic classification and nomenclature of impact-derived and impact-related materials. Approximately equivalent units (for cartographic purposes) are shown on the same line. Fundamental units in each hierarchy are marked with an asterisk (*). Derived from the North American Stratigraphic Code and International Stratigraphic Guide, 2nd edition (NACSN 1983; Salvador 1994), as discussed in the text.

Lithostratigraphic units	Lithodemic units	Allostratigraphic units
Supergroup	*Supersuite*[a]	---
Group	*Suite*[a]	*Allogroup*
Formation *	***Lithodeme*** *	***Alloformation*** *
Member, lens, lentil	---	*Allomember*
Bed, flow	---	---
Lithostratigraphic horizon or lithohorizon	---	---

[a]*Supersuite* and *Suite* may also be named *Complex*, see text.

may be obscured, and the individual rocks or rock sequence cannot be readily mapped." Thus, in a formal name, "complex" would be a valid rank term, according to the Guide (Salvador 1994). A lithostratigraphic horizon (or lithohorizon) is "the surface of lithostratigraphic change, commonly the boundary of a lithostratigraphic unit, or a lithologically distinctive very thin marker bed within a lithostratigraphic unit" (Salvador 1994). A lithostratigraphic horizon (or lithohorizon), although small and usually not mappable, is a formal unit, according to the Guide (Table 2).

Informal lithostratigraphic units are not discussed in much detail within the International Stratigraphic Guide, 2nd edition. However, it is worth noting that the Guide specifically assigns all types of rock bodies "that can be recognized by their lithological properties and to which casual reference is made but for which there is inappropriate basis to justify designation as a formal unit" to the category of "informal" (Salvador 1994). We are of the opinion that if an impact-derived or impact-related unit can be identified on a lithological basis, it likely justifies designation as a formal unit. For this reason, we focus upon nomenclature of formal units in this report.

In our view, formal lithostratigraphic terminology, as in the North American Stratigraphic Code and the International Stratigraphic Guide, 2nd edition, may be applied successfully in those instances where impact-derived and impact-related materials fulfill the following criteria:
(1) the units can be mapped or are "traceable in the subsurface;"

(2) the units are "bodies of rocks, bedded and unbedded, that are defined and characterized on the basis of their observable lithologic properties," and
(3) the units are essentially independent from "inferred geologic history" and "time concepts" (i.e., they are *primarily* objectively defined units).

As noted above, many of the previously suggested terms related to impact-derived and impact-related materials, as presented by Stöffler and Grieve (1994a; 1994b), are basically petrologic terms (e.g., *proximal* and *distal impactites*, *impact breccias and melt rocks*, *impact breccia dikes*, etc.). These terms would be *inappropriate* as lithic rank terms for any formal unit. However, units comprising one or more of these petrologically defined types could be mapped as one or more of the various ranks of lithostratigraphic units described in the Code and Guide, as long as those units are primarily objectively defined.

For example, a typical "prevailingly tabular" proximal impactite (*sensu* Stöffler and Grieve 1994a; 1994b) could be mapped as a formation in most instances. This is the case in the vicinity of Ries crater, Germany, where ejecta of that crater were mapped as a continuous and discontinuous unit termed the Bunte Breccia (formalized by Hüttner 1969). Similarly, at Albion Island in northern Belize, and in adjacent Quintana Roo, México, comparable ejecta, derived from Chicxulub impact structure, México, have been mapped as the Albion Formation (Ocampo et al. 1996).

In southern Nevada, the Alamo Breccia Member (commonly referred to as the "Alamo Breccia;" but formally named Alamo Breccia Member by Sandberg et al. 1997) is a mappable unit in the Upper Devonian section. This unit, which contains impact ejecta and sedimentary structures due to seismic disturbance triggered by impact energy (Warme and Kuehner 1998), is a member within both the Guilmette Formation and the Devil's Gate Limestone (two laterally equivalent Upper Devonian shelfal carbonate units (Sandberg et al. 1997)).

There are many places where distinctive impact-derived or impact-related layers are informally named, but are *in effect* members or beds within host formations. For example, at the Cretaceous-Tertiary boundary, an informal, ejecta-bearing tsunami unit called the "basal Clayton sands" within the Paleocene Clayton Formation of Alabama (Smit et al. 1996) could easily (and appropriately) be mapped as a named member or bed of the overlying Clayton Formation. Similarly, in northeastern México, informally delineated ejecta-bearing tsunami sand deposits at the base of the Paleocene Velasco Formation (Smit et al. 1996) could be mapped as a named member or bed of the overlying Velasco.

Use of lithostratigraphic concepts is appropriate in the examples above, and in similar situations where the goal of stratigraphic nomenclature is to delineate an *objective* lithologic unit for mapping purposes. Stratigraphic studies of impact-derived and impact-related materials that emphasize a unit's chronostratigraphic characteristics (i.e., the precise synchroneity or isochroneity of such a unit) should employ the more appropriate chronostratigraphic terminology (see section 3.5 below).

Crater-filling units, which include monomict and polymict impact breccias, impact melt, melt rock, and melt breccia, suevite breccia, and perhaps other impact-related materials, could comprise a formation (including various members), or perhaps a lithostratigraphic group (including various formations). An example of such a crater-filling unit is the Chicxulub Breccia within the Chicxulub impact structure, México (Hildebrand et al. 1991; Sharpton et al. 1996). Another example is the Exmore Breccia (Poag et al. 1994), which fills part of the Chesapeake Bay impact structure, but also forms a continuous layer outside the crater in the subsurface of southeastern Virginia. Even though the formation concept has been applied in these two noteworthy examples, we think that the diversity and complexity of rocks in other crater-filling units could justify usage of different terminology. Specifically, the International Stratigraphic Guide (2nd edition; Salvador 1994) concept of "complex" (or the Code's concept of a lithodemic unit -- see section 3.3 below) could be also applied.

At a finer scale, distal impactite, or more specifically an impactoclastic air-fall bed (*sensu* Stöffler and Grieve 1994a; 1994b), could be viewed as a lithostratigraphic horizon (or lithohorizon), as described in the Guide (Salvador 1994). However, we caution against such classification because the Guide's lithostratigraphic horizon (or lithohorizon) need *not* be a synchronous or isochronous unit (Salvador 1994). We think that distal impactites and impactoclastic air-fall beds are inherently synchronous and isochronous units. However, if age relations of some distal impactites and impactoclastic air-fall beds were unclear (e.g., due to reworking or tectonic modification), perhaps the terms lithostratigraphic horizon or lithohorizon would be appropriate. Otherwise, we recommend chronostratigraphic terminology for these deposits (see section 3.5 below).

Without reference to the Code, Guide, or any comparable scheme, Stöffler and Grieve (1994a; 1994b) introduced the concept of "impact formation," which they defined as a "geological formation produced by impact; includes various lithological and structural units inside and beneath an impact crater (inner impact formations), the continuous ejecta blanket (outer impact formations) and distal ejecta such as tektites and impactoclastic air fall beds." We think it important to note here that this concept of a "impact formation" is not the same as lithostratigraphic formation as described in the North American Stratigraphic Code and International Stratigraphic Guide, 2nd edition (NACSN 1983; Salvador 1994). In the Stöffler and Grieve (1994a; 1994b) concept of an "impact formation," many potentially disparate units would be included within an interpreted geological framework, which is genetically related to a short-duration impact event. To be equivalent to a formation of the lithostratigraphic hierarchy, any designated unit must be essentially independent from "inferred geologic history" and "time concepts" (NACSN 1983). The "impact formation" is closely tied to "inferred geologic history" and "time concepts," and, thus, is intentionally more genetic in concept than a lithostratigraphic formation. The "impact formation" could be viewed as in informal lithostratigraphic unit, however.

The "impact formation" is more closely akin to the genetic "geologic units" used in planetary mapping schemes (*sensu* Wilhelms, 1990), than any currently defined lithostratigraphic, lithodemic, or chronostratigraphic units (NASCN 1983; Salvador 1994). "Geologic units" of planetary mapping have been defined as "discrete three-dimensional bodies of rock ... formed, relative to those neighboring units, (a) by a discrete process or related processes, and (b) in a discrete time span" (Wilhelms, 1990). Both the "impact formations" and the "geologic units" of planetary mapping are hybrid units, combining aspects of both informal lithostratigraphic units and chronostratigraphic units (see section 3.5 below) in their definitions.

3.3
Formal Lithodemic Units, their Hierarchy, and Applicability

According to the North American Stratigraphic Code (NACSN 1983), a formal *lithodemic unit* (from Greek *demas* meaning living body or frame) is defined as "a body of predominantly intrusive, highly deformed, and/or highly metamorphosed rock, distinguished and delimited on the basis of rock characteristics." In contrast to formal lithostratigraphic units, formal *lithodemic units* "generally do not conform to the Law of Superposition" (Table 1), and contacts of such units may be "sedimentary, extrusive, intrusive, tectonic, or metamorphic" (NACSN 1983). The only similar concept of a rock unit that is discussed in the International Stratigraphic Guide, 2nd edition (Salvador 1994) is the "complex" (see section 3.2 above). (Note: The term "complex" is also used by the North American Stratigraphic Code (NACSN 1983), but with much more restrictive meaning, as noted below in this section.)

Lithodemic units are intended to be entirely objective mapping units. The Code notes that "concepts based on inferred geologic history properly play no part in the definition of a lithodemic unit." However, the Code notes that "where two rock masses are lithically similar but display objective structural relations that preclude the possibility of their being even broadly of the same age, they should be assigned to different lithodemic units."

The *lithodeme* is "the fundamental unit in lithodemic classification" (Table 2) and is defined as "a body of intrusive, pervasively deformed, or highly metamorphosed rock, generally non-tabular and lacking primary depositional structures and characterized by lithic homogeneity." The lithodeme is mappable and also "traceable in the subsurface" (NACSN 1983). The lithodeme is the fundamental unit within the lithodemic classification and, thus, occupies the conceptual position of the formation within lithostratigraphic hierarchy.

Suite and *supersuite* are, respectively, the two higher ranked categories of units in formal lithodemic classification (NACSN 1983). The *suite* is defined as a formal unit comprising two or more associated lithodemes of the same class. By *class of suite* is meant "metamorphic, intrusive, *or* plutonic," (NACSN 1983). A *supersuite* is a unit comprised of "two or more suites … having a degree of natural relationship to one another, either in a vertical or a lateral sense" (NACSN 1983).

We see no reason why the lithodemic units just mentioned could not be expanded by definition to include impact-related and impact-derived materials under certain circumstances. This may require that impact-derived and impact-related rocks are recognized as a "class" of rocks, like the others (igneous, sedimentary, and metamorphic), and that perhaps is not likely to happen soon within the general geological community. Alternatively, the concept of "metamorphic rock" could be broadened to include shocked and other rocks modified by impact heat and pressure.

We see no problem with using lithodemic hierarchy to help describe mappable units comprised of impactites (including impact breccias), impact melts, melt rocks, and melt breccias, impact pseudotachylites, suevite breccias, and potentially other proximal impact-derived and impact-related materials (terminology from Stöffler and Grieve 1994a; 1994b). We feel that the caveat restricting lithodemic usage to rocks that "generally do not conform to the Law of Superposition" probably excludes most impact ejecta deposits.

One potential example of mappable lithodemic units in impact-derived and impact-related materials exists in the large pseudotachylite dikes seated within the Proterozoic Witwatersrand Supergroup cropping out in the Vredefort impact structure, Republic of South Africa (Reimold and Colliston 1994). While some of these dikes may or may not be entirely impact-related (Reimold and Colliston 1994; Reimold 1995), they (1) are mappable, (2) do not conform to the Law of Superposition, and (3) have "tectonic" contacts, and (4) are entirely objectively defined. Thus, a case could be made for application of lithodemic concepts in this instance. If both impact and tectonic pseudotachylites are present in the area, as has been strongly suggested (Reimold 1995), they would be separately named lithodemic entities, according to the Code.

A "complex" in lithodemic classification is defined as "an assemblage or mixture of rock of two or more genetic classes, i.e., igneous, sedimentary, or metamorphic, with or without highly complicated structure" (NACSN 1983). The term "complex" can be used in lieu of a rank term or lithic rank term (e.g., breccia) and is an unranked or non-hierarchical concept, which is "commonly ... comparable to suite or supersuite" and is "named in the same manner" (NACSN 1983). The International Stratigraphic Guide, 2nd edition (Salvador 1994) notes that use of "complex" (which is viewed in the Guide as a *lithostratigraphic* unit) may be useful when working with an "assemblage of diverse rocks" wherein each individual "lithic component is impractical (to map) at ordinary scales." Among the potential types of complexes noted, the "structural complex" is described as "heterogeneous mixtures of disrupted bodies of rock in which some individual components are too small to be mapped ... (and) where there is no doubt that the mixing or disruption is due to tectonic processes." This concept could be applied in some instances to impact-derived and impact-related materials within and near impact structures especially in large impact structures and those modified by tectonism. Alternatively, the concept of a lithodemic complex within the Code itself could be easily modified to include most impact geologic settings (i.e., an "impact complex"). Complexes are named using an appropriate geographic term and the word "Complex" only (NACSN 1983; Salvador 1994).

At Sudbury impact structure, Ontario, a unit named "Sudbury Igneous Complex" (*sic*) has been used in mapping since the work of Dressler (1984). This unit, which represents "a differentiated melt sheet, a hybrid endogenic magma that intruded in several pulses, or an endogenic magma topped by impact-generated granophyres" (Ariskin et al. 1999), is a body of largely impact-related material to which the Law of Superposition *does not apply* unto its constituent units. In contrast, the originally overlying Onaping Formation, which represents a succession of various impact melt layers, melt breccias, impact breccias, and suevites (not in that order; Avermann 1994), comprises an adjacent body of impact-related materials to which the Law of Superposition *does apply* unto its constituent units. Thus, whereas the Sudbury Igneous Complex *(sic)* could be interpreted as a lithodemic unit, the overlying Onaping Formation is clearly a lithostratigraphic entity.

3.4
Formal Allostratigraphic Units, their Hierarchy, and Applicability

According to the North American Stratigraphic Code (NACSN 1983), there is a category of units called "allostratigraphic units," which are "mappable stratiform bodies of sedimentary rock that (are) defined and identified on the basis of (their) bounding discontinuities" (Table 2). Formal allostratigraphic units may be created to distinguish between:
(1) "superposed discontinuity-bounded deposits of similar lithology;"
(2) "contiguous discontinuity-bounded deposits of similar lithology;" or
(3) "geographically separated discontinuity-bounded units of similar lithology or to distinguish as single units discontinuity-bounded deposits characterized by lithic homogeneity."

Internal characteristics of these units "may vary laterally and vertically throughout the unit" (NACSN 1983).

The Code makes clear that allostratigraphic units are to be objective units, but allows that (1) genetic interpretation, (2) well-documented geologic history, and (3) age relationships "may influence the choice of the unit's boundaries." There is a ranked hierarchy of units which mimic the lithostratigraphic rank hierarchy: allogroup; alloformation (fundamental unit); and allomember (Table 2). Allostratigraphic units are named like lithostratigraphic and lithodemic units, i.e., they have an appropriate local geographic (i.e., proper) name and a rank term that is capitalized. The International Stratigraphic Guide, 2nd edition (Salvador 1994) does not list any category of unit that is directly comparable to the allostratigraphic units of the Code.

Recognizing that the Code authors intended for allostratigraphic units to be "stratiform bodies of sedimentary rock" perhaps limits somewhat the applicability of this concept in the present discussion. However, many impact-derived and impact-related materials accumulate as a result of sedimentation processes and many such materials are contained within sedimentary strata. In particular, distal impact breccias and distal impactites, including impactoclastic air-fall beds (terminology of Stöffler and Grieve 1994a; 1994b), and other impact-related units bearing impact-produced components such as shocked minerals, microkrystites, impact spherules and spheroids, tektites and microtektites, and other impact materials (e.g., iridium-bearing materials) could be classified as allostratigraphic units. One caveat here is that any unit so conceived should have discontinuities at its upper and lower boundaries. This may limit the potential for application of allostratigraphic classification to impact-derived and impact-related materials in such a way that only selected types of ejecta could be treated with this terminology. With proximal ejecta, the bounding discontinuities may be more petrographically distinct (e.g., bedrock-breccia contacts or upper-bounding paleosols), but with Phanerozoic distal ejecta, bounding discontinuities may be more commonly of a biostratigraphic nature.

A possible example of impact-derived and impact-related materials that could be mapped as an allostratigraphic unit occurs in central Belize, Central America. In this area, an informal unit, the Teakettle diamictite, which contains Chicxulub ejecta mixed with locally derived materials (Pope and Ocampo 2000), is found within depressions upon a karst surface. The Teakettle diamictite has a discontinuity, marked by a paleosol, upon its upper surface. Another nearby example is the Albion Formation of northern Belize (mentioned in section 3.2 as a lithostratigraphic unit). The Albion Formation contains a higher proportion of Chicxulub ejecta and rests upon a ballistically eroded Maastrichtian bedrock surface. Further, it has as a paleosol on its upper surface (Ocampo et al. 1996) and, thus, could be considered an alloformation as well as a formation, depending upon the emphasis of the investigator.

The Peñalver Formation of western Cuba, a 200-m thick mass-movement deposit containing Cretaceous-Tertiary boundary impact ejecta from the Chicxulub impact, is also a discontinuity-bounded unit (as described by Takayama et al. 2000). The basal conglomerate facies rests upon a scour surface and the unit's top is marked by a sedimentological break with sediments of the overlying Paleocene Apolo Formation (Takayama et al. 2000).

3.5
Formal Chronostratigraphic Units, their Hierarchy, and Applicability

The North American Stratigraphic Code and the International Stratigraphic Guide, 2nd edition (Salvador 1994) both recognize the well-established concept of the chronostratigraphic units. A chronostratigraphic unit, as defined by the Code (NACSN 1983) is "a body of rock established to serve as the material reference for all rocks formed during the same span of time." A key, distinguishing characteristic of all chronostratigraphic units is that "each of its boundaries (upper and lower) is synchronous" (Table 1). Thus, the chronostratigraphic unit "serves as the basis for defining (a) specific interval of time" (NACSN 1983). In the Guide, a chronostratigraphic unit is quite similar, being defined as a body of rock, "layered or unlayered, that (was) formed during a specific interval of geologic time" (Salvador 1994).

In both the Code and Guide, the stated goal in supporting the erection of chronostratigraphic units was to "establish a standard global chronostratigraphic scale" for the purposes of enhanced temporal classification (Salvador 1994). To this end, both the Code and Guide recognize a hierarchical structure of chronostratigraphic units (i.e., the eonothem, erathem, system, series, stage, and substage; NACSN 1983; Salvador 1994). These concepts were originally heavily dependent upon biostratigraphic data, but are now extensively supported by radiometric and stable isotopic data as well.

The Code's and the Guide's hierarchical scheme of chronostratigraphy does *not* have particular relevance to the present discussion, except that impact-derived and impact-related materials are present at the "event-boundaries" of some systems and stages (some examples are recounted by Montanari and Koeberl 2000). The International Stratigraphic Guide, 2nd edition (Salvador 1994) recognizes a non-hierarchical concept that attends this type of "event-bounding stratigraphy," namely the chronostratigraphic horizon (or chronohorizon). The Guide defines this kind of feature as "a stratigraphic surface or interface that is isochronous (i.e., of equal duration, and) everywhere of the same age." Theoretically without readily measurable thickness, the chronostratigraphic surface (or chronohorizon) has been "commonly applied" (outside of impact stratigraphy, of course) to "very thin and distinctive intervals that are essentially isochronous over their whole geographic extent and thus constitute excellent time-reference or time-correlation horizons" (Salvador 1994). Certainly, these chronostratigraphic surfaces (or chronohorizons) need not be the same as boundaries for hierarchical units, and can occur within hierarchical chronostratigraphic units. The geochronologic equivalent of a chronostratigraphic horizon (or chronohorizon) is "a *moment* (or an *instant*, if it has no resolvable time duration on a geologic scale)" (Salvador 1994).

Perhaps the best-documented example of impact-related chronostratigraphic horizons (or chronozones) are in the two upper Eocene impactoclastic air-fall (beds) horizons found globally in marine sediment cores and within the Eocene/Oligocene Global Stratotype Section and Point (GSSP) at Massignano, Italy (as noted in Farley et al. 1998; Montanari and Koeberl 2000). These two chronostratigraphic horizons (or chronozones) are separated by approximately 25 cm (10 to 20 Ka; Wei 1995) in most places. The older layer, rich in microkrystites, shocked-quartz grains, nickel-rich spinels, and iridium-bearing components (Pierrard et al. 1998), is global in extent and likely represents ejecta from the Popigai impact structure in Siberia (Montanari and Koeberl 2000). In contrast, the younger layer, rich in microtektites, shocked-mineral phases, and high-pressure polymorphs (Glass 1989), seems more restricted to the eastern U.S. coastal area, the Caribbean, and the Gulf of México. This layer is the same as the "North American strewn field" of tektites (Glass 1989), which likely represents impact ejecta of the Chesapeake Bay crater, eastern U.S. (Montanari and Koeberl 2000).

The North American Stratigraphic Code (NACSN 1983) and the International Stratigraphic Guide, 2nd edition (Salvador 1994) both recognize another sort of non-hierarchical chronostratigraphic unit, namely the *chronozone*. The Code contains a more detailed view of this concept than the Guide and, thus, the Code's view is presented here.

A chronozone is "a non-hierarchical, but commonly small, formal chronostratigraphic unit, and its boundaries may be independent of ... ranked (hierarchical) units" (NACSN 1983). The chronozone is an isochronous unit, which may be "based upon a biostratigraphic unit ... , a lithostratigraphic unit ... , or a magnetopolarity unit." In the instance of a lithostratigraphic unit being used as a chronozone, the prefix *litho-* would be applied to the formal rank term, i.e., *lithochronozone*. The *scope* of a chronozone "may range markedly, depending upon the purpose for which it was defined." Therefore, if a lithochronozone were defined as, for example, all distal and proximal ejecta of an impact event, its *scope* would be delimited by this definition.

We think lithochronozones would have their greatest usefulness and application to impact-derived and impact-related materials if these materials were in a measurably thick ejecta blanket composed of distal and/or proximal impactites, including distal impactoclastic air-fall beds (terminology of Stöffler and Grieve 1994a; 1994b). However, a group, suite, or complex of rocks comprising a small crater-filling impactite unit could also fit into the definition of lithochronozone in some instances. Also, the "impact formation" (*sensu* Stöffler and Grieve 1994a; 1994b; discussed above) could be considered as a type of lithochronozone in some instances.

Lithochronozones are not the same as the chronostratigraphic horizons (or chronohorizons) discussed in the Guide (Salvador 1994). Lithochronozones are meant to have readily measurable thickness (but not necessarily have mappability; NACSN 1983). Whereas, chronostratigraphic horizons (or chronohorizons) may be an immeasurably thin stratigraphic surface strewn with impact-derived and impact-related materials (e.g., tektites, microtektites, microkrystites, shocked mineral grains, nickel-bearing spinels, impact-related carbon phases, mineral phases bearing cosmic platinum-group element signatures, etc.).

Lithochronozones may be "either formally or informally" defined (NACSN 1983). The Code views formally defined lithochronozones as those having a material referent that is also a formal unit in some other category (e.g., lithostratigraphic, biostratigraphic, or magnetopolarity). The Code implies that all other lithochronozones are, by default, informally defined, including those "key beds and markers" used in "industry investigations" and other units that are "useful to define geographic distributions of lithofacies."

We suggest that a useful distinction between formally defined and informally defined lithochronozones in the present instance could be made upon the basis of direct relation to an established impact structure. For example, we think good examples of a *formally defined* lithochronozone comprising impact-derived and impact-related materials (including tsunami deposits), which have already been named as lithostratigraphic units, are:

(1) the Campanian Crow Creek Member of the Pierre Shale of South Dakota and Nebraska (described by Izett et al. 1998), and

(2) the Lower Cretaceous Myklegardfjellet Bed at the base of the Rurikfjelleet Formation (= base of the Berriasian Stage, i.e., the Jurassic-Cretaceous boundary) on Oppdalsåta Mountain in central Svalbard (Dypvik et al. 1996; Johnsen et al. 2001).

These units are related, respectively, to the Manson impact structure in Iowa (Izett et al. 1998) and the Mjølnir impact structure on the floor of the Barents Sea (Dypvik et al. 1996). Likewise, distal impactoclastic air-fall beds (ejecta) within the Proterozoic Bunyeroo Formation of South Australia, which have been informally named for their petrology (Wallace et al. 1996), could be formally named as well, because they have been directly identified with the Acraman impact structure, South Australia (Wallace et al. 1996).

We think good examples of *informally defined* lithochronozones comprising impact-derived and impact-related materials are the several Archean spherule layers within the Wittenoon Formation in the Hamersley Basin of Western Australia (described by Simonson 1992) and within the Monteville Formation of the Transvaal Supergroup in South Africa (described by Simonson et al. 1997). The globally distributed, impactoclastic air-fall bed (layer) generated by the Chicxulub impact in México poses a peculiar problem of nomenclature, because it is a lithochronozone of such great lateral extent (see occurrences plotted in Smit 1999). Originally described as a distal impactoclastic air-fall bed within the Scaglia Rossa Formation in the Umbria-Marche region of Italy (Alvarez et al. 1980), the unit has been traced worldwide (Smit 1999). Therefore, no single local name for this lithochronozone seems entirely appropriate. The distal impactoclastic air-fall bed at the Cretaceous-Tertiary Global Stratotype Section and Point, located a few kilometers west of the town of El Kef, Tunisia, which is part of the "ejecta layer and boundary clay" layer, has no formal name. At the base of the stratotype section for the Danian, at Stevns Klint, Denmark, the distal impactoclastic air-fall bed is called "Fiskeler." Thus, at Stevns Klint and vicinity, we could refer to this layer uniformly as the Fiskeler Lithochronozone.

Names of lithochronozones, according to the Code, are proper nouns (geographic terms) that were previously assigned to the lithostratigraphic unit upon which the lithochronozone is based (NACSN 1983). In many instances, while working with impact-derived and impact-related materials, this source for names would probably not work well. Therefore, we recommend seeking appropriate local geographic names for impact-derived and impact-related lithochronozones and generally avoiding using the same name as the derivative impact structure.

In both the Code and Guide, the corresponding geochronologic unit of the chronozone is the *chron*. For the purposes of the present discussion, the chron for a unit comprised of impact-derived and impact-related material would be a relatively short-duration interval related to cosmic impact (i.e., an "impact chron"). Such an interval includes the time of contact through early modification stages of the impact-cratering event (described by Melosh 1989) and the time represented by any geologic record of direct environmental aftermath (whether local or global), potentially including impact-related air-fall deposits, tsunami sediments, mass-movement deposits, seismites, etc.

Proper names for chrons, according to the Code, are usually "identical with those of the corresponding chronostratigraphic units" (NACSN 1983). The Code makes no provision for "independently formed" names of chrons. We think that the local geographic name should be used first with the name of the derivative impact structure, if known, in parentheses. For example, if we were to classify the Myklegardfjellet Bed on Svalbard as the Myklegardfjellet Lithochronozone, the latter would be the material referent for the Myklegardfjellet (Mjølnir) Chron (data from Dypvik et al. 1996; Johnsen et al. 2001).

4
Conclusions and Comments

The impact geology community should not ignore extant stratigraphic nomenclature because it can be quite helpful in organizing and classifying impact-derived and impact-related Earth materials. Where mapping of objectively defined units within such materials is the primary objective of the study, formal lithostratigraphic and lithodemic units would be most useful. Where emphasis is placed upon bounding discontinuities of mappable impact-derived and impact-related materials (especially regarding impact ejecta), allostratigraphic units may be used. Where emphasis is placed upon temporal aspects of impact-derived and impact-related materials (including temporal correlation), lithochronozones and chronostratigraphic horizons (or chronohorizons) may be used.

We hope that more common adoption of extant stratigraphic nomenclature in impact geology studies will improve scientific communication through standardization of terminology. Further, we think usage of such stratigraphic nomenclature will promote better understanding between impact geologists and non-specialists in this area. In the future, we hope changes in the wording of the Code and Guide will better accommodate impact geology examples in their texts.

Acknowledgements

We thank Professor Uwe Reimold for his many valuable suggestions that improved this manuscript.

References

Alvarez LW, Alvarez W, Asaro F, Michel HV (1980) Extraterrestrial cause for the Cretaceous-Tertiary extinction. Science 208: 1095-1108

Ariskin AA, Deutsch A, Ostermann M (1999) Sudbury Igneous Complex: Simulating phase equilibria and in situ differentiation for two proposed parental magmas. In: Dressler BO, Sharpton VL (eds) Large Meteorite Impacts and Planetary Evolution II. Geological Society of America, Special Paper 339, pp 373-388

Avermann M (1994) Origin of the polymict, allochthonous breccias of the Onaping Formation, Sudbury Structure, Ontario, Canada. In: Dressler BO, Grieve RAF, Sharpton VL (eds) Large Meteorite Impacts and Planetary Evolution. Geological Society of America, Special Paper 293, pp 265-274

Cowie JW (1986) Guidelines for boundary stratotypes. Episodes 9: 78-82

Deutsch A, Langenhorst F (1994) Geological formations in and around impact structures. In Marfunin AS (ed), Mineral Matter in Space, Mantle, Ocean Floor, Biosphere, Environmental Management, and Jewelry, Springer-Verlag, Berlin, pp 89-95

Dressler BO (1984) General geology of the Sudbury area. In: Pye EG, Naldrett AJ, Giblin PE (eds) The Geology and Ore Deposits of the Sudbury Structure. Ministry of Natural Resources, Ontario Geological Survey, Toronto, Special Volume 1, pp 57-82

Dypvik H, Gudlaugsson ST, Tsikalas F, Attrep Jr. M, Ferrell Jr. RE, Krinsley DH, Mørk A, Faleide JI, Nagy J (1996) Mjølnir structure: An impact crater in the Barents Sea. Geology 24: 779-782

Farley KA, Montanari A, Shoemaker EM, Shoemaker CS (1998) Geochemical evidence for a comet shower in the late Eocene. Science 280: 1250-1253

French BM (1999) Traces of Catastrophe – a Handbook of Shock-Metamorphic Effects in Terrestrial Meteorite Impact Structures. Lunar and Planetary Institute Contribution 954, Lunar and Planetary Institute, Houston, 120pp

Glass BP (1989) North American tektite debris and impact ejecta from DSDP site 612. Meteoritics 24: 209-218

Hildebrand AR, Penfield GT, Kring DA, Pilkington M, Carmargo ZA, Jacobsen SB, Boynton WV (1991) Chicxulub crater: A possible Cretaceous-Tertiary boundary impact crater on the Yucatan Peninsula, Mexico. Geology 19: 867-871

Hüttner R (1969) Bunte Trümmermassen und Suevit. Geologica Bavarica 61: 142-200

Izett GA, Cobban WA, Dalrymple GB, Obradovich JD (1998) ^{40}Ar/^{39}Ar age of the Manson impact structure, Iowa, and correlative impact ejecta in the Crow Creek Member of the Pierre Shale (Upper Cretaceous), South Dakota and Nebraska. Geological Society of America Bulletin 110: 361-376

Johnsen SO, Mørk A, Dypvik H, Nagy J (2001) Outline of the geology of Svalbard. Short geological review and guidebook for 7th ESF IMPACT workshop. In: Smelror M, Dypvik H, Tsikalas F (eds) Submarine Craters and Ejecta-Crater Correlation and Icy Impacts and Icy Targets (Abstracts and Proceedings of the Norwegian Geological Society 1), pp 91-112

Laznicka P (1988) Breccias and Coarse Fragmentites – Petrology, Environments, Associations, and Ores. Elsevier, Amsterdam, pp 691-698

Melosh HJ (1989) Impact Cratering: A Geologic Process. Oxford University Press, New York, 245pp

Montanari A, Koeberl C (2000) Impact Stratigraphy: The Italian Record. Lecture Notes in Earth Sciences, Vol. 93, Springer, Berlin-Heidelberg, 364pp

NACSN (North American Commission on Stratigraphic Nomenclature) (1983) North American Stratigraphic Code. American Association of Petroleum Geologists Bulletin 67: 841-875

Ocampo AC, Pope KO, Fischer AG (1996) Ejecta blanket deposits of the Chicxulub crater from Albion Island, Belize. In: Ryder G, Fastovsky D, Gartner S (eds) The Cretaceous-Tertiary Event and other Catastrophes in Earth History. Geological Society of America, Special Paper 307, pp 75-88

Pierrard O, Robin E, Rocchia R, Montanari A (1998) Extraterrestrial Ni-rich spinels in upper Eocene sediments from Massignano, Italy. Geology 26: 307-310

Poag CW, Powars DS, Poppe LJ, Mixon RB (1994) Meteoroid mayhem in Ole Virginny: Source of the North American tektite field. Geology 22: 691-694

Pope KO, Ocampo AC (2000) Chicxulub high-altitude ballistic ejecta from central Belize [abs.]. Lunar and Planetary Science 31: abstract no. 1419, CD-ROM, 2 pp

Raikhlin AI, Selivanovskaya TV, Masaitis VL (1980) Rocks of terrestrial impact craters: problems of classification [abs.]. Lunar and Planetary Science 11: 911-913

Reimold WU (1995) Pseudotachylite in impact structures – generation by friction melting and shock brecciation?: A review and discussion. Earth-Science Reviews 39: 24-265

Reimold WU (1998) Exogenic and endogenic breccias: a discussion of major problematics. Earth-Science Reviews 43: 25-47

Reimold WU, Colliston WP (1994) Pseudotachylites of the Vredefort Dome and the surrounding Witwatersrand Basin, South Africa. In: Dressler BO, Grieve RAF, Sharpton VL (eds) Large Meteorite Impacts and Planetary Evolution. Geological Society of America, Special Paper 293, pp 177-196

Salvador A (ed) (1994) International Stratigraphic Guide, 2nd ed. International Union of Geological Sciences and Geological Society of America, 214pp

Sandberg CA, Morrow JR, Warme JE (1997) Late Devonian Alamo impact event, global kellwasser events, and major eustatic events, eastern Great Basin, Nevada and Utah. In: Link PK, Kowallis BJ (eds) Proterozoic to Recent Stratigraphy, Tectonics, and Volcanology, Utah, Nevada, southern Idaho, and central Mexico (Brigham Young University Geology Studies 42(I)) pp 129-160

Sharpton VL, Marín LE, Carney JL, Lee S, Ryder G, Schuraytz BC, Sikora P, Spudis PD (1996) A model of the Chicxulub impact basin based on evaluation of geophysical data, well logs and drill core samples. In: Ryder G, Fastovsky D, Gartner S (eds) The Cretaceous-Tertiary Event and other Catastrophes in Earth History. Geological Society of America, Special Paper 307, pp 55-74

Simonson BM (1992) Geological evidence for a strewn field of impact spherules in the early Precambrian Hamersley Basin of Western Australia. Geological Society of America Bulletin 104: 829-839

Simonson BM, Beukes NJ, Hassler S (1997) Discovery of a Neoarchean impact spherule horizon in the Transvaal Supergroup of South Africa and possible correlations to the Hamersley Basin of Western Australia [abs.]. Lunar and Planetary Science 28: 1323-1324

Smit J (1999) The global stratigraphy of the Cretaceous-Tertiary boundary impact ejecta. Annual Reviews Earth and Planetary Science 27: 75-113

Smit J, Roep TB, Alvarez W, Montanari A, Claeys P, Grajales-Nishimura JM, Bermudez J (1996) Coarse-grained, clastic sandstone complex at the K/T boundary around the Gulf of Mexico: Deposition by tsunami waves induced by the Chicxulub impact? In: Ryder G, Fastovsky D, Gartner S (eds) New Developments Regarding the KT Event and Other Catastrophes in Earth History. Geological Society of America, Special Paper 307, pp 151-182

Stöffler D, Grieve RAF (1994a) Classification and nomenclature of impact metamorphic rocks: A proposal to the IUGS subcommission of the systematics of metamorphic rocks. Lunar and Planetary Science 25: 1347-8

Stöffler D, Grieve RAF (1994b) Classification and nomenclature of impact metamorphic rocks: A proposal to the IUGS subcommission of the systematics of metamorphic rocks. In: Montanari A, Smit J (eds) Post-Östersund Newsletter, European Science Foundation (ESF) Network on Impact Cratering and Evolution of Planet Earth, Strasbourg, pp 9-15

Takayama H, Tada R, Matsui T, Iturralde-Vinent M, Oji T, Tajika E, Kiyokawa S, García D, Okada H, Hasagawa T, Toyoda K (2000) Origin of the Peñalver Formation in northwestern Cuba and its relation to the K/T boundary impact event. Sedimentary Geology 135: 295-320

Wallace MW, Gostin VA, Keays RR (1996) Sedimentology of the Neoproterozoic Acraman impact ejecta horizon, South Australia. Australian Geological Survey Organization Journal, Australian Geology and Geophysics 16: 443-451

Warme JE, Kuehner H-C (1998) Anatomy of an anomaly: the Devonian catastrophic Alamo breccia of southern Nevada. International Geology Review 40: 189-216

Wei W (1995) How many impact-generated microspherule layers in the upper Eocene? Palaeogeography, Palaeoclimatology, Palaeoecology 114: 101-110

Wilhelms DH (1990) Geologic mapping. In: Greeley R, Batson RM (eds) Planetary Mapping, Cambridge University Press, New York, pp 208-260

Appendix: Requirements for the Definition of Formal Stratigraphic Units

The North American Stratigraphic Code (NACSN 1983) and the International Stratigraphic Guide, 2nd edition (Salvador 1994) present detailed descriptions of what is necessary to formally define and name stratigraphic units and what constitutes a stratotype (or type section) for such units. The reader is referred to these publications for all details of their suggested requirements, however we will present a summary below for the purposes of introducing readers not familiar with such practice.

North American Stratigraphic Code Requirements for Formally Named Geologic Units

The Code says that "naming, establishing, revising, redefining, and abandoning formal geologic units requires publication in a recognized scientific medium of a comprehensive statement which includes: (1) intent to designate or modify a formal unit; (2) designation of category and rank of unit; (3) selection and derivation of name; (4) specification of stratotype (where applicable); (5) description of unit; (6) definition of boundaries; (7) historical background; (8) dimensions, shape, and other regional aspects; (9) geologic age; (10) correlations; and (11) possibly its genesis."

The Code specifies that the requisite "publication in a recognized scientific medium," when first issued, must be: (1) reproduced in ink on paper by some method that assures numerous identical copies and wide distribution; (2) issued for the purpose of scientific, public, permanent record; and (3) readily obtainable by purchase or free distribution." Further, the Code notes that the following types of publication do *not* meet the requirements as above: microfilms, microcards, notations on an illustration distributed to persons on a field trip, proof or galley sheets, open-file reports, theses, dissertations, dissertation abstracts, media used in a scientific presentation, scientific abstract, a legend of a geologic map or figure caption, labels on rock specimens, a document archived in a library or museum, an anonymous publication, a report in the popular press, a report in a legal document, and guidebooks that are distributed only to field-trip participants.

To be valid, the Code notes, a unit "must serve a clear purpose and be duly proposed and duly described and the intent to establish (the unit) must be specified." The Code specifically notes: "casual mention of a unit ... does not establish a new formal unit, nor does mere use in a table, columnar section, or map."

The Code also notes "category and rank of new or revised units must be specified." For many reasons, the Code prescribes that "specification and unambiguous description of the category is of paramount importance" in each instance. Further, "selection and designation of an appropriate rank from the distinctive terminology developed for each category help serve this function."

For proper derivation of names, the Code requires compound nomenclature. For most categories listed in the Code, a "geographic name combined with an appropriate rank or descriptive term" is appropriate. Geographic names for units "are derived from permanent natural or artificial features at or near which the unit is present." This is preferable to names that might be derived from "impermanent features such as farms, schools, stores, churches, crossroads, and small communities." The Code recommends that names be selected from "those shown on topographic, state, provincial, county, forest service, hydrographic, or comparable maps, particularly those names approved by a national board for geographic names." Disappearance of a permanent or impermanent feature after derivation of a name from it does not affect the unit name. For geographic terms with two parts to the name, the Code lets stand well-established names with both parts, but recommends in the instance of new names omission of the generic part of a place name (e.g., river, lake, etc.) "unless it is required to distinguish between two otherwise identical names." (However, "two names should not be derived from this same feature" using this rule). The Code recommends, "a unit should not be named for the source of its contents." The example used in the Code is that of a glacier and its till, but we should note that a crater and its ejecta would be another example. Finally, the initial letters of *all* parts of a formal name are capitalized (see examples in the main text above).

According to the Code, stability of names is maintained by the "rule of priority" and "by preservation of well-established names." Stratigraphic names, once assigned, should not be changed without "explaining the need" via publication. Priority in naming units should be respected, but "priority alone does not justify displacing a well-established name by one neither well-known nor commonly used." However, an "inadequately established name" should not be preserved only because of its priority. The Code notes that "redefinition (of assigned names) is preferable to abandonment of the names of well-established units, which may have been defined imprecisely" during an earlier time of lesser standards. Spelling of a geographic name "commonly conforms to the usage of the country and linguistic group involved" in the selection. To avoid duplication of names, clearinghouses are maintained in the U.S., México, and Canada, and readers are referred to the details of how to check for duplication of names as described in the Code.

The Code recommends designation of a stratotype of a unit or a boundary stratotype (also called type sections and type localities) because "it is essential in the definition of most formal geologic (material) units." A stratotype is "the standard (whether original or subsequently designated) for a named geologic unit or boundary and constitutes the basis for definition or recognition of that unit or boundary." Therefore, the Code requires a stratotype to be "illustrative and representative of the concept of the unit or boundary being defined." For additional remarks on detailed aspects of the stratotype, the reader is referred to the Code.

In the published description of any new stratigraphic unit, the Code requires a detailed "unit description." This description should be so clear that "any subsequent investigator can recognize that unit unequivocally." The following, at minimum, should be in the published description: composition, texture, primary structures, structural attitude(s), biologic remains, geochemistry, geophysical properties, geomorphic expression, unconformable or cross-cutting relations, and age. Further, criteria should be specified about the unit's boundaries and how to distinguish between the new unit and adjacent units. A proposed new stratigraphic unit must have "a nomenclatorial (*sic*) history of (the) rocks assigned" to the unit. This is considered part of the justification for definition of a new unit.

If major changes are needed in an established formal unit, the Code recognizes redefinition, revision, and abandonment as ways to accomplish this. Revision and abandonment require "as much justification as establishment of an new unit," in keeping with the Code view that priority is important. Redefinition is "a correction or change in the descriptive term applied to a stratigraphic or lithodemic unit." Redefinition does not require a new geographic term. Revision is different and "involves either minor changes in the definition of one or both boundaries of a unit, or in the unit's rank." Abandonment is different from revision and redefinition and formal abandonment involves "improperly defined or obsolete stratigraphic, lithodemic, or temporal units." Abandonment is allowed only if there is "sufficient justification" and "recommendations are made for the classification and nomenclature to be used" in place of the abandoned unit. For more information about redefinition, revision, and abandonment, the reader is referred to the Code.

International Stratigraphic Guide, 2nd edition, Requirements for Formally Named Geologic Units

The Guide says, "the proposal of a new formal stratigraphic unit requires a statement of intent to introduce the new unit and the reasons for doing so." Such a proposal should include a:
(1) "clear and complete definition, characterization, and description of the unit so that any subsequent investigator can recognize it unequivocally;"
(2) "proposal of the kind, name, and rank of the unit;" and
(3) "designation of a stratotype (type section) or type locality on which the definition and description of the unit is based."

Further, the proposal should be "published in a recognized scientific medium." Similarly, revision and redefinition of an established unit must be so published.

In general, the Guide says, a new unit description should include "a clear account if its boundaries, diagnostic properties, and attributes. For lithostratigraphic units, the emphasis should be placed upon "lithologic properties," whereas with chronostratigraphic units, "emphasis should be placed on features bearing on age and time-correlation."

As in the Code, names of formal stratigraphic units are "compound." For most unit categories, names should consist of "a geographic name combined with an appropriate term indicating the kind and rank of the unit." The name of a new stratigraphic unit "should be unique," therefore, "before attempting to establish a new formal stratigraphic name, authors should refer to national, state, or provincial records of stratigraphic names to determine whether a name has been used previously." The geographic component of a name derives "from permanent natural or artificial features at or near which the stratigraphic unit is present." Recommendations in the Guide parallel those in the Code on the matter of name selection, spelling, possible reasons for change in name, rule of priority, and duplication of names. Further, the Guide notes, "use of a geographic name for a stratigraphic unit should be subject to approval by the (appropriate) national organization for place-names."

The Guide comments extensively, more so than the Code, on the matter of stratotypes and type localities and requirements for their selection. The following are pertinent definitions related to this concept.

(1) *Stratotype* (or type section) is "the original or a subsequently designated standard of reference of a named layered stratigraphic unit or of a stratigraphic boundary." A stratotype is "a specific interval or point in a specific sequence of rock strata and constitutes the standard for the definition and characterization of the stratigraphic unit or boundary being defined."

(2) *Unit-stratotype* is "the type section of a layered unit that serves as the standard of reference for the definition and characterization of the unit." In the instance of a "complete, well-exposed layered unit," the "upper and lower limits of the unit-stratotype are its boundary-stratotypes."

(3) *Boundary-stratotype* is "a specified sequence of rock strata in which a specific point is selected that serves as the standard for definition and recognition of a stratigraphic boundary."

(4) *Composite-stratotype* is "a unit-stratotype formed by the combination of several specified type intervals of strata, called *component-stratotypes*." This kind of stratotype may be useful where "a certain lithostratigraphic unit may not be entirely exposed in any single section."

(5) *Type locality* is the "specific geographic locality" in which "the unit-stratotype or the boundary-stratotype" of a "layered stratigraphic unit or a boundary between layered units" is situated. Alternatively, the type locality is the general geographic "locality where the unit or boundary was originally defined or named" for units lacking a properly designated stratotype.

(6) *Type area* is "the geographic area or region that encompasses the stratotype or type locality of a stratigraphic unit or stratigraphic boundary.

The Guide's requirements for stratotypes (and type sections) vary for each category, but in general "the most important requisite of a stratotype is that it adequately represents the concept for which it is the material type." The ideal unit-stratotype would be "a complete exposure of all rocks in a unit from bottom to top and throughout its entire lateral extent." In the case of global chronostratigraphic units, the emphasis is placed upon marking the lower boundary at various "Global boundary stratotype section and point" (GSSP) locations. Descriptions of stratotypes should be both geographic and geologic; therefore appropriately detailed topographic and geologic maps (so that others may find the site(s)) are needed to convey this information. A stratotype "should offer reasonable assurance of long-range preservation" and be "geographically accessible to all." Boundaries of global chronostratigraphic units must be approved (see Cowie 1986). Requirements for type localities for nonlayered igneous and metamorphic rock bodies are "similar to those that apply to the selection of stratotypes (or type sections) of layered stratigraphic units." Namely, they should be selected carefully to fully represent the "concept of the unit … both geographically and geologically … and should be easily accessible."

Main Geochemical Signatures Related to Meteoritic Impacts in Terrestrial Rocks: A Review

María-Jesús Muñoz-Espadas[1], Jesús Martínez-Frías[2] and Rosario Lunar[3]

[1] Departamento de Geología, Museo Nacional de Ciencias Naturales, CSIC, José Gutiérrez Abascal 2, 28006 Madrid, Spain. (majem@mncn.csic.es)
[2] Centro de Astrobiología, CSIC-INTA, Carretera de Torrejón a Ajalvir, 28850 Torrejón de Ardoz, Madrid, Spain. (martinezfrias@mncn.csic.es)
[3] Departamento de Cristalografía y Mineralogía, Facultad de Ciencias Geológicas, Universidad Complutense de Madrid, Avenida Complutense s/n, 28040 Madrid, Spain. (lunar@geo.ucm.es)

Abstract. The chemical composition of impact melt rocks, breccias and ejecta layers is dominated mainly by the proportions and composition of target rocks. However, small quantities of vaporized and molten meteorite material mixed with them significantly alter the concentrations and ratios of certain elements and isotopes. The identification of this meteoritic signature is used to propose an impact formation for structures of uncertain origin, as well as a possible criterion for inferring the impactor type. The most common criteria applied to these studies are the detection of a positive siderophile element anomaly, the Re-Os isotopic system, and the Mn-Cr isotopic system. An enrichment in Cr, Ni and Co, at the ppm level, and Os, Re, Ir, Ru, Rh, Pd and Au, at the ppb level, usually indicates the presence of a component containing high abundances of those elements. These ratios help identify the projectile as either a chondrite or an iron meteorite, but do not detect an achondritic projectile. The application of the Re-Os system method is based in the low ^{187}Os/^{188}Os ratio of chondritic and iron meteorites, in comparison with present day higher ratio of normal upper crust, and the high Os contents in meteorites compared to crustal rocks. The admixture of a meteoritic component to crustal rocks produces a ^{187}Os/^{188}Os anomaly. The Mn-Cr system considers the deviations of the ^{53}Cr/^{52}Cr ratios from the standard terrestrial ^{53}Cr/^{52}Cr ratio as a result of the addition of an extraterrestrial component. This isotopic study often allows an accurate determination of the impactor type, especially when combined with PGE ratios. The applications and limitations of each method are reviewed.

1
Introduction

The presence of a meteoritic component in impact-derived rocks, either in the crater or in distal ejecta, can be decisive in assigning an impact origin for a certain structure or ejecta layer (Koeberl and Shirey 1997). Some projectiles can be identified from the presence of physical remnants of the meteorite (e.g., Meteor Crater, Arizona). However, the high temperatures during impact usually vaporize impactors larger than around 40 m in diameter (which hardly undergo atmospheric retardation, and encounter the target with relatively undiminished force) (Grieve 1997; Montanari and Koeberl 2000). Also, meteorites are not very resistant towards erosion, and last only a few thousand years (in the case of stony meteorites) or a few tens to hundreds of thousand years (in the case of iron meteorites) on the surface of the Earth. Thus, meteorite fragments usually only survive in young ($\sim$0.1 Ma) craters smaller than 1.5 km in diameter. There are some rare exceptions; for example, abundant meteorite fragments have been found on the ocean floor in deposits from the Eltanin impact event, which hit the South Pacific Ocean about 2.1 million years ago (Kyte and Brownlee 1985), believed to be from a 1-4 km-sized projectile (Gersonde et al. 1997; Flores et al. 2002). Probable meteorite material (severely altered and replaced) has also been found in K-T boundary sediments (Kyte 1998; Robin et al. 1993).

A number or recent hydrocode simulations of impacts at various angles indicate that the shock pressure and the amounts of solid, molten and vaporized projectile can vary considerably with the angle of impact (e.g., Pierazzo et al. 1997; Pierazzo and Melosh 1999, 2000; Schnabel et al. 1999). A downrange focusing of projectile material was observed in oblique impacts, especially in low impact angles ($\leq$30°), accompanied by the ejection of most of the projectile from the opening crater in the early stages of the impact. Shock melting and vaporization occur inside the projectile, mainly in its leading part, for all impacts except for the most oblique considered, a 15° impact, decreasing with angle of impact. The shock is weakest in the trailing half of the projectile, where any component surviving the impact in the solid state has its origin. According to Pierazzo and Melosh (1999), over 50% of the projectile is entrained in the expansion plume in the early phases of the 15° impact (these authors regret that their simulations do not continue long enough to see how this early entrainment could influence the distribution of the projectile material worldwide).

Schnabel et al. 1999 considered velocity, but not angle of fall, in their simulations of an hydrocode for the Canyon Diablo impact. Their choice of a 15 m radius and an impact velocity of 20 km s^{-1} for the impactor yields an estimation of $\sim$15% solid material as a 1.5 to 2 m thick shell, covering the trailing hemisphere and about one third of the leading hemisphere. If an impact velocity of 15 km s^{-1} is assumed, the solid shell thickens to $\sim$5 m and constitutes $\sim$63% of the mass. In both cases the rest of the proyectile melts, but does not vaporize. Vaporization would require higher but atypical velocities for Earth-crossing asteroids.

The quantity of the recondensed meteoritic vapor that may be mixed with the vaporized, molten, or shocked and brecciated target rocks is generally ≤1%. Schmidt et al. (1997) determined a meteoritic component of 0.5% of a nominal CI component for Sääksjärvi crater, and about 0.1% for Mien and Dellen. French et al. (1997) detected a minor extraterrestrial component (≤0.15%) in the melt–bearing breccias of the Gardnos impact structure. Nevertheless, the abundance of a meteoritic component is sometimes found to be much higher than these values. The amounts found in samples from the Clearwater East impact structure, Canada, corresponds to 4 to 7% of a nominal CI component, according to Palme et al. 1979 and Schmidt 1997. Recently, McDonald (2002), after re-evaluating the existing PGE data, proposed that up to 8% ordinary (possibly type-L) chondrite component is present in the impact melt. For the Morokweng impact melt rocks, McDonald et al. (2001) calculated an ordinary chondrite component at levels of up to 7.5%, in agreement with the earlier assessment by Koeberl et al. (1997). Such high abundances of meteoritic contamination may be explained by a higher impact angle, or lower impact velocity. In some cases it is possible to differentiate the chemical signature of the usually minute meteoritic contamination from the compositional signature of the normal terrestrial target rocks (Grieve 1991; Koeberl and Shirey 1997). For this purpose, detailed trace element and/or isotopic studies are necessary, as reviewed in the following chapters.

2
Siderophile Trace Element Analysis

Some siderophile and related elements are more abundant in meteorites than in terrestrial crustal rocks (Table 1). Therefore, melt rock siderophile and highly siderophile element (HSE) abundances and their interelement ratios in the impact rocks are compared to the average continental crust composition of these elements (Tables 1 and 2). An enrichment in Cr, Ni and Co, at ppm level, and Os, Re, Ir, Ru, Rh, Pd and Au, at ppb level, usually indicates the presence of either a chondritic or an iron meteoritic component. An achondritic signature is more difficult to discern, because these meteorites have significantly lower contents of the key siderophile elements (Koeberl and Shirey 1997; Schmidt 1997; Koeberl 1997, 1998; Table 1). Chondrites have high abundances of Cr (typically about 0.26 wt%; Anders and Grevesse 1989), whereas iron meteorites have more variable Cr contents that are typically around 100 times lower than in chondrites (Buchwald 1975). Enrichments in Cr and low Ni/Cr or Co/Cr ratios can be used to distinguish between chondritic and iron (Ni/Cr ~ 40000 and Co/Cr ~ 100 in the latter) projectiles (Evans et al. 1993). However, as the Co, Cr, and Ni contents are common on the upper crust (average 8, 37, and 45 ppm respectively; Schmidt et al. 1997), their elemental enrichments may be ambiguous.

Platinum group elements (PGE) are better suited for identifying a meteoritic component. The abundances of the PGE (Ru, Rh, Pd, Os, Ir, Pt) and Au in chondrites and iron meteorites are several orders of magnitude higher than those detected in terrestrial crustal rocks (e.g., Morgan et al. 1975; Palme et al. 1978, 1979; Morgan and Wandless 1983; Evans et al. 1993). Chondrites typically contain about 400-800 ppb Ir and Os (depending on the chondrite type), whereas the concentration of Ir and Os in the continental crust is approximately 0.02 ppb, according to Taylor and McLennan (1985). A more recent determination by Peucker-Ehrenbrink and Jahn (2001) sets a similar proxy for the values of Ir and Os in the upper continental crust: 22 and 31 pg/g, respectively. This means that the signal to background ratio is very high for PGE in impact rocks. The abundance of platinum-group elements and their interelemental ratios have been used to determine the type or class of meteorite. However, it has to be considered that some PGE enrichment is normal in certain terrestrial rocks; for example, gold mineralizations near the Bosumtwi crater were suggested to be responsible for the Ir enrichment in Ivory Coast tektites (Jones, 1985). Although the degree of PGE fractionation in various types of mantle rocks has been recognized to be quite large (McDonald et al. 1994; Gueddari et al 1999; Rehkämper et al. 1999; Schmidt et al. 2000) the PGE patterns of the mantle and in some mantle-derived rocks may be similar to those of chondrites (Table 1; Koeberl and Shirey 1997).

Table 1. Siderophile elements composition of several terrestrial (basalt, granite, peridotite and Earth´s upper crust) meteoritic (CI, eucrite), and impact (Clearwater East structure) materials.

	Basalt JB-1A (1)	Granite G-1 (2)	Peridotite JP (1)	Upper continental crust (3)	CI (4)	Eucrite (Juvinas) (5)	Clearwater East Impact melt rocks (6)
ppm							
Co	38.6	2.3	116	8[#]	508[a]	3.3	-
Cr	392	20	2807	37[#]	2650[a]	2330	-
Ni	139	3.4	2458	45[#]	10700[a]	1.1	-
ppb							
Au	0.71	3.2	2300	0.40	148	7.1	4.90
Ir	0.023	2	20	0.03	480	0.028	25.19
Os	0.018	0.11	79	0.03	492	0.018	26.94
Pd	0.6	1.9	13	2.00	560[a]	4	32.20
Pt	1.6	8.2	49	-	982	-	153*
Re	0.18	0.63	0.15	0.08	39	0.01	0.58
Rh	-	< 5	-	0.38	140	-	9.58
Ru	-	< 400	65	1.06	683	-	38.12

Data sources: (1) Terashima et al. (1994). (2) Gladney et al. (1991). (3) Schmidt (1997) except # Schmidt et al. (1997). (4) Jochum (1996), except a Wasson and Kallemeyn (1988), average. (5) Morgan et al. (1978). (6) Average of 5 samples by Schmidt (1997), except *Evans et al. (1993), one sample.

It would, therefore, be difficult to distinguish a meteoritic component if the entire target was – for example - an ophiolite mantle section. However, impact melt rocks are silica normative when a CIPW norm is calculated, because of the normally high volume of siliceous crustal rocks they incorporate. Mantle rocks are generally not silica saturated, and conventional geochemistry would reveal this (McDonald, personal communication, 2002).

Knowledge of the amount of siderophile elements that are provided by the basement rocks is a prerequisite for the identification of extraterrestrial material in impact melts. This indigenous component has to be subtracted from the melt content of the siderophile elements in the impact melt or breccia to obtain the net meteoritic contribution (e.g., Palme 1980). Mixing calculations can be carried out to determine the relative proportion of the various target rocks types that are known or suspected to contribute to breccias or melt rocks (e.g., French et al. 1997; Koeberl et al. 1998b), for example, rock components found included in a suevite. The harmonic least-squares (HMX) mixing calculation program (Stöckelmann and Reimold 1989) can be employed for this purpose. Unfortunately, mixing calculations are sometimes complicated by the uncertainties surrounding the exact type and composition of target rocks. Erosion or partial burial by younger rock sequences may make it difficult to confidently estimate the proportions of different target rocks, especially if they are large or more than a few million years old (McDonald, personal communication, 2001), or the indigenous PGE concentrations very low or highly variable (e.g., Schmidt and Pernicka 1994). The analytical procedure is also difficult and time consuming, and this method only gives reliable results if the target stratigraphy is simple.

Before making calculations, the average compositions are scrutinized to determine which parameters show variations between the target rock groups large enough to be useful for distinguishing the resulting mixtures. Element concentrations likely to have undergone post-impact changes, such as alkalies, are of little use in calculating target rock proportions, as they may give unreliable results. Once a model of the bulk composition of the impact rocks has been built up, the siderophile element composition of this model is compared to the siderophile element abundances observed in the impactites. Schmidt (1997) used only one analysis (a quartz monzonite from the basement) to infer the entire target suite PGE contribution to the Clearwater East impact melt rocks, which is not very representative of the target area petrology. The composition of that quartz monzonite was then subtracted from the average PGE concentrations of the melt rocks to produce a series of "net PGE ratios". Schmidt's results appeared to match corresponding ratios in CI chondrites. However, PGE-bearing particles in the Clearwater East melt are heterogeneously distributed, yielding a broad range of Ir concentration in the melt rocks (e.g., Palme et al. 1981; Evans et al. 1993). As a result, the geochemical significance of the average concentrations used by Schmidt (1997) is questionable.

The correlation method is a better alternative. PGE-Ir regressions are frequently used for this purpose (e.g., Morgan and Petrie 1979; Palme 1980; McDonald et al. 2001). Among the HSE, Ir and Os have the lowest CI-normalized abundance in crustal rocks (6.5 and 6.2 x 10^{-5}, respectively; Evans et al. 1993). Consequently, their contribution to a melt rock contaminated by a meteoritic component is lower that for other elements. If the meteoritic component is homogeneously distributed and indigenous Ir is negligible, then correlations and their slopes (controlled by the PGE/Ir ratios of the dominant PGE component) can be determined. The y-axis intercepts at Ir equals zero is used to constrain the indigenous PGE contribution (Fig. 1). The correlation method is limited by the fact that small variations in the absolute concentrations of HSE in impact melt rock samples with high concentrations of strongly siderophile elements have a large influence on the y-intercept (at Ir = 0) (Schmidt 1997). Fitting a tightly constrained regression line, through many samples of impact melt from a crater, averages out deviations for individual samples and permits extrapolation back to a y-axis intercept that is independent of assumptions over the target rock composition (McDonald et al. 2001). McDonald (2002) reviewed and re-interpreted the available PGE data from samples from the Clearwater East impact melt with this approach and concluded that the most consistent projectile was an ordinary (possibly type-L) chondrite. This agrees with Cr isotope data of Shukolyukov and Lugmair (2000b).

Table 2. Comparison of element ratios of different meteorites and impact melt rocks

| | Meteorites | | | | | | Impact Melt Rocks | |
	CI	H	L	LL	IIIAB	IAB	ClearwaterEast	Morokweng
Os/Ir	1.06[#]	-	-	-	-	-	0.98	-
Ru/Ir	1.42	1.42	1.47	1.49	1.38	2.04	1.53	1.50
Rh/Ir	0.29	0.30	0.33	0.34	-	-	0.36	0.53
Pd/Ir	1.21[#]	1.11	1.35	1.64	0.29	1.01	1.31[#]	1.54
Pt/Ir	1.97	2.08	2.17	2.22	1.9	2.2	8.6*	2.10
Ni/Ir	23000*	-	-	-	1900	7800	-	13400
Cr/Ir	5.5*	-	-	-	0.02	-	-	8020
Ir/Au	3.22	3.45	3.03	2.70	5.55	1.05	5.48[#]	3.22
Pt/Au	7.07*	-	-	-	10.7	2.3	2.95*	-
Pd/Au	4.0*	-	-	-	6.6	1.2	1.97*	-
Ru/Rh	4.95	4.63	4.40	4.33	-	-	3.04	4.14
Ir/Re	12[#]	-	-	-	-	-	57[#]	-
Pt/Pd	-	1.88	1.16	1.35	-	-	-	1.29
Pt/Ru	1.39*	-	-	-	1.4	1.1	8.6*	-
Pd/Ru	0.79*	-	-	-	0.12	3.55	2.95*	1.09

Data sources: Averages: CI: Jochum (1996), 2 analyses, except [#]Schmidt (1997) and *Evans et al. (1993) and references therein. H, L and LL: McDonald (2002) Number of analyses: H, 2; L, 3; LL, 2. IIIAB and IAB: Evans et al. (1993) and references therein. Clearwater East: Impact melt rocks, McDonald (2002), except [#] Schmidt (1997), 5 analyses, and *Evans et al. (1993), 2 analyses. Morokweng: Impact melt rocks, McDonald et al. (2001), 16-18 analyses, depending on ratio.

Different processes occurring during the formation of impact glasses and melts cause problems with meteoritic component identification. The main factors controlling the incorporation of a meteoritic component into impact rocks are impact angle and projectile velocity (see introduction). However, this energy scaling relationship does not explain the fractionation detected within a single crater (e.g., Meteor Crater, Wabar, and some Australian craters). Locations at different radial distances from the impact site have been found to show different siderophile element and PGE signatures, which do not coincide exactly with the interelement ratios of the impactor when it is partly preserved. Differences in vaporization and/or condensation temperatures could cause the PGE to fractionate among themselves, but no obvious correlation with any physical or chemical properties has jet been detected (Attrep et al. 1991; Mittlefehldt et al. 1992 a, b; Koeberl and Shirey 1997). Fractionation effects have also been detected for distal ejecta from the K-T boundary impact in different localities (Evans et al. 1993; Evans and Chai 1997).

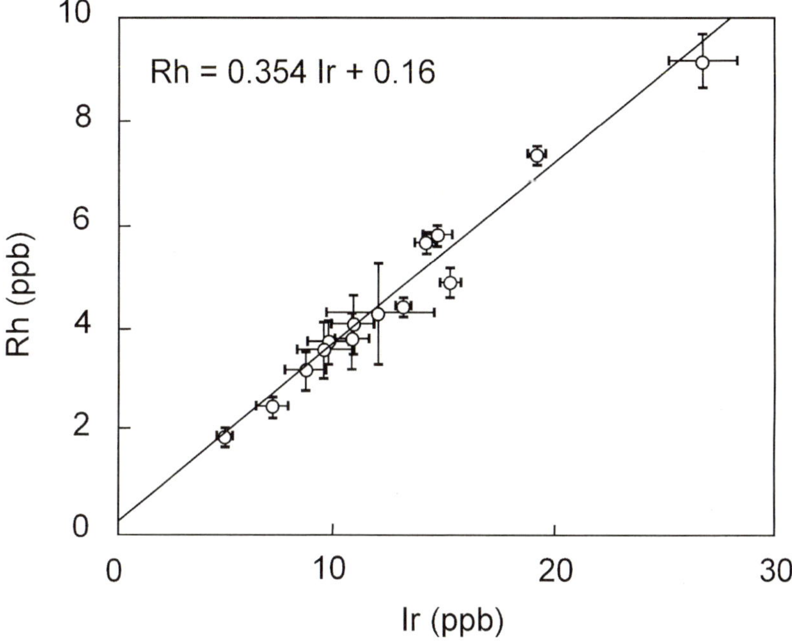

Fig. 1. Regression between contents of Ir and Rh in impact melt rocks from the Morokweng structure, South Africa. Error bars show one standard deviation of the mean concentration. From McDonald et al. (2001).

Siderophile element fractionation in the impact melt while still molten is also possible. In large craters, where the melt can remain hot for several thousand years, different mineral phases, such as pyroxenes, magnetite and chromite, may take up various proportions of the siderophile elements (Ni, Co, and Cr, but not PGE), leading to an irregular distribution of these elements and, possibly, fractionated interelement ratios and patterns (Palme et al. 1979; Koeberl et al. 1994a, Koeberl 1998). This can render ratios such as Cr/Ni, Ni/Ir and Cr/Ir of limited use. The fractionation of PGE observed in Clearwater East and Morokweng was attributed to a small-scale redistribution in sulphide and oxide phases (e.g., magnetite, chromite; Koeberl 1997). However, other authors (McDonald, personal communication, 2001) consider that the evidence for vaporization or condensation fractionation of the PGEs in thick melt sheets or global ejecta is not strong, and possibly explained by the small sizes of the sample taken for analysis. When larger samples are considered, the PGEs appear less fractionated.

Hydrothermal processes associated with the hot impact melt may also change the PGE concentrations in impact rocks (Gostin et al. 1989; Wallace et al. 1990), although distant ejecta deposits are not affected by this process. The relative mobility of these elements has to be considered before interpreting the ratios involving PGE, as this may lead to non-chondritic ratios even if a meteoritic component is present. For example, the Au/Ir ratio has been used to distinguish between cosmic and terrestrial signatures (Palme 1982). However, Au is much more mobile than Ir under terrestrial conditions, making Au/Ir less useful than other PGE ratios. Remobilization from ejecta deposits may also differ between marine (underwater at the time of the impact) and continental (on land at the time of the impact) sites. In marine K-T sites, the mobility sequence was found to be Rh > Au > Pd > Pt > Ru > Ir, while in continental K-T sites an Au > Pd > Pt ≈ Rh > Ru > Ir sequence was determined (Evans et al. 1993). As Ru and Ir are the least mobile of the PGE (Cousins 1973; Cousins and Vermaak 1976; Westland 1981), the use of the Ru/Ir ratio minimizes problems in impactor identification created by differential remobilization. Evans et al. (1993) concluded that the incomplete database of PGE contents in meteorites, and effects of fractionation during melt sheet cooling or the vaporization/condensation of ejecta, limit the application of PGE abundances and interelemental ratios to impactor identification. Nevertheless, more complete datasets on PGE in impactites and in different meteorite types are now available, and the quality of the data has been improved by appropriate replicate analysis, overcoming the problem of the precision levels (McDonald 1998). A large number of samples is necessary in order to perform statistical analysis of the PGE ratios and reliable regression analysis. When these conditions are fulfilled, PGE ratios can help to distinguish among different types of projectiles (McDonald et al. 2001; McDonald 2002).

3
The Re-Os Isotopic System

The abundance of Re and Os and the $^{188}Os/^{187}Os$ isotopic ratios may allow a confirmation of the presence of a cosmic component in terrestrial rocks, although they do not permit the determination of the meteorite type. This was first shown for the K-T boundary by Turekian (1982) and Luck and Turekian (1983), and for rocks at impact craters by Fehn et al. (1986). The Re-Os isotopic system is based on the ^{187}Re decay via β-decay into ^{187}Os, with a halflife of 42.3 ± 1.3 Ga (Lindner et al. 1989). All seven naturally occurring osmium isotopes are stable. Regarding the geochemical behavior of Re and Os, the former is moderately incompatible and is, therefore, enriched in the melt, whereas the latter is strongly retained in the mantle during partial melting of mantle rocks and remains in the residue. This behavior results in high Re, but low Os, concentrations in most crustal rocks, and their Re/Os ratio is usually no less than 10 (Koeberl 1998). The Earth's mantle has Re and Os concentrations, which are much lower than those of meteorites. However its Re/Os ratio is indistinguishable from that of meteorites (Faure 1986), especially chondrites and irons, which have relatively high Re and Os contents, with Os more abundant than Re (ca. 600 and 50 ppb, respectively, in chondrites), resulting in Re/Os ratios ≤ 0.1. Moreover, achondrites are an exception, as they have low Re and Os concentrations (contents of 0.06 and 0.44 ppb, respectively, have been measured in Moore County eucrite; Mason 1979; Table 3). Consequently, meteoritic and mantle rock $^{187}Os/^{188}Os$ ratios experience only small changes with time, in contrast to crustal rocks (Fig. 2). Present-day mantle has a low $^{187}Os/^{188}Os$ ratio of 0.12-0.13 (e.g., Smoliar et al. 1996, Meisel et al. 1996). Meteorites also have low $^{187}Os/^{188}Os$ ratios. For example, a suite of data for 12 ordinary chondrites define a narrow range of present day $^{187}Os/^{186}Os$ of 0.1289 $\pm$ 0.0011. Data for 10 enstatite chondrites define a similar average – present day $^{187}Os/^{186}Os$ of 0.1283 $\pm$ 0.0005. In contrast, a suite of four carbonaceous chondrites define a 1-2% lower $^{187}Os/^{186}Os$ of 0.1259 $\pm$ 0.0005 (Meisel et al. 1996). $^{187}Os/^{188}Os$ ratios of about 0.67-1.61 are representative of upper continental crust (Koeberl and Shirey 1997; Koeberl et al. 1998a).

Table 3. Average abundances of Re and Os in terrestrial and extraterrestrial materials.

Rock	Re (ppb)	Os (ppb)	Re/Os
CI chondrites (1)	36.5	486	0.075
Ordinary chondrites (1)	57	660	0.086
IIAB (1)	1.39-4799	12.7-65270	0.073-0.11
IIIAB (1)	2.83-1444	17.1-18430	0.078-0.202
Eucrites (1)	0.01-0.06	0.018-0.44	0.081-0.136
Basalt JB-1A (2)	0.18	0.018	10
Granite G-1 (3)	0.63	0.11	5.727
Peridotite JP (2)	0.15	79	0.002
Average Upper Continental Crust (4)	0.8	0.03	2.66
Average Upper Continental Crust (5)	0.198	0.031	6.387

Data sources: (1) Koeberl and Shirey (1997) and references therein; (2) Terashima et al. (1994); (3) Gladney et al. (1991); (4) Schmidt (1997); (5) Peucker-Ehrenbrink and Jahn (2001).

Due to the high Os and Re abundances in chondrites, the admixture of a small meteoritic component to crustal target material significantly alters the Os isotopic characteristics of the resulting impact melt or breccias (Table 4, Fig. 3). While a high abundance of Os could be due to crustal enrichment processes and incorporation of ore minerals, non-crustal Os isotopic compositions clearly indicate the presence of a meteoritic or mantle component. To distinguish between mantle and meteoritic signals, the differences in total abundance of Os in both materials is considered (Koeberl and Shirey 1997).

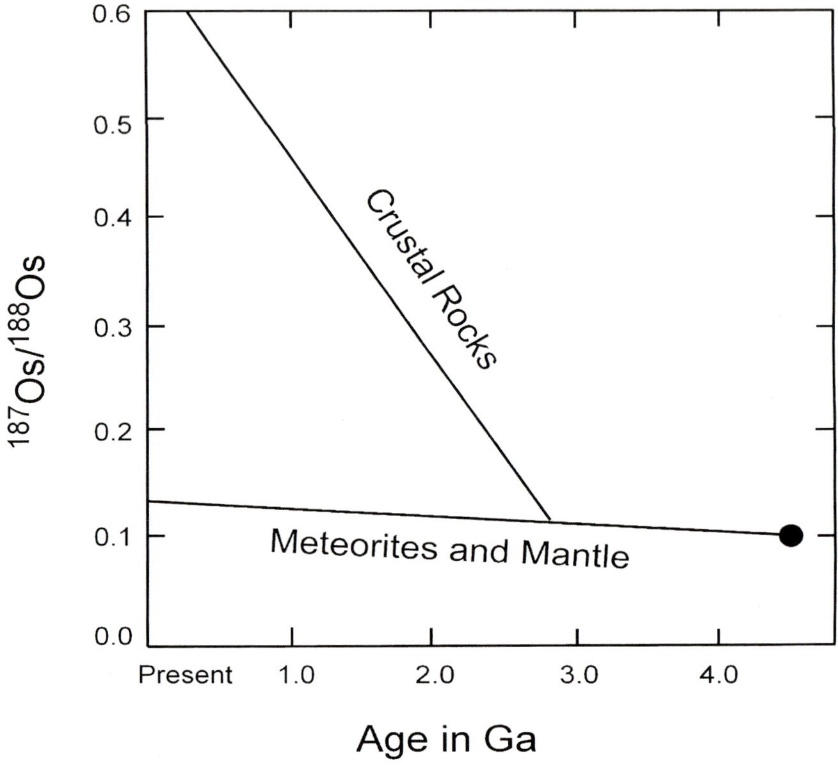

Fig. 2. Schematic evolution of the ^{187}Os/^{188}Os ratio of meteorites and the Earth's mantle and crust. From Faure (1986).

Mantle rocks have about 1-4 ppb Os, and typical chondrites have about 400-800 ppb Os. Thus, about 100 times more mantle material than meteoritic material would need to be added to normal crustal rocks to give the same Os isotopic ratio as the bulk rock. Ultramafic precursor rocks may be discerned by detailed geological field investigation, petrographic study of the rocks, trace element and, if necessary, supplementary Rb-Sr and/or Sm-Nd isotopic analysis (Koeberl and Shirey 1997).

The method has been applied to the K-T boundary clay in various locations, in order to test its impact versus volcanic origin, e.g., Stevns Klint, Denmark (Luck and Turekian, 1983), Woodside Creek, New Zealand (Lichte et al. 1986), and Sumbar, Turkmenistan (Meisel et al. 1995). In the latter study, the variation of the $^{187}Os/^{188}Os$ ratio was measured across a complete boundary section. A significant sudden decrease of the $^{187}Os/^{188}Os$ ratio from the end-Cretaceous rock layers to the actual K-T boundary clay was observed, in correlation with the maximum Ir and Os concentration; in the early Tertiary rocks, the $^{187}Os/^{188}Os$ ratio increases to higher values. The osmium isotopic system has also proved useful to confirm the impact origin of suspicious structures. The origin of Vredefort structure in South Africa, for example, has been controversial. Granophyric rock dykes exposed in its basement and along the boundary between the core and the supracrustal rocks have been suggested to represent either an igneous intrusion (e.g., Bisschoff 1972), or impact melt injected into fractures in the floor of the impact structure (French et al. 1989; French and Nielsen 1990). Koeberl et al. (1996a) found that most Vredefort Granophyre samples have considerably higher Os contents than the country rocks from which the granophyre is likely to have been formed. The $^{187}Os/^{188}Os$ ratios overlap the meteoritic data range and indicate that the granophyre samples contain some meteoritic Os, suggesting up to 0.2% of a chondritic component. The Vredefort Granophyre was, therefore, confirmed to represent an impact melt rock. Other case studies are the Ivory Coast tektites and

Table 4. Os abundances and isotopic data for K-T boundary samples, and a variety of impact glasses and impact breccias (Koeberl and Shirey 1997, and references therein).

Sample	Os (ppb)	$^{187}Os/^{188}Os$	$^{187}Os/^{186}Os$
K-T boundary localities			
Starkville South, CO, U.S.A.	25	0.14	1.2
Madrid, CO, U.S.A.	12.2	0.140	1.167
Woodside Creek, New Zealand	60	0.135	1.12
Stevns Klint, Denmark	110	0.1668	1.386
Ivory Coast tektites			
IVC 8902	0.0889	0.2087	1.734
IVC 2069	0.129	0.1528	1.270
Bosumtwi crater impact glasses			
BI 9201	0.125	0.9009	7.49
Kalkkop crater impact breccias			
Br-2 (100.4)	0.0354	0.487	4.049
Br-3 (112.7)	0.1886	0.2149	1.790
Chicxulub impact melt rocks			
CI-N10-1A	25.2	0.113	0.941
CI-N10-2	0.056	0.505	4.200

Morokweng crater (Koeberl and Shirey 1993a, b; Koeberl et al. 2002), Chicxulub, Mexico (Koeberl et al. 1994a), Saltpan crater, South Africa (Koeberl et al. 1994b), Kalkkop crater, South Africa (Koeberl et al. 1994c), or the Manson structure, Iowa, U.S.A. (Koeberl and Shirey 1996; Koeberl et al. 1996b).

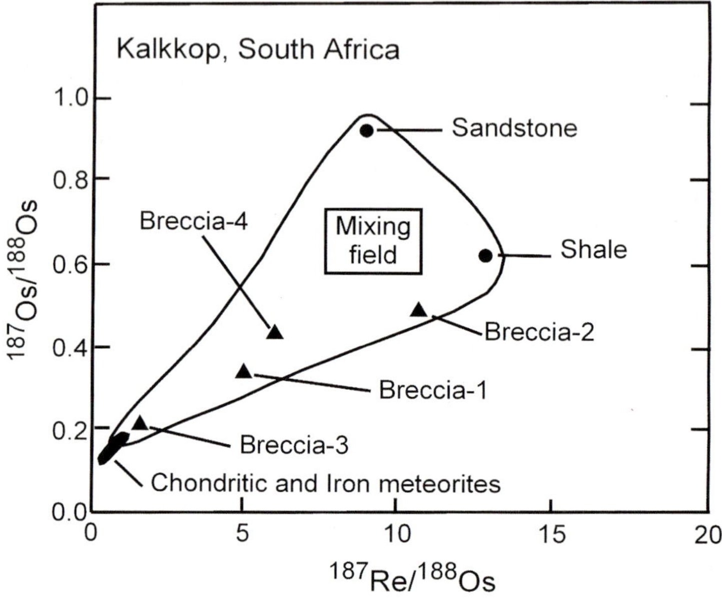

Fig. 3. Ratios of ^{187}Os/^{188}Os versus ^{187}Re/^{188}Os for target rocks (shale and sandstone) of the Kalkkop impact crater, South Africa, in comparison with data for four impact breccias and the data array for chondritic and iron meteorites (small solid dots). From Koeberl and Shirey (1997).

4
The Cr-Mn Isotopic System

The chromium isotope systematics were recently explored for impact studies. The radioactive nuclide ^{53}Mn decays to stable ^{53}Cr with a half-life of 3.7 Ma. 53Mn is now extinct in the Solar System, but was present when the early planetesimals were forming, as indicated by variations in the relative abundance of the radiogenic daughter ^{53}Cr in various ancient objects in the Solar System (Birck and Allègre 1988; Hutcheon et al. 1992; Nyquist et al. 1997). Lugmair and Shukolyukov (1998) performed high-precision mass spectrometric analysis of chromium, and developed a technique that allows to measure small ^{53}Cr/^{52}Cr variations of less than 1 ε with an uncertainty of 0.05 to 0.10 ε units (1 ε is one part in 10^4). These isotopic variations are measured as the deviations of the

$^{53}Cr/^{52}Cr$ ratios from the standard terrestrial $^{53}Cr/^{52}Cr$ ratio. Terrestrial samples exhibit $^{53}Cr/^{52}Cr$ ~0 ε regardless of their origin, because the Earth homogenized after ^{53}Mn had fully decayed (Table 5). Samples from the Moon give the same result as the Earth, because of their close genetic relationship (e.g., Hartmann and Davis 1975; Hartmann 1986; Stevenson 1987; Melosh 1989). Most classes of meteorite have excess ^{53}Cr relative to the terrestrial value. The ordinary chondrites show a characteristic $^{53}Cr/^{52}Cr$ of ~0.48 ε. Although the $^{53}Cr/^{52}Cr$ ratios of individual eucrites and diogenites vary because of an early planet-wide Mn/Cr fractionation, their parent body (Vesta) is characterized by the close-to-chondritic ^{53}Cr excess of ~0.57 ε. The Mn-Cr isotope systematics of the angrites, primitive achondrites, and pallasites are also consistent with $^{53}Cr/^{52}Cr$ ratios of ~0.5 ε in their bulk parent bodies. The ^{53}Cr excess is ~0.22 ε for the Martian meteorites, and ~0.17 ε for the EH-chondrites (Fig. 4; Shukolyukov and Lugmair 1998). Carbonaceous chondrites, however, show an apparent deficit of ^{53}Cr of ~-0.40 ε. This results from the use of $^{54}Cr/^{52}Cr$ ratio for a second order fractionation correction. They were found to contain Cr of presolar origin, characterized by mostly elevated but sometimes lower than normal $^{54}Cr/^{52}Cr$ ratios (Podosek et al. 1997). The actual, unnormalized $^{53}Cr/^{52}Cr$ ratio is similar to that of other undifferentiated meteorites, and the apparent ^{53}Cr deficit in the carbonaceous chondrites is actually due to an excess of ^{54}Cr.

Since the measured excesses of radiogenic ^{53}Cr are very small for the samples with relatively low Cr concentrations, such as olivines and those with long exposure ages, a potential contribution from spallation reactions has to be taken into account. The use of the $^{54}Cr/^{52}Cr$ ratio for a second order fractionation correction introduces an additional uncertainty due to the presence of spallogenic ^{54}Cr. Birck and Allègre (1985) determined production rates for Cr in the iron meteorite Grant. The production rates for ^{53}Cr and ^{54}Cr turned out to be approximately the same: ~2.9 x 10^{11} atoms/g per Ma. These values agree reasonably well with the less precise values for the production rate of ^{53}Cr in Grant of ~2.3 x 10^{11} atoms/g per Ma by Shima and Honda (1966). Thus, with known exposure age and Cr and Fe concentrations (Fe is the main target for Cr production), a spallation correction can be performed (Lugmair and Shukolyukov 1998).

The observed distribution of radiogenic ^{53}Cr may not be due to differences in the bulk Mn/Cr ratios of the parent bodies. Instead, this distribution may reflect an original spatial heterogeneity of ^{53}Mn in the early Solar System, which is now evident as a radial gradient in the radiogenic ^{53}Cr abundances (Lugmair and Shukolyukov 1998; Shukolyukov and Lugmair 2000a). Regardless of the scenario, this observed difference allows us to distinguish extraterrestrial material on the basis of high-precision measurements of the Cr isotopic composition.

This method was successfully applied by Shukolyukov et al. (1999) and Koeberl et al. (2002) for the Morokweng structure impact melt. A substantial portion of the Cr in their samples was found to be of cosmic origin ($^{53}Cr/^{52}Cr$ = 0.24-0.27 ε), and the isotopic ratio was clearly different from those of the carbonaceous and enstatitic chondrites. Thus, they concluded that the most probable projectile is an ordinary chondrite type material. The authors

Table 5. $^{53}Cr/^{52}Cr$ ratios and Cr concentrations in terrestrial and extraterrestrial samples (from Shukolyukov and Lugmair 1998).

Sample	Cr (ppm)	$^{53}Cr/^{52}Cr$ (ε*)
Terrestrial minerals, rock, and sediment		
Laboratory standard - terrestrial normal	-	$\equiv 0$
KH-1 Px, Kilbourne Hole, USA (px)	2500	-0.01 ± 0.08
JAG 89-9, Jagersfontein, S. Africa (gte)	-	0.04 ± 0.08
SC Ol, San Carlos Volcanic Field, USA (ol)	202	-0.03 ± 0.11
MB 81-14, Deccan Traps, India (basalt)	112	-0.04 ± 0.06
ODP 31-302-5-5, Western Pacific (clay)	34	-0.02 ± 0.09
Bulk meteorites and their parent bodies		
Allende (CV3)	3540	-0.38 ± 0.11
Orgueil (CI)	2530	-0.46 ± 0.07
OC (H, L) and their parent bodies	~ 3900	~0.48
Eucrites	1600-3200	0.7 - 1.3
Diogenites	6000-12700	0.4 - 0.6
HED parent body (Vesta)	-	~0.57†
Primitive achondrites and their parent bodies	-	~0.53
Angrites	800-1800	0.4 - 0.7
Angrite parent body	-	~0.48†
Pallasite parent body	-	~0.52†
SNC-meteorites (Mars)	-	~0.22
EC (EH) and their parent body	2900-3900	~0.17
Lunar anorthosite 60025	24	0.00 ± 0.09

*1 ε unit is 1 part in 10^4. †Calculation based on Mn-Cr isotope systematics. px = pyroxene, gte = garnet, ol = olivine.

complemented the Cr isotope study with the PGE ratios (Cr/Ir) to focus their result to a L chondrite. Shukolyukov and Lugmair (1998) measured chromium concentrations in K-T boundary sediments commonly 20 to 30 times higher than those in background sediments (e.g., Kyte et al. 1980, 1985). In samples from Caravaca, Spain, Shukolyukov and Lugmair (1998) measured Cr isotopic characteristics of the K-T boundary clays, finding contents of 991 ppm Cr, in comparison with 40 and 69 ppm measured below and above the impact layer (Table 6), and the 185 ppm of the average Cr concentration of the bulk continental crust (Taylor and McLennan 1985). The Cr isotopic system can then be used to test if a considerable part of the Cr in a supposed impact layer is indeed of cosmic origin (rather than mostly from terrestrial ejecta material or volcanic ash), and it can provide direct isotopic evidence for the impact hypothesis, and in particular the impactor type. However, Shukolyukov and Lugmair (1998) noted that none of the meteorite classes they studied have a Cr isotopic composition similar to that of the K-T samples (which have a deficit of ^{53}Cr). Furthermore, they tentatively indicated that the obtained deficit of ^{53}Cr could be the result of an elevated ^{54}Cr/^{52}Cr ratio in the samples (as discussed in the first paragraph of this section). In consequence, a carbonaceous chondrite was considered to be the best candidate for the impactor type. Their calculations showed 80% of the Cr in the K/T sediments could have originated from a meteorite of this class.

Table 6. ^{53}Cr/^{52}Cr ratios and Cr concentrations in the K-T boundary samples and their backgrounds, and impact rocks.

Sample	Cr (ppm)	^{53}Cr/^{52}Cr (ε*)
K-T boundary and background clays (1)		
SK503, Caravaca, Spain	991	-0.40 ± 0.08
CA10b, Caravaca, Spain, 10-15 cm below K-T (clay)	40	-0.01 ± 0.07
CA16a, Caravaca, Spain, 16-20 cm above K-T (clay)	69	-0.02 ± 0.12
Barberton impact deposits and background material (2)		
D-4, Barberton spherule bed, South Africa	-	-0.32 ± 0.06
Background sample, 3 m below	-	0.01 ± 0.06
Morokweng impact melt (3)		
MO15, Morokweng, South Africa	359	0.24 ± 0.04
MO48, Morokweng, South Africa	408	0.27 ± 0.03

Data sources: (1) Shukolyukov and Lugmair (1998). (2) Shukolyukov and Lugmair (2000b). (3) Shukolyukov et al. (1999). *1ε unit is 1 part in 10^4.

Chromium isotopic studies were necessary to confirm the presence of an extraterrestrial component in the spherule bed (S4) from the Barberton Greenstone Belt, South Africa, supporting the hypothesis of an impact origin, and making this the oldest impact deposit on Earth (Shukolyukov et al. 2000). Previous PGE analyses showing relative abundances of Ir, Pt, and Os nearly similar to those in chondritic meteorites (Kyte et al. 1992) were not considered conclusive by some

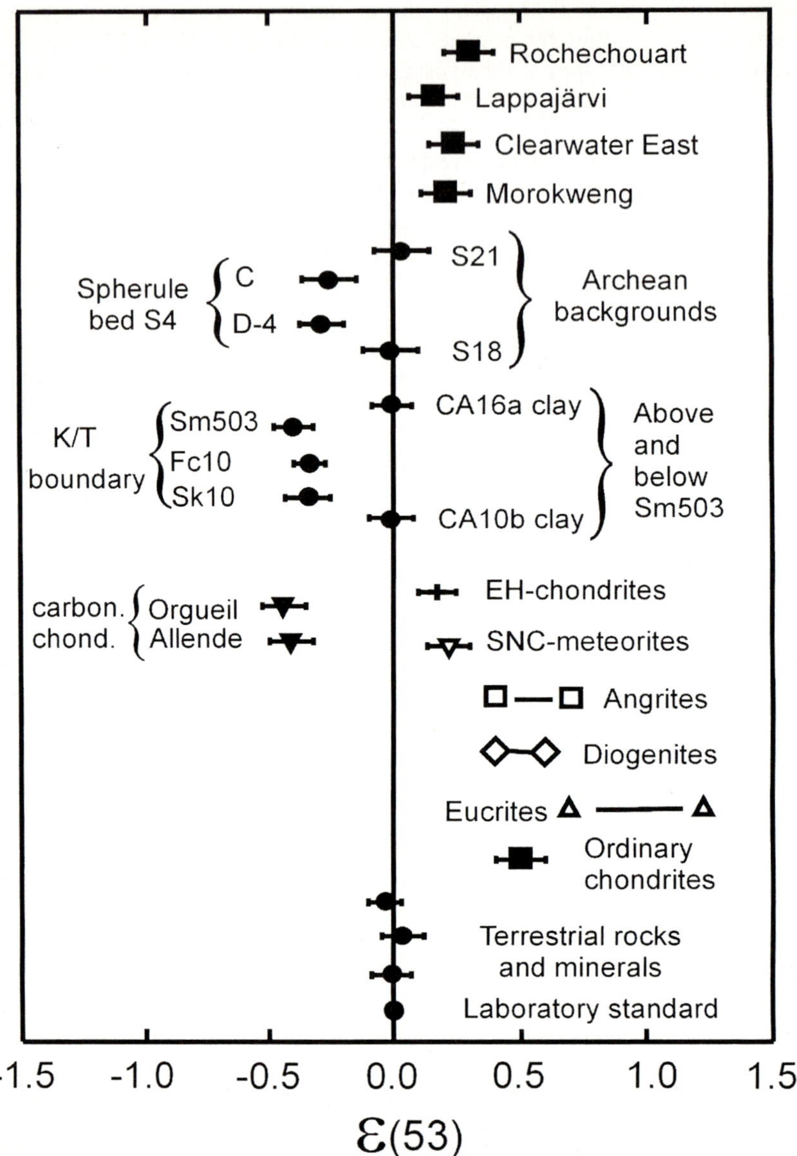

Fig. 4. ^{53}Cr/^{52}Cr ratios in various terrestrial, impactite, K-T boundary and meteoritic samples. After Shukolyukov and Lugmair (2000b) and Shukolyukov et al. (2000). K-T boundary samples: SM503: Caravaca, Spain; FC10 and SK10: Stevens Klint, Denmark. Archean samples: Spherule bed S4: Barberton Greenstone Belt, South Africa.

authors (Koeberl et al. 1993; Koeberl and Reimold 1995), who argued that the extreme enrichments of siderophile elements (up to 5 times the chondritic abundances), much higher than in any other known impact deposit (or even in meteorites), could be explained by secondary mineralization. However, samples from the spherule bed were found to have a $^{53}Cr/^{54}Cr$ ratio between ~0.26 ± 0.11 ε. and ~0.32 ± 0.06 ε, where as background sediments yielded normal terrestrial $^{53}Cr/^{54}Cr$ ratio (Shukolyukov et al. 2000).

A drawback with the chromium isotopic method is that a substantial amount of the chromium has to be of extraterrestrial origin to show an effect in the Cr isotopic composition of terrestrial rocks. For example, in rocks with crustal Cr abundances, only meteoritic components <1.2% can be detected (Koeberl et al. 2002). Therefore, the Cr isotopic method is much less sensitive than the Os isotopic method. Additionally, analytical procedure is complicated and requires extreme precision, thus limiting the number of samples that can be analyzed. Nevertheless, the Cr isotopic method, where applicable, can provide additional information regarding the nature of the impactor (Koeberl et al. 2002). Although this determination is not possible for iron meteorites, which do not carry significant amounts of Cr, it allows to discern different types of ordinary chondrites, and it is the only procedure that can discriminate achondritic impactors, which have low concentrations of Os, but higher abundances of Cr.

5
Conclusions

Meteoritic components have been identified in about a quarter of the craters known on Earth (Table 7; Koeberl 1998). The estimates of impacting body types are dominated by chondrites (41% of the inferred impactors), probably because their geochemical signature is the most easily identifiable due to the relative high abundance of siderophile elements and Cr. Non-chondritic impacting bodies are often identified simply because the body was not chondritic, rather than through any firm indication that it was an achondritic or an iron body (Grieve 1991, 1997). Iron meteorites have been proposed for 34% of the impact structures, mainly on the basis of the presence of meteorite fragments, or metallic spherules, at the crater site. Conflicting identifications have been made for a number of impact structures. Both iron and chondritic projectiles have been proposed for Bosumtwi (Palme et al. 1978; Koeberl and Shirey 1993b), Zhamanshin (Palme et al. 1978; Glass et al. 1983), and Rochechouart (Wolf et al. 1980; Lambert 1982) impact structures. Many identifications based only on Ni, Co, and Cr studies are highly uncertain, as no PGE or isotopic data are available.

A search for siderophile anomalies at some impact structures may be unsuccessful due to a variety of reasons already mentioned above. This may include sampling problems, the impact of a differentiated projectile with little or no siderophile element enrichment with respect to terrestrial materials, or uncertainties in making corrections for indigenous siderophile elements because of the many different types of target rocks. Projectile type determination is also complicated by fractionation processes that take place during the formation of impact melts and glasses, vaporization/condensation of the ejecta, and the post-impact alteration and mobilization of the key elements (Evans et al. 1993; Schmidt 1997; Koeberl 1998). However, if the previous drawbacks can be excluded, PGE ratios can identify projectiles if samples of sufficient quality are carefully analyzed (e.g., McDonald et al. 2001); these studies can be complemented with Cr isotopic studies (e.g., Shukolyukov et al. 1999; Koeberl et al. 2002).

Isotopic studies of Os and Cr are very efficient in the detection of meteoritic component in a number of impact melt rocks and breccias. The Os isotopic method is a more sensitive tool for detecting an extraterrestrial component, although laboratory procedures (sample preparation, digestion, and measurement) are complex, and the Os isotopic method does not give information regarding the projectile type. The Cr isotopic method has the disadvantage of involving a complicated and time consuming analytical procedure. Also, a significant proportion of the Cr in an impactite, compared to the abundance in the target, has to be of extraterrestrial origin. However, where applicable, it can discriminate the Cr source (terrestrial or extraterrestrial) and provide additional information on the nature of the impactor (Koeberl et al. 2002).

Table 7. Terrestrial impact structures with inferred impactor types. Modified from Koeberl (1998) (see footnotes).

Name	Location	Diameter	Impactor	Evidence
Kaalijärvi	Estonia	0.11*	Iron (IAB)	M
Wabar	Saudi Arabia	0.116	Iron (IIIAB)	M, S
Henbury	Northern Territory, Australia	0.157*	Iron (IIIAB)	M, S
Odessa	Texas, USA	0.168*	Iron (IIIAB)	M
Boxhole	Northern Territory, Australia	0.17	Iron (IIIAB)	M
Macha	Russia	0.3*	Iron	MS
Aouelloul	Mauritania	0.39	Iron (IIIB, IIID?)	S, Os
Monturaqui	Chile	0.46	Iron (IAB)	M, S
Kalkkop	South Africa	0.64	Chondrite?	S
Wolfe Creek	Western Australia, Australia	0.875	Iron (IIIAB)	M, S
Barringer (Meteor)	Arizona, USA	1.186	Iron (IAB)	M, S
Tswaing (Saltpan)	South Africa	1.2	Chondrite	S, Os
New Quebec	Quebec, Canada	3.44	Chondrite	S
Brent	Ontario, Canada	3.8	Chondrite (L?)	S

Table 7. (cont.)

Name	Location	Diameter	Impactor	Evidence
Rio Cuarto	Argentina	4.5*	Chondrite (H)	M, S, Os
Gow Lake	Saskatchewan, Canada	5	Iron?	S
Gardnos	Norway	5	Chondrite	S, Os
Sääksjärvi	Finland	6	Stony-iron?	S
Wanapitei Lake	Ontario, Canada	7.5	Chondrite	S
Ilyinets	Ukraine	8.5	Iron?	S
Mien	Sweden	9	Stone?	S
Bosumtwi	Ghana	10.5	Chondrite? Iron?	S, Os
Ternovka	Ukraine	11	Chondrite?	S
Nicholson Lake	Northwest Territories, Canada	12.5	Achondrite	S
Zhamanshin	Kazakhstan	14	Chondrite (Iron?)	S
El'Gygytgyn	Russia	18	Achondrite?	S
Dellen	Sweden	19	Stone?	S
Obolon	Ukraine	20	Iron?	S
Lappajärvi	Finland	23	Chondrite	S, Cr
Rochechouart	France	23	Chondrite? Iron?	S, Cr
Ries	Germany	24	Achondrite?	S
Boltysh	Ukraine	24	Chondrite?	S
Strangways	Northern Territory, Australia	25	Achondrite	S
Clearwater East	Quebec, Canada	26	Chondrite (L?)	S, Cr
Mistastin	Labrador, Canada	28	Iron?	S
Manson	Iowa, USA	35	Chondrite	S, Os
Mjølnir	Norway	40	Iron	S
Kara	Russia	65	Chondrite?	S
Acraman	South Australia, Australia	90	Chondrite	Es
Morokweng	South Africa	70-80	Chondrite (L or LL)	S, Cr, Os
Popigai	Russia	100	Chondrite	S, Es

Table 7. (cont.)

Name	Location	Diameter	Impactor	Evidence
Chicxulub	Yucatan, Mexico	170	Chondrite	S, Os, Es
Vredefort	South Africa	300	Chondrite	S, Cr, Os

Diameters (in kilometers) from: Earth Impact Database (2002), except Morokweng (Reimold et al. 2000) *Crater field; diameter corresponds to the largest dimension of largest structure. Evidence: S, siderophile element enrichment and/or pattern; Os, Os isotopic ratio; M, meteorite fragments; MS, metallic spherules; Es, siderophile element enrichment in ejecta. References in Koeberl (1998), and later works: Kalkkop (Reimold et al. 1998), S (PGE) in Morokweng (McDonald et al. 2001) and Clearwater Lake East (McDonald 2002), Cr in Lappajärvi, Rochechouart and Clearwater Lake East (Shukolyukov and Lugmair 2000b), S in Boltysh (Lorenz 1999) and Mjølnir (Dypvik and Attrep 1999), M in Lappajärvi (Badjukov and Raitala 2000) and Cr and Os in Morokweng and Vredefort (Koeberl et al. 2001, 2002).

Acknowledgements

Frank Kyte and Iain McDonald provided constructive reviews of the manuscript; and Christian Koeberl added helpful comments and editorial advice. We thank Paul Giblin for the revision of the manuscript. This work forms part of the PhD of MJME and is also related to the main research guidelines of the Laboratory of Planetary Geology (Centro de Astrobiologia) in the framework of the IMPACT program of the ESF. The first author acknowledges a predoctoral grant (FP98-20250393) from the Ministerio de Ciencia y Tecnología.

References

Anders E, Grevesse N (1989) Abundances of the elements: Meteoritic and solar. Geochimica et Cosmochimica Acta 53: 197-214

Attrep M, Orth CJ, Quintana LR, Shoemaker CS, Shoemaker EM, Taylor SR (1991) Chemical fractionation of siderophile elements in impactites from Australian meteorite craters [abs.] Lunar and Planetary Science 22: 39-40

Badjukov DD, Raitala J (2000) Schock-reworked remnants of a projectile matter in impact melts of the Lappäjarvi crater. [abs.] Lunar and Planetary Science 31: abstract no. 1591

Birck JL, Allègre CJ (1985) Isotopes produced by galactic cosmic rays in iron meteorites. In: Isotopic Ratios in the Solar System, Cepadues-Editions, Paris, pp 21-25

Birck JL, Allègre CJ (1988) Manganese-chromium isotope systematics and development of the early solar system. Nature 331: 579-584

Bisschoff AA (1972) The dioritic rocks at the Vredefort Dome. Transactions of the Geological Society of South Africa 75: 31-45

Buchwald VF (1975) Handbook of Iron Meteorites. University of California Press, Berkeley, 1418 pp

Cousins CA (1973) Notes on the geochemistry of the platinum group elements. Transactions of the Geological Society of South Africa 76: 77-81

Cousins CA, Vermaak CF (1976) The contribution of Southern African ore deposits to the geochemistry of the platinum group metals. Economic Geology 71: 287-305

Dypvik H, Attrep M (1999) Geochemical signals of the Late Jurasic, Mjølnir marine impact. Meteoritics and Planetary Science 34: 393-406

Earth Impact Database (2002) <http://www.unb.ca/passc/ImpactDatabase/> (Accessed: 4/October/2002)

Evans NJ, Chai CF (1997) The distribution and geochemistry of platinum-group elements as event markers in the Phanerozoic. Palaeogeography, Palaeoclimatology, Palaeoecology 132: 373-390

Evans NJ, Gregoire DC, Grieve RAF, Goodfellow WD, Veizer J (1993) Use of platinum-group elements for impactor identification: Terrestrial impact craters and Cretaceous-Tertiary boundary. Geochimica et Cosmochimica Acta 57: 3737-3748

Faure G (1986) Principles of Isotope Geology. 2nd Edition. J. Wiley and Sons, New York, 589 pp

Fehn U, Teng R, Elmore D, Kubik PW (1986) Isotopic composition of osmium in extraterrestrial samples determined by accelerator mass spectrometry. Nature 323: 707-710

Flores JA, Sierro FJ, Gersonde R (2002) Calcareous plankton stratigraphy around the Pliocene Eltanin asteroid impact area (SE Pacific): documentation and application for geological and paleoceanographic reconstruction. Deep Sea Research. II 49: 1011-1028

French BM, Nielsen RL (1990) Vredefort bronzite granophyre: chemical evidence for an origin as meteorite impact melt. Tectonophysics 171; 119-138

French BM, Orth CJ, Quintana CR (1989) Iridium in the Vredefort bronzite granophyre: impact melting and limits on a possible extraterrestrial component. In: Proceedings of the 19th Lunar and Planetary Science Conference, Cambridge University Press and Lunar and Planetary Institute, Houston, pp 733-744

French BM, Koeberl C, Gilmour I, Shirey SB, Dons JA, Naterstad J (1997) The Gardnos impact structure, Norway: Petrology and geochemistry of target rocks and impactites. Geochimica et Cosmochimica Acta 61: 873-904

Gersonde R, Kyte FT, Bleil U, Diekmann B, Flores JA, Gohl K, Grahl G, Hagen R, Kuhn G, Sierro FJ, Völker D, Abelmann A, Bostwick JA (1997) Geological record and reconstruction of the late Pliocene impact of the Eltanin asteroid in the Southern Ocean. Nature 390: 357-363

Gladney ES, Jones EA, Nickell EJ, Roelandts I (1991) 1988 compilation of elemental concentration data for USGS DTS-1, G-1, PCC-1 and W-1. Geostandards Newsletter 15: 199-396

Glass BP, Fredriksson K, Florensky PV (1983) Microirghizites recovered from a sediment sample from the Zhamanshin impact structure. Proceedings of the 14th Lunar and Planetary Science Conference, Journal of Geophysical Research 88: B319-B330

Gostin VA, Keays RR, Wallace MW (1989) Iridium anomaly from the Acraman impact ejecta horizon: Impacts can produce sedimentary iridium peaks. Nature 340: 542-544

Grieve RAF (1991) Terrestrial impact - The record in the rocks. Meteoritics 26: 175-194

Grieve RAF (1997) Extraterrestrial impact events: the record in the rocks and the stratigraphic column. Palaeogeography, Palaeoclimatology, Palaeoecology 132: 5-23

Gueddari K, La Flèche MR, Camiré G (1999) First data on platinum group elements (PGE) geochemistry of the Mont Albert peridotites (Quebec) Comptes Rendus de l'Académie des Sciences - Series IIA - Earth and Planetary Science 29: 479-486

Hartmann WK (1986) Moon origin: The impact trigger hypothesis. In: Hartmann WK, Phillips RJ Taylor GJ (eds) Origin of the Moon. Houston, Lunar and Planetary Institute, pp 579-608

Hartmann WK, Davis DT (1975) Satellite-sized planetesimals and the lunar origin. Icarus 24: 504-515

Hutcheon ID, Olsen E, Zipfel J, Wasserburg GJ (1992) Chromium isotopes in differentiated meteorites: Evidence for ^{53}Mn [abs.] Lunar and Planetary Science 23: 565-566

Jochum K (1996) Rhodium and other platinum-group elements in carbonaceous chondrites. Geochimica et Cosmochimica Acta 60: 3353-3357

Jones WB (1985) Chemical analysis of Bosumtwi crater target rocks compared with Ivory Coast tektites. Geochimica et Cosmochimica Acta 49: 2569-2576

Koeberl C (1997) Impact cratering: the mineralogical and geochemical evidence. In: Johnson KS, Campbell JA (eds) Proceedings, "The Ames structure and similar features". Oklahoma Geological Survey Circular 100, pp 30-54

Koeberl C (1998) Identification of meteoritical components in impactites. In: Grady MM, Hutchison R, McCall GJH, Rothery DA (eds) Meteorites: Flux with time and impact effects. Geological Society, London, Special Publication 140, pp 133-152

Koeberl C, Reimold WU (1995) Early Archean spherule beds in the Barberton Mountain Land, South Africa: no evidence for impact origin. Precambrian Research 74: 1-33

Koeberl C, Shirey SB (1993a) Osmium isotopes in Ivory Coast tektites: Confirmation of a meteoritic component and rhenium depletion [abs.] Lunar and Planetary Science 24: 809-810

Koeberl C, Shirey SB (1993b) Detection of a meteoritic component in Ivory Coast tektites with rhenium-osmium isotopes. Science 261: 595-598

Koeberl C, Shirey SB (1996) Re-Os isotope study of rocks from the Manson impact structure. In: Koeberl C, Anderson RR (eds) The Manson Impact Structure, Iowa: Anatomy of an Impact Crater. Geological Society of America, Special Paper 302: 311-339

Koeberl C, Shirey SB (1997) Re-Os isotope systematics as a diagnostic tool for the study of impact craters and distal ejecta. Palaeogeography, Palaeoclimatology, Palaeoecology 132: 25-46

Koeberl C, Reimold WU, Boer RH (1993) Geochemistry and mineralogy of early Archean spherule beds, Barberton Mountain Land, South Africa: evidence for origin by impacts doubtful. Earth and Planetary Science Letters 119: 441-452

Koeberl C, Sharpton VL, Schuraytz BC, Shirey SB, Blum JD, Marin LE (1994a) Evidence for a meteoritic component in impact melt rock from the Chicxulub structure. Geochimica et Cosmochimica Acta 58: 1679-1684

Koeberl C, Reimold WU, Shirey SB (1994b) Saltpan impact crater, South Africa: geochemistry of target rocks, breccias, and impact glasses, and osmium isotope systematics. Geochimica et Cosmochimica Acta 58: 2893-2910

Koeberl C, Reimold WU, Shirey SB, Le Roux FG (1994c) Kalkkop crater, Cape Province, South Africa: Confirmation of impact origin using osmium isotope systematics. Geochimica et Cosmochimica Acta 58: 1229-1234

Koeberl C, Reimold WU, Shirey SB (1996a) A Re-Os isotope and geochemical study of the Vredefort Granophyre: clues to the origin of the Vredefort structure, South Africa. Geology 24: 913-916

Koeberl C, Reimold WU, Kracher A, Träxler B, Vormaier A, Körner W (1996b) Mineralogical, petrological, and geochemical studies of drill cores from the Manson impact structure, Iowa. In: Koeberl C, Anderson RR (eds) The Manson Impact Structure, Iowa: Anatomy of an Impact Crater. Geological Society of America, Special Paper 302, pp 145-219

Koeberl C, Armstrong RA, Reimold, WU (1997) Morokweng, South Africa: A large impact structure of Jurassic-Cretaceous boundary age. Geology 25: 731-734

Koeberl C, Reimold WU, Shirey SB (1998a) The Aouelloul crater, Mauritania: On the problem of confirming the impact origin of a small crater. Meteoritics and Planetary Science 33: 513-517

Koeberl C, Reimold WU, Blum JB, Chamberlain CP (1998b) Petrology and geochemistry of target rocks from the Bosumtwi impact structure, Ghana, and comparison with Ivory Coast tektites. Geochimica et Cosmochimica Acta 62: 2179-2196

Koeberl C, Peucker-Ehrenbrink B, Reimold WU, Shukolyukov A, Lugmair GW (2001) Comparison of Os and Cr isotopic methods for the detection of meteoritic components in impact melt rocks from the Morokweng and Vredefort impact structures, South Africa. [abs.] Catastrophic Events and Mass Extinctions: Impacts and Beyond. Abstract no. 3048

Koeberl C, Peucker-Ehrenbrink B, Reimold WU, Shukolyukov A, Lugmair GW (2002) A comparison of the osmium and chromium isotopic methods for the detection of meteoritic components in impactites: Examples from the Morokweng and Vredefort impact structures, South Africa. In: Koeberl C, MacLeod, K (eds) Catastrophic Events and Mass Extinctions: Impacts and Beyond, Geological Society of America, Special Paper 356: 607-617

Kyte FT (1998) A meteorite from the Cretaceous/Tertiary boundary. Nature 396: 237-239

Kyte FT, Brownlee DE (1985) Unmelted meteoritic debris in the Late Pliocene iridium anomaly - Evidence for the ocean impact of a nonchondritic asteroid. Geochimica et Cosmochimica Acta 49: 1095-1108

Kyte FT, Zhou Z, Wasson JT (1980) Siderophile-enriched sediments from the Cretaceous-Tertiary boundary. Nature 288: 651-656

Kyte FT, Smit J, Wasson JT (1985) Siderophile interelement variations in the Cretaceous-Tertiary boundary sediments from Caravaca, Spain. Earth and Planetary Science Letters 73: 183-195

Kyte FT, Zhou L, Lowe DR (1992) Noble metal abundances in an early Archean impact deposit. Geochimica et Cosmochimica Acta 56:1365-1372

Lambert P (1982) Anomalies within the system: Rochechouart target rock meteorite. In: Silver LT, Schultz PH (eds). Geological Implications of Impacts of Large Asteroids and Comets on the Earth. Geological Society of America, Special Paper 190, pp 57-68

Lichte FE, Wilson, SM, Brooks RR, Reeves RD, Holzbecher J, Ryan DE (1986) New method for the measurement of osmium isotopes applied to a New Zealand Cretaceous/Tertiary boundary shale. Nature 322: 130-132

Lindner M, Leich DA, Russ GP, Bazan JM, Borg RJ (1989) Direct determination of the half-life of ^{187}Re. Geochimica et Cosmochimica Acta 53: 1597-1606

Lorenz CA (1999) Trace elements geochemistry in impact melts of the Boltysh crater, Ukraine. [abs.] Lunar and Planetary Science 30: abstract no. 1597

Luck JM, Turekian KK (1983) Osmium-187/Osmium-186 in manganese nodules and the Cretaceous-Tertiary boundary. Science 222: 613-615

Lugmair GW, Shukolyukov A (1998) Early solar system timescales according to ^{53}Mn-^{53}Cr systematics. Geochimica et Cosmochimica Acta 62: 2863-2886

Mason B (1979) Meteorites. In: Data of Geochemistry, United States Geological Survey Professional Paper 440-B-1, B117-120

McDonald I (1998) The need for a common framework for collection and interpretation of data in platinum-group element geochemistry. Geostandards Newsletter 22: 85-91

McDonald I (2002) Clearwater East impact structure: A re-interpretation of the projectile type using new platinum-group element data from meteorites. Meteoritics and Planetary Science 37: 459-464

McDonald I, Hart RJ, Tredoux M (1994) Determination of the platinum-group elements in South African kimberlites by nickel sulphide fire-assay and neutron activation analysis. Analytica Chimica Acta 289: 237–247

McDonald I, Andreoli MAG, Hart RJ, Tredoux M (2001) Platinum-group elements in the Morokweng impact structure, South Africa: evidence for the impact of large ordinary chondrite projectile at the Jurassic-Cretaceous boundary. Geochimica et Cosmochimica Acta 65: 299-309

Meisel T, Krähenbühl U, Nazarov MA (1995) Combined osmium and strontium isotopic study of the Cretaceous-Tertiary boundary at Sumbar, Turkmenistan: A test for impact vs. volcanic hypothesis. Geology 23, 313-316

Meisel T, Walker RJ, Morgan JW (1996) The osmium isotopic composition of the Earth´s primitive upper mantle. Nature 383: 517-520

Melosh HJ (1989) Impact Cratering: A Geologic Process. Oxford, Oxford University Press, 245 pp

Mittlefehldt DW, See TH, Hörz F (1992a) Projectile dissemination in impact melts from Meteor crater, Arizona [abs.] Lunar and Planetary Science 23, 919-920

Mittlefehldt DW, See TH, Hörz F (1992b) Dissemination and fractionation of projectile materials in the impact melts from Wabar crater, Saudi Arabia. Meteoritics 27: 361-370

Montanari A, Koeberl C (2000) Impact Stratigraphy – The Italian Record. Lecture Notes in Earth Sciences, vol. 93, Springer Verlag, Heidelberg-Berlin, 364 pp

Morgan JW, Petrie KW (1979) The distribution of volatile and siderophile elements in the impact melt of East Clearwater (Quebec). Proceedings of the 10th Lunar and Planetary Science Conference, pp 2465-2492

Morgan JW, Wandless GA (1983) Strangways Crater, Northern Territory, Australia: Siderophile element enrichment and lithophile element fractionation. Journal of Geophysical Research 88: A819-A829

Morgan JW, Higuchi H, Ganapathy R, Anders E (1975) Meteoritic material in four terrestrial meteorite craters. Proceedings of the 6th Lunar Science Conference, pp 1609-1623

Morgan JW, Higuchi H, Takahashi H, Hertogen J (1978) A "chondritic" eucrite parent body; inference from trace elements. Geochimica et Cosmochimica Acta 42: 27-38

Nyquist L, Lindstrom D, Shih CY, Wiesmann H, Mittlefehldt DW, Wentworth S, Martinez R (1997) Mn-Cr isotopic systematics of chondrules from the Bishunpur and Chainpur meteorites [abs.] Lunar and Planetary Science 28: 1033-1034

Palme H (1980) The meteoritic contamination of terrestrial and lunar impact melts and the problems of indigenous siderophiles in the lunar highland. Proceedings of the 11th Lunar and Planetary Science Conference, pp 481-506

Palme H (1982) Identification of projectiles of large terrestrial impact craters and some implications for the interpretation of Ir-rich Cretaceous/Tertiary boundary layers. In: Silver LT, Schultz PH (eds) Geological Implications of Impacts of Large Asteroids and Comets on Earth. Geological Society of America Special Paper 190, pp 223-233

Palme H, Janssens MJ, Takahasi H, Anders E, Hertogen J (1978) Meteorite material at five large impact craters. Geochimica et Cosmochimica Acta 42: 313-323

Palme H, Göbel E, Grieve RAF (1979) The distribution of volatile and siderophile elements in the impact melt of East Clearwater (Quebec). Proceedings of the 10th Lunar and Planetary Science Conference, pp 2465-2492

Palme H, Grieve RAF, Wolf R (1981) Identification of the projectile at Brent crater, and further considerations of projectile types at terrestrial craters. Geochimica et Cosmochimica Acta 45: 2417-2424

Peucker-Ehrenbrink B, Jahn B (2001) Rhenium-osmium isotope systematics and platinum group element concentrations: Loess and the upper continental crust. Geochemistry, Geophysics, Geosystems. 2: 33-59

Pierazzo E, Melosh HJ (1999) Hydrocode modeling of Chicxulub as an oblique impact event. Earth and Planetary Science Letters 165: 163-176

Pierazzo E, Melosh HJ (2000) Hydrocode modeling of oblique impacts: The fate of the projectile. Meteoritics and Planetary Science 35: 117-130

Pierazzo E, Vickery AM, Melosh HJ (1997) A reevaluation of impact melt production. Icarus 127: 408-423

Podosek FA, Ott U, Brannon JC, Neal CR, Bernatowicz TJ, Swan P, Mahan SE (1997) Thoroughly anomalous chromium in Orgueil. Meteoritics and Planetary Science 32: 617-627

Rehkämper M, Halliday AN, Alt J, Fitton JG, Zipfel J, Takazawa E (1999) Non-chondritic platinum- group element ratios in oceanic mantle lithosphere: petrogenetic signature of melt percolation? Earth and Planetary Science Letters 172: 65-81

Reimold WU, Koeberl C, Reddering, JSV (1998) The 1992 drill core from the Kalkkop impact crater, Eastern Cape Province, South Africa: stratigraphy, petrography, geochemistry and age. Journal of African Earth Earth Sciences, 26: 573-592

Reimold WU, Armstrong RA, Koeberl C (2000) New results from the deep borehole at Morokweng, North West Province, South Africa: Constraints on the size of the J/K boundary age impact structure [abs.] Lunar and Planetary Science 31: abstract no. 2074, CD-ROM

Robin E, Froget L, Jehanno C, Rocchia R (1993) Evidence for a K/T impact event in the Pacific Ocean. Nature 363: 615-617

Schmidt G (1997) Clues to the nature of the impacting bodies from platinum-group elements (rhenium and gold) in borehole samples from the Clearwater East crater (Canada) and the Boltysh impact crater (Ukraine). Meteoritics and Planetary Science 32: 761-767

Schmidt G, Pernicka E (1994) The determination of platinum group elements (PGE) in target rocks and fall-back material of the Nördlinger Ries impact crater (Germany). Geochimica et Cosmochimica Acta 58: 5083-5090

Schmidt G, Palme H, Kratz KL (1997) Highly siderophile elements (Re, Os, Ir, Ru, Rh, Pd, Au) in impact melts from three European impact craters (Sääksjärvi, Mien, and

Dellen): Clues to the nature of the impacting bodies. Geochimica et Cosmochimica Acta 61: 2977-2987

Schmidt G, Palme H, Kratz KL, Kurat G (2000) Are highly siderophile elements (PGE, Re and Au) fractionated in the upper mantle of the earth? New results on peridotites from Zabargad. Chemical Geology 163: 167-188

Schnabel C, Pierazzo E, Xue S, Herzog GF, Masarik J, Cresswell RG, di Tada ML, Liu K, Fifield LK (1999) Shock melting of the Canyon Diablo Impactor: Constraints from Nickel-59 Contents and Numerical Modeling. Science 285: 85-88

Shima M, Honda M (1966) Distribution of spallation produced chromium between alloys in iron meteorites. Earth and Planetary Science Letters 1: 65-74

Shukolyukov A, Lugmair GW (1998) Isotopic evidence for the Cretaceous-Tertiary impactor and its type. Science 282: 927-929

Shukolyukov A, Lugmair GW (2000a) On the ^{53}Mn heterogeneity in the early Solar System. Space Science Reviews 92: 225-236

Shukolyukov A, Lugmair GW (2000b) Extraterrestrial matter on Earth: Evidence from the Cr isotopes [abs.] In: Catastrophic Events and Mass Extinctions: Impacts and Beyond, Houston, Lunar and Planetary Institute Contribution no. 1053, pp 197-198

Shukolyukov A, Lugmair GW, Koeberl C, Reimold WU (1999) Chromium in the Morokweng impact melt: isotopic evidence for extra-terrestrial component and type of the impactor [abs.]. Meteoritics and Planetary Science 34: A107-A108

Shukolyukov A, Kyte FT, Lugmair GW, Lowe DR, Byerly GR (2000) The oldest impact deposits on Earth – First confirmation of an extraterrestrial component. In: Gilmour I, Koeberl C (eds) Impacts and the early Earth, Lecture Notes in Earth Sciences, vol. 91, Springer Verlag, Berlin, pp 99-116

Smoliar MI, Walker RJ, Morgan JW (1996) Re-Os ages of group IIA, IIIA, IVA, and IVB iron meteorites. Science 271: 1099-1102

Stevenson DJ (1987) Origin of the Moon - the collision hypothesis. Annual Review of Earth and Planetary Sciences 15: 271-315

Stöckelmann D, Reimold WU (1989) The HMX mixing calculation program. Mathematical Geology. 21: 853-860

Taylor SR, McLennan SM (1985) The Continental Crust: Its Composition and Evolution. Blackwell Scientific Publications, Oxford, 312 pp

Terashima S, Imai N, Itoh S, Ando A, Mita N (1994) 1993 Compilation of analytical data for major elements in seventeen GSJ geochemical reference samples, Igneous Rock Series, Bulletin of the Geological Survey (Society) of Japan 45: 305-381

Turekian KK (1982) Potential of ^{187}Os/^{186}Os as a cosmic versus terrestrial indicator in high iridium layers of sedimentary strata: In: Silver LT, Schultz PH (eds) Geological Implications of Impacts of Large Asteroids and Comets on the Earth. Geological Society of America, Special Paper 190, pp 243-249

Wallace MW, Gostin VA, Keays RR (1990) Acraman impact ejecta and host shales: Evidence for low-temperature mobilization of iridium and other platinoids. Geology 18: 132-135

Wasson JT, Kallemeyn GW (1988) Compositions of chondrites. Philosophical Transactions of the Royal Society of London A325: 535-544

Westland AD (1981) Inorganic chemistry of the platinum-group elements. Canadian Institute of Mining and Metallurgy 23: 5-18

Wolf R, Woodrow A, Grieve RAF (1980) Meteoritic material at four Canadian impact craters. Geochimica et Cosmochimica Acta 44: 1015-1022

Impact Decompression Melting: A Possible Trigger for Impact Induced Volcanism and Mantle Hotspots ?

Adrian P. Jones[1], David G. Price[1], Paul S. DeCarli[1,2], Neville Price[1] and Richard Clegg[3]

[1]Department of Geological Sciences, University College London, Gower Street, London, WC1E 6BT, United Kingdom. (adrian.jones@ucl.ac.uk, d.price@ucl.ac.uk, paul.decarli@ sri.com)
[2] SRI International, Menlo Park, CA 94025, USA.
[3] Dynamics House, Hurst Road, Horsham, W Sussex, RH12 2DT, United Kingdom. (all@centdyn.demon.co.uk)

Abstract. We examine the potential for decompression melting beneath a large terrestrial impact crater, as a mechanism for generating sufficent quantity of melt to auto-obliterate the crater. Decompression melting of the sub-crater mantle may initiate almost instantaneously, but the effects of such a massive melting event may trigger long-lived mantle up-welling or an impact plume (I-plume) that could potentially resemble a mantle hotspot. The energy released is largely derived from gravitational energy and is outside (but additive to) the conventional calculations of impact modelling, where energy is derived solely from the kinetic energy of the impacting projectile, be it comet or asteroid; therefore the empirical correlation between total melt volume and crater size will no longer apply, but instead be non-linear above some threshold size, depending strongly on the thermal structure of the lithosphere. We use indicative hydrocode simulations (AUTODYNE-2D) to identify regions of decompression beneath a dynamic large impact crater, (calculated as P-Lithostatic P) using SPH and Lagrangian solvers. The volume of melting due to decompression is then estimated from comparison with experimental phase relations for the upper mantle and depends on the geotherm. We suggest that the volume of melt produced by a 20 km iron projectile travelling at 10 km/s into hot oceanic lithosphere may be comparable to a Large Igneous Province (LIP ~10^6 km^3). The mantle melts will have plume-like geochemical signatures, and rapid mixing of melts from sub-horizontal sub-crater reservoirs to depths where garnet and/or diamond is stable is possible. Direct coupling between impacts and volcanism is therefore a possibility that should be considered with respect to global stratigraphic events in the geological record. Maximum melting would be produced in young oceanic lithosphere and could produce oceanic plateaus, such as the Ontong Java plateau at ~120 Ma. The end-Permian Siberian Traps, are also proposed to be the result of volcanism triggered by a major impact at ~250 Ma, onto continental or oceanic crust. Auto-obliteration by volcanism of all craters larger than ~200 km would explain their anomalous absence on Earth compared with other terrestrial planets in the solar system. This model provides a potential explanation for the formation of komatiites and other high degree partial melts. Impact reprocessing of parts of the upper mantle via impact plumes is consistent with models of planetary accretion after the late heavy bombardment and provides an alternative explanation for most primitive geochemical signatures currently attributed to plumes as originating from the deep mantle or outer core.

1
Introduction

Researchers have already suggested that several larger geological features had an impact origin, but have auto-obliterated the traditional evidence of impact by subsequent large-scale igneous activity. Examples of such suggestions include the Bushveld Complex (Hamilton 1970, Rhodes 1975), the Deccan Traps (Rampino 1987; Negi et al. 1993), the break up of tectonic plates (Seyfert and Sirkin 1979; Price 2001), the formation of oceanic plateaus (Rogers 1982) and catastrophic mantle degassing from volcanism triggered by oceanic impact (Kaiho et al. 2001, but see Koeberl et al. 2002). An alternative mechanism relating volcanism to giant impacts proposed by Boslough et al. (1986) concerned the potential for antipodal focussing of energy transmitted through the Earth to trigger volcanism on the other side of the Earth to the impact itself, although the physics of this specific mechanism have recently been questioned by Melosh (2000). These suggestions have usually been rejected on the grounds that an impact model is less plausible than the widely accepted plume model (Mahoney and Coffin 1997; Richards et al. 1989). In the case of the Deccan traps, an iridium-rich layer between flows is taken by Bhandari et al. (1995) to indicate that this volcanism was already active before the K/T bolide event, as an argument against impact volcanism. Similarly, convincing evidence for impact in rocks from the Bushveld Complex have not been found (e.g., Buchannan and Reimold 1998). The present paper is an attempt to demonstrate more rigorously the plausibility of an impact model for the initiation of a large-scale igneous event. In addition, Glikson (1999) pointed to the planetary-scale role of mega-impacts in the history of development of the Earth's crust, and drew attention to the likely preferential melting efficiency of mega-impacts in oceanic lithosphere due to their higher geothermal gradients and thinner crust. Many of Glikson's ideas and fundamental implications are substantiated by our results for decompression melting, as predicted both by Glikson and ourselves (Price 2001).

Central to this paper is our contention that the phenomenon of pressure-release melting, or decompression melting, described in detail later, is the key to understanding the volumes of melt generated during large impacts and that in part this process has been overlooked or wrongly de-emphasised (Melosh 1989; Pierazzo et al. 1997). Melosh (2000) contends that there is no firm evidence that impacts can induce volcanic activity in the impact crater region, and he presents strong arguments, based on the amount of energy available, against the proposal that an impact could trigger volcanism at a distance. He notes that suggestions of impact-induced volcanism have often been based on observations of the large basalt-filled basins on the lunar nearside, but these are undermined by the discovery of large unfilled farside basins, and by the evidence that nearside volcanism apparently postdated basin formation by as much as 1 Gyr. Melosh concluded that pressure-release melting was highly unlikely on the Moon and he discounted the possibility of presure-release melting on the Earth. We agree with Melosh that pressure-release melting is unlikely on the Moon. If the temperature profile (temperature vs. pressure) were similar to the Earth's oceanic profile, excavation to a depth of approximately 500 km would be necessary to trigger pressure-release melting on the Moon. The largest verified terrestrial craters (Vredefort, Sudbury, Chicxulub; all ~200 km crater size) are all continental, and may be too small to have triggered pressure-release melting in a continental shield with low geothermal gradient; or if they did generate decompression melts, these have not yet been recognized. However, we do not agree that decompression melting can be ignored; our indicative simulations imply that a Sudbury-scale impact crater (~200 km diameter crater) would trigger instantaneous pressure-release melting if it occurred on oceanic lithosphere where geothermal gradients are high. Somewhat larger impacts on continental lithosphere would be required to trigger volcanism, which we propose to be the case for the Siberian Traps. The potential energy range available from very large impactors is vast. To put this into context, the largest conjectured terrestrial impact was the Moon-forming event, when an impactor 10% of the mass of the Earth (a true 'mega-impact') apparently ripped away part of the entire mantle possibly briefly exposing the Earths already-differentiated core (Canup and Asphaug 2001). We focus on less extreme large impacts likely to generate terrestrial craters in the range of ~>200 km, which may have been relatively common during the early part of the Earth's history, and are still dwarfed by potential projectiles available in the (upper) size range of known near-earth crossing objects.

In this paper we address the traditional objections to an impact-related origin of major terrestrial igneous features and will conclude (1) that the plume hypothesis may not explain all of the features to which it is currently applied, (2) the generally dismissed process of pressure-release melting does provide a mechanism for larger impacts to generate large volumes ($\sim 10^6 \mathrm{km}^3$) of melt and (3) the flux of larger impactorsis sufficient to explain the number of large igneous provinces (LIPs; $\sim 10^6 \mathrm{km}^3$ of melt) seen on Earth. We propose that a candidate oceanic LIP generated by impact volcanism might be the Ontong Java Plateau and a candidate continental LIP might be the Siberian Traps; we suggest a range of features by which this hypothesis may be tested. We propose that mantle hotpots triggered by large impacts offer a plausible upper mantle alternative to deep rooted lower mantle plumes, and will be associated with a comparable array of igneous, geochemical and metasomatic features. We recognize that this concept of reducing very large energetic geological processes to very short timescales and extraterrestrial triggers will require a substantial shift in approach by many traditional Earth scientists, but we believe that the underlying arguments are unavoidable.

2
Large Igneous Provinces (LIPS)

Large igneous provinces are widely thought to be produced by mantle melting resulting from a plume. Two main hypotheses have been proposed to explain the relationship between mantle plumes and flood basalts. In the plume-head hypothesis, Campbell and Griffiths (1990) consider that a large plume head, with a diameter of ~1000 km, originates at the core-mantle boundary and rises to form, beneath the lithosphere, an oblate circular disk with a diameter of ~2000 km. This leads to uplift of the overlying lithosphere of 0.5-1.0 km, and the development of volcanic activity. Plume-head melting occurs as the consequence of adiabatic decompression when the top of the plume reaches the top of the asthenosphere. Melting, they contend, will start at the hot leading edge of the top of the plume, where the plume can melt to produce high MgO magmas. As the plume head continues to rise and flatten, the cooler entrained-mantle edge of the plume may start to melt if it rises to sufficiently low pressures at shallow depths. In the second model, White and McKenzie (1989) assumed a much smaller plume, with an unspecified origin. They emphasized that it is the production of melt material, which is of paramount importance, and note that the potential temperature of the plume is only 100-300°C higher than the surrounding mantle. Only in the low velocity zone (LVZ) are the P and T conditions such that the mantle is close to melting. As the increase in temperature caused by the plume is modest, the plume will only give rise to melting in a relatively narrow depth zone immediately beneath the LVZ. Consequently, they conclude that the depth of the stem of the plume is immaterial. But vital to their model is the coincident development of lithospheric thinning, which determines the volume of melt produced.

Although the role of plumes and hot spots in the development of volcanic chains such as Hawaii is widely accepted, there are some, however, who question whether such plumes can be responsible for all LIPs. Thus, for example, Saunders et al. (1992) maintain that the relatively short period between the initial contact from below, to the generation of melt is likely to be less than 10 Ma. Despite the heat transfer that may take place between plume and continental lithosphere, they argue that large volumes of melt material are unlikely to be generated, and even that the melt that occurs may freeze in-situ as heat is lost to the lithosphere. Campbell and Griffiths (1990) point out the shortcomings in the White and McKenzie hypothesis, while Anderson (1998) questions both plume models, and suggests that sources of geochemical anomalies and melting processes may occur instead at shallow depths in the mantle (we agree). Other authors suggest detailed field evidence in some large igneous provinces does not support either mantle plume model. Thus, in a recent review Sharma (1997) observes; "Collectively the [cited field] observations suggest that the Siberian Traps eruption cannot be linked directly either to lithospheric stretching in the absence of a plume or to hotspot initiation. Yet there appears to be consensus supporting a plume origin among those working on the Siberian Traps. Two pieces of evidence have engendered such a confluence of opinion: (i) the large volume ($>2 \times 10^6$ km^3) of magma emplaced and (ii) the short duration ~1 Ma of eruption." Subsequent geological and geophysical papers are also incompatible with a conventional mantle plume as the cause of the Siberian Traps (Czamanske et al. 1998; Elkins-Tanton and Hager 2000).

In the following section, we indicate how decompression melting resulting from a large impact might generate large volumes of melt, which could be emplaced very rapidly, and might offer an alternative explanation to the mantle plume model for the Ontong Java Plateau, the Siberian Traps, and perhaps other LIP's.

3
Impact Melting

There is a well-established correlation between observed terrestrial crater size and the total volume of impact melt (Fig. 1, after Cintala and Grieve 1994). However, the observed craters are all in continental crust and perhaps the largest, Sudbury, is ~200 km diameter, with a lower bound estimated melt volume of ~ 8000 km^3. Studies of the largest known terrestrial craters, Sudbury, Vredefort, and Chicxulub, indicate similar rim diameters (~200-250 km). We concentrate on a detailed discussion of the well-studied Sudbury crater. Stöffler et al, (1994) summarized the results of an eight-year research project on the Sudbury structure. On the basis of textural, chemical, and isotopic evidence, they concluded that the Sudbury Igneous Complex (SIC) represents a differentiated impact melt with no significant deep-sourced magmatic or volcanic contribution. They also cite geophysical evidence that the SIC is not funnel-shaped with an extension to deeper levels of the crust. Their revised estimate of the original crater rim diameter is 220 km, and they estimate that the original volume of Sudbury impact melt was about 12,500 km^3. The depth of the transient cavity, the maximum depth of excavation and the maximum depth of melting are estimated to be in the ranges of 28 to 37 km, 15 to 21 km, and 25 to 35 km, respectively. These latter estimates are derived from a combination of constraining field observational data with the heuristic scaling relations presented in Melosh (1989). Assuming a 20 km/s impact of a projectile with density of 3 g/cm^3, Pi-group scaling relations predict a projectile diameter of about 14 km, corresponding to an impact energy of 8.6 x 10^{23} J. As noted by Melosh (1989), depending on one's choice among the proposed scaling relations, the uncertainty in prediction of impact energy from crater diameter could be as high as a factor of forty, for very large impact craters. We, therefore, have no qualms in comparing Stöffler's results with hydrocode calculations of impact events that differ by as much as a factor of three in impact energy. Pierazzo et al. (1997) calculated melt production for the 20 km/s impact of a 10 km diameter dunite projectile (3.3 x 10^{23} J) on various targets. In this context, the calculated volume of granitic melt, 8900 km^3, is in good agreement with Stöffler's estimated volume. The cited calculation was concerned only with melt production and did not run long enough to determine other crater parameters, such as transient crater diameter. Roddy et al. (1987) calculated the 20 km/s impact of a 10 km diameter quartz projectile (2.6 x 10^{23} J) on a layered continental site. The maximum transient crater depth was ~37 km, the upper limit of Stöffler's estimate, but the maximum transient crater diameter was ~80 km, substantially less than the 110 km estimated by Stöffler. One may note that the transient crater calculated by Roddy et al. (1987) has a much greater depth-to-diameter ratio than predicted by more generic calculations of O'Keefe and Ahrens (1993). These generic calculations generally supported Pi-group scaling. However, they do not appear to have modelled silicate phase transitions accurately and they were restricted to an impact velocity of 12 km/s. Roddy et al. (1987) explicitly modelled the effects of silicate phase transformations, which are known to have a major effect on wave propagation (Swegle 1990).

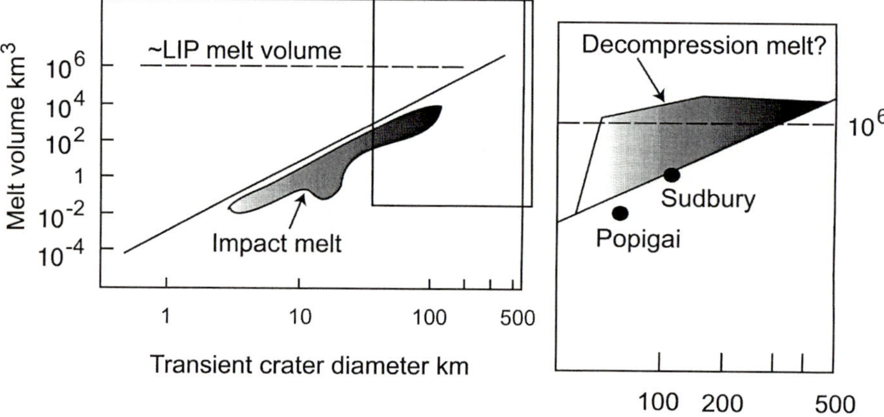

Fig. 1.   Correlation of observed volume of impact melt versus crater diameter for terrestrial impact craters (e.g., Cintala and Grieve 1994) compared with melt volume required for a Large Igneous Province (LIP ~10^6 km^3). Enlargement schematic shows the hypothetical increase in melt volume, due to decompression melting of lithospheric mantle, resulting in a non-linear relationship with crater diameter. Decompression melting triggered by impact might produce sufficient magma to feed a LIP.

We note that the estimated original volume of Sudbury melt, 12,500 km^3, is substantially less than the ~10^6 km^3 volume of a large igneous province. The calculations of Pierazzo et al. (1997) indicate that production of 10^5 km^3 of melt corresponds to a 20 km/s vertical impact of a 22.4 km diameter dunite projectile (4 x 10^{24} J). Using Pi-group scaling, the predicted transient crater diameter is 145 km, leading to a final diameter of about 300 km. For the 12 km/s vertical impacts modelled by O'Keefe and Ahrens (1993), the maximum depth of excavation is a constant fraction, about 0.05, of the final diameter. This implies that the predicted depth of excavation would be only 15 km, the lower bound of Stöffler's estimate for Sudbury (derived from Lakomy's (1990) geological study of the footwall breccia). The contradictions between Sudbury ground truth and the results of applying generic scaling relations imply that additional detailed modelling is needed. We are particularly interested in the behaviour of a heated target, such as the Earth, whose geothermal gradient is well understood.

4
Decompression Melting

Partial melting of the mantle occurs wherever the ambient temperature exceeds the mantle solidus temperature. Under adiabatic conditions in the upper mantle this situation arises during uplift or decompression of hot mantle, since the melting temperature for mantle peridotite increases with pressure (positive dT/dP). The mantle potential temperature is the temperature at any depth on the mantle solidus intersected by the adiabatic ascent path of a known melt temperature at the surface; this is adjusted for the additional thermal loss associated with latent heat of melting. McKenzie and Bickle (1988) correlated the total 2-D thickness of melt that can be extracted with the mantle potential temperature and degree of lithospheric thinning. Thus, the uniform thickness of oceanic crust (~7 km) is consistent with the volume of melt produced if the mantle has a potential temperature of ~1280°C. We now consider how decompression melting may be induced by a large impact, where lithospheric thinning is effectively instantaneous, as required by McKenzie and Bickle (1988).

Decompression melting has not been encountered in laboratory shock experiments, nor is it expected, since it is a phenomenon restricted to large-scale impacts. It is well understood however, and is the main process, advocated by geophysicists for melting on Earth. It is seen in mantle xenoliths rapidly decompressed by rising volcanic magmas (Jones et al. 1983), and can be simulated in sacrificial solid-media experiments (Langenhorst et al. 1998). Therefore, it should be seriously considered whenever an impact is sufficiently large to cause the transient crater depth to excavate a substantial fraction of the local crustal thickness, and thereby cause a sudden drop in lithostatic pressure beneath the crater. This is because the temperature interval between ambient geotherm and lithological melting closes rapidly with increasing depth. By contrast, decompression of most crustal melts, causes freezing, since these generally have negative melting curves at low pressures (Wyllie 1979). There is thus an increasingly likelihood for decompression melting with increasing transient crater depth (Ht). Terrestrial geotherms are fixed at depths of approximately 400 and 660 km by the olivine to ß-phase and spinel to perovskite phase transitions respectively (Poirier 2000). At much shallower depths, geotherms are superadiabatic and vary according to lithospheric structure. Variations in the shallow geotherms represent exactly the region of interest for impacts. For oceanic lithosphere, geotherms vary with age from hot and young to cold and old. Geotherms for continental crust extend from the coolest gradients typical of stable cratons to those that overlap with lower oceanic values during active regional metamorphism.

The volume of decompression melt can be estimated by combining calculations of the pressure drop beneath an impact crater with mantle melting behaviour from published experimental data (as recently compiled, for example, by Thompson and Gibson 2000). For mantle peridotite, the degree of partial melting is, to a first order, related directly to the excess temperature above the solidus, for any given pressure. For example (Fig. 2), a pressure reduction of 15 kbar (1.5 GPa) is equivalent to raising the temperature by up to ~150°C and, in peridotite previously at solidus temperature (T), leads to 20-40% melting. This simple observation is the

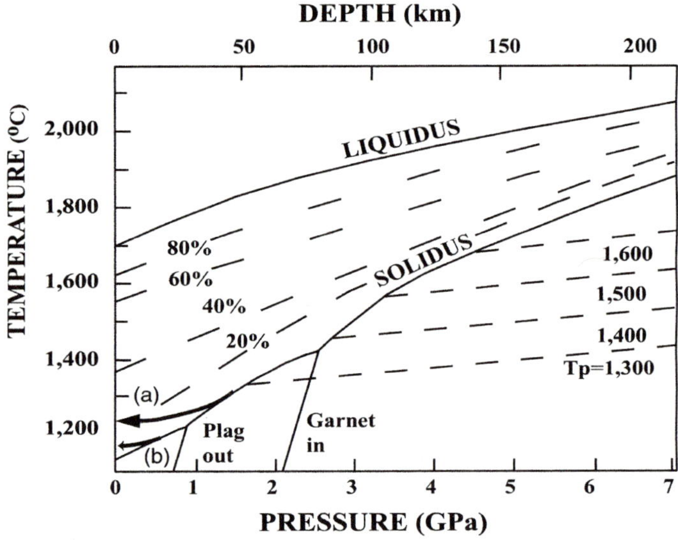

Fig. 2. Phase relations for mantle peridotite, showing degrees of melting at temperatures above the solidus, and curves for mantle potential temperatures (T_p) in upper mantle peridotite (after Thompson and Gibson, 2000). Melt compositions vary with the degree of melting and correspond to basalt (~10-20% melting), picrite (up to 30% melting) and komatiite (>30%). Two examples of decompression melting are shown, (corrected for latent heat of melting, but uncorrected for impact heating, or adiabatic uprise). A pressure decrease of (a) - 1.5 GPa is similar to raising the temperature by up to ~150°C and, in peridotite previously at solidus temperature, leads to 20-30% melting (picritic). **(b)** -0.5 GPa causes ~10% melting (basaltic). Any contribution to heating from impact would increase the degree of melting. We propose that decompression melting is important for hot target lithosphere (Earth) and may trigger large-scale volcanism. The mantle thermal anomaly could be long-lived and may superficially resemble a hotspot, but with no lower mantle root.

crux of our argument, it represents an enormous potential for substantially melting the mantle <u>beneath</u> an impact crater, and has profound consequences for the geological history of the Earth. Melt compositions will vary according to the degree of melting and correspond approximately to komatiite (>30%), picrite (up to 30% melting) and basalt (circa 10-20% melting) respectively. At pre-impact depths shallower than ~75 km and for lower degrees of partial melting, there is also a compositional dependance on pressure for various varieties of basalt. In the most favourable case, thermally active oceanic lithosphere is already in a partially molten state at shallow depth.

Melting is not a kinetically hindered process because it is entropically so favourable, and so decompression melting will occur virtually instantaneously in hot mantle wherever there is sufficient reduction in pressure beneath a large impact, including reduction of lithostatic load by excavation of crater material, massive central uplift or lithostatic modification during formation of multi-ring structures.

5
Hydrocode Model

To quantify the instantaneous stress drop resulting from impact crater formation, we have performed indicative hydrodynamic simulations using the hydrocode AUTODYNE-2D (version 4.1) similar to that described by Hayhurst and Clegg (1997). The AUTODYNE-2D code has been well validated by data from small-scale hypervelocity experiments with a variety of target and impactor materials (Hayhurst et al. 1995). The impact parameters were not intended to represent the complexities of a real impact, but were chosen so that most of the calculation would take place in a regime where Hugoniot uncertainties were small. The model "lithosphere" has a pre-determined pressure gradient to simulate the effects of lithostatic load, similar to the global geophysical model for the Earth called PREM (Primitive Earth Reference Model, Poirier 2000). There was no pre-impact thermal gradient employed in this simulation, but the self-compression density and thermal effects of gravitational and shock compression, were included. Lithostatic pressure and total pressure were calculated separately and integrated at the end of each run to quantify the pressure change, and specifically to determine regions of negative pressure, or decompression. The target dimensions are a 2-D box 300 km by 300 km, mirrored along the vertical axis of the crater to give a model space 600 km by 300 km. The lower boundary (300 km depth) was chosen to avoid back reflections in the model, but still caused noise in the data at the end of each run; this could be extended in future models or amended using a different solver, to a boundary transparent to shock. The target material selected was basalt, (SESAME EOS number 7530) using a no-strength model. Obviously future models could incorporate layers to represent crust, and peridotite to represent mantle. The pure iron impactor (SESAME EOS number 2410), was modelled as a sphere of 10 km radius with initial contact velocityof 10 km/s. The model symmetry used normal incidence, with idealised cylindrical symmetry. The SPH solver (Smoothed Particle Hydrodynamics) was filled with 22,500 particles, and progress in terms of (for example) velocity, density, temperature and pressure of each, being calculated in each step of the run.

After 40 seconds, the simulation produces a transient crater which has proportions of depth to diameter close to 1:1, which greatly excedes the 1:3 ratio of conventional impact crater assumptions (Melosh 1989); however, high aspect transient craters have been found in previous simulations. Thus, Roddy et al. (1987) using a 10 km quartz, 10% porosity impactor at 20 km/s find at maximum depth of 39 km ~ 30 seconds after impact, the diameter of the cavity is only 62 km (aspect ratio ~1:2). Also, Pierazzo et al. (1997) calculate for a similar impact with 10 km dunite moving at 20 km/s at the same time after impact, a crater diameter of ~60 km and depth of ~35 km (aspect ratio again ~1:2). Our calculation shows that, as expected on the basis of simple analytical considerations, the calculated depth-to-diameter ratio does indeed depend on relative shock and release properties of both impactor and target. We used a larger impactor-target density ratio (iron:basalt) that the simulations referred to above, and this led to the 1:1 depth-to-

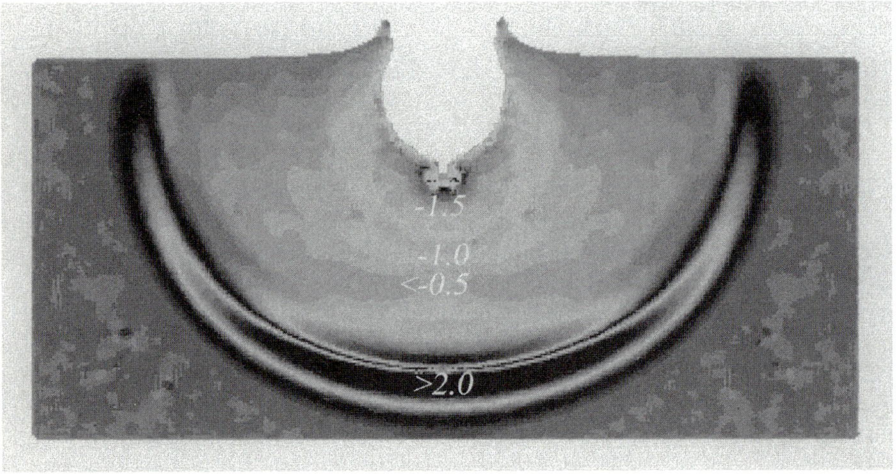

Fig. 3. Indicative hydrocode model of a simulated impact designed to show regions where decompression melting should occur. Model conditions: 300 x 300 km cell, impactor = 10 km radius iron, velocity 10 kms^{-1}, orthogonal impact, target = basalt (homogeneous), pressure gradient = PREM (Poirier 2000). Labelled are pressure zones relative to lithostatic load, for −1.5, -1.0, <-0.5 GPa. If these zones occured in hot (young) oceanic lithosphere, decompression partial melting should occur. In the short term, this could ~instantaneously generate the volume of melt required, (basaltic, picritic or komatiitic) to form a large igneous province (LIP ~10^6 km^3); in the longer term, the thermal signature and could resemble a mantle hotspot, or impact plume (I-plume).

diameter ratio in our model. The calculations and analyses are validated by data from small-scale hypervelocity impact experiments with a variety of target and impactor materials (Pond and Glass 1970). The aspect ratio will directly influence the depth of mantle impact.

The results for the simulation show that after 40 seconds, there is a virtually spherical transient crater ~ 100 km in diameter (Fig 3), below which there are clearly identified zones of decompression. Figure 4 plots pressure versus depth below the transient crater, and shows three curves, for (a) lithostatic load (starting condition), (b) pressure induced by impact and (c) pressure difference (b-a). It can be seen that at 40 seconds after impact, a zone of decompression with magnitude ~ 1.0 GPa extends over a large interval from 120 to 180 km depth. Comparison of this information with Fig 2 shows that in the Earth's mantle, this decompression would occur in garnet peridotite and overlap with the stability field for diamond (pressures higher than ~5 GPa); if melting were initiated at this depth, and erupted, the geochemical signature of any resultant volcanic lavas should reflect the influence of garnet.

There are two causes of decompression – one long term, the other more transient. The latter transient effect is due to the interaction of rarefaction waves originating at the free surface. A longer-lasting zone of decompression occurs directly beneath the crater produced by the excavation of the crater material and the resultant loss of lithostatic load. The amount of melting generated by these processes can be estimated by direct comparison of the decompression values calculated in the simulation (as in Fig. 3) with the mantle melting relations shown in Fig. 2. Melting will occur virtually instantaneously over a range of depths during the course of the impact. We calculate that, for young oceanic lithosphere, the integrated volume of rock to experience super-solidus conditions is ~ 2 x 10^7 km^3 during the course of the shock event. This would lead to the production of ~ 3 x 10^6 km^3 of melt as the depressurised volume of mantle experiences an average of 15% partial melting. In a real impact, the melt extraction process would be complicated by, for example, gravitational instability of newly formed low-density melts beneath the impact crater, melt viscosity, foundering of crustal rocks, variations in porosity and permeability in shattered rocks, and explosive interaction with water. Withdrawal of a large volume of melt from the mantle, previously unsupported by, for example, a deep rising conventional plume, could lead to further mass up-flow of the upper mantle during a secondary stage of dynamic flow or collapse into the vacated "space" with resultant further melting (Price 2001). For simplicity, we therefore assume delivery of only ~30% of the melt to the surface. Thus, our results provide an estimate of ~ 1 x 10^6 km^3 of basaltic melt, comparable to the characteristic volume of LIP's.

There are two main caveats, which we point out about our simulation. Firstly, our simulation uses materials with no inherent strength, and treats the target as a fluidized material. Justification for this is provided by the observation of asthenospheric doming beneath large lunar craters (Neumann et al. 1996), where lunar mantle (thought to be broadly similar to Earth's silicate mantle) may have flowed as a liquid due to shock (Elkins-Tanton, pers. comm., 2002).

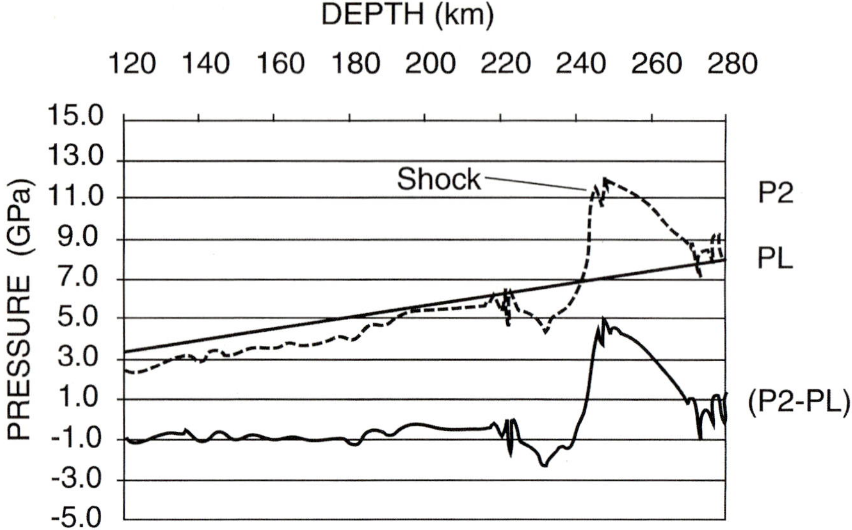

Fig. 4. Pressure versus depth at 40 seconds after impact for indicative model (AUTODYNE-2D). PL = lithostatic load, P2 = impact pressure, (P2-PL) = pressure difference (negative values = decompression). Decompression of ~ -1.0 GPa extends from a depth of ~120 to 180 km, and at lesser values to ~200 km depth, where some melt may exist everywhere in the Earth's mantle (the low velocity zone). The model indicates that decompression melting might be a significant process triggered by large impact craters on Earth, and is expected to be most effective in oceanic lithosphere, where geothermal gradients are high.

However, if friction in the model is increased, and the material treated as a cold brittle solid, then the zone of decompression beneath the crater attenuates much more quickly with depth, diminishing the potential for decompression melting. High-friction models may be more appropriate for shallow crustal impacts, where the rocks may fail under shear and tensile loading, but even there, friction during impact is apparently dramatically reduced during impacts to very low levels, perhaps due to acoustic fluidization (Ivanov 1998). Secondly, the time slice we have chosen at 40 seconds coincides approximately with the maximum depth dimension of the transient crater. This time may not represent the state of the mantle after crater formation is completed, although it apparently does for strength-free or friction-free mantle. We expect the full crater to develop in these simulations in about ~200-400 seconds, and we have run decompression volume versus depth profiles as a function of time. Our simulation shows similar results at 60 and 90 seconds, after which the model degraded due to undesirable interaction with the 300 km depth limit.

We have compared our results from AUTODYNE-2D with calculations provided by Boris Ivanov (pers. comm. 2002) using a different hydrocode (SALES; see also Ivanov et al. 1997, Ivanov and Deutsch 1999) and dunite in place of basalt target, but using similar impactor dimensions. He estimates a volume of 2 to 4 x 10^6 km^3 mantle decompressed to >~ 0.1 GPa in a fluidized mantle, which is less than about half the amount in our model. Given the differences in hydrocodes, number of data points, geothermal gradients (he uses continental) and materials, this is actually rather similar. Both models show that the decompressed mantle volume is more than an order of magnitude larger than the total excavated crater volume (~3 x 10^5 km^3). Our initial model has therefore succeeded in demonstrating the potential for melting due to decompression, in contrast to previous impact melt studies which have concentrated on comparing shock heating with geothermal gradients (e.g., Pierazzo et al. 1997; Turtle and Pierazzo 2000). We are currently refining the model towards a more complex (e.g., layered) lithosphere target, including the melt extraction process.

6
Flux of Impactors

Having shown that a large impact into hot lithosphere could potentially generate large volumes of melt, we need to consider whether the probability of this occurring is large enough to be significant in the Phanerozoic history of the Earth. These arguments have been well rehearsed in discussion of the striking coincidence of timing between emplacement of flood lavas (LIPs) and at least 5 major extinction events at stratigraphic boundaries throughout the Phanerozoic (Rampino 1987, Rampino and Stothers 1988, Courtillot 1992). Recent calculations imply formation of >450 terrestrial craters of D> 100 km since the late heavy bombardment, and cratering rate estimates solely for oceanic impacts (crater >30km) suggest that a large 200 km crater may occur every 150 Ma, and a 500 km crater every 450 Ma (Glikson 1999; Shoemaker et al. 1990; Koulouris et al. 1999). Examination of the terrestrial impact record over the last ~100 Ma shows that a crater with diameter ~100 km or more has occurred on average once every 35 Ma (Popigai 100 km, 35 Ma; Chesapeake Bay 85 km, 35 Ma; Chicxulub 180 km, 65 Ma). Similar impact rates are inferred independently from studies of comets and for the combined probabilities of comets and asteroids; Weissman (1997) indicated that the impact probability of long period comets large enough to produce craters >10 km is about 1 per million years, and estimates an interval of 1.7 x 10^7 yrs between potentially catastrophic long period comet impacts. Both comets and asteroids cause impacts, but comets can have much higher velocities. If one assumes that this flux has remained constant since the end of the late heavy bombardment (at ~3.8-4 Ga), then the derived flux is very similar to previous recent estimates (Grady et al. 1998). There are perhaps ~1000 craters of diameter >10 km "missing " from the geological record in the last 3000 Ma. More significantly, the expected number of craters > 200 km diameter is ~25 and there should also be 1 to 5 craters of diameter > 500 km; these have not yet been identified. Our contention is that the larger craters would have been auto-obliterated by impact volcanism, now represented by some LIPs, and that they will appear very different to conventional craters.

7
Impact Signatures

Oceanic impacts would generally be devoid of mineralogical indicators like shocked quartz, and although the oceanic crust contains other minerals potentially susceptible to shock effects, these are generally dominated by plagioclase feldspar, which transforms to glass (maskelynite) and is unlikely to survive even modest hydrothermal alteration. Potential mineral indicators of oceanic impact derived from the target oceanic crust, could include spinels, perhaps also including nickel-bearing and chromium-bearing varieties (Robin et al. 2000), and spherules. Most mantle minerals (olivine, pyroxenes, garnet) when shocked, transform to metastable phases or glass, or more likely just melt due to their higher initial temperatures (Fig. 5). Any hot minerals and glass would then be susceptible to seawater alteration, to secondary hydrous minerals. Thus, the likelihood of resistant minerals with distinctive shock features to survive over geological time from an oceanic impact, is substantially lower than for a continental impact. Geologically old fractured impacted oceanic terrains might show extensive hydrous mineral development. In general, we agree with the overview presented by Elston (1992), that in large impacts, "smoking gun" shock phenomena are likely to be lacking, largely because heat effects overwhelm shock effects.

In the hypothetical scenario of an impact sufficiently large to auto-obliterate, by definition the traditional proximal indicators will be obscured except or until erosion or other geological processes remove the lava deposits. Distal deposits of glassy materials should be lower in silica compared to continental impact glasses. Large craters might still be identified in plan view by circular structures reflecting crustal or tectonic deformation (Price 2001) sometimes mirrored in remote sensing geophysical data, or by radial distribution of the igneous sequence. Thus, the Sudbury Igneous Complex (conventional impact melt) is located centrally within the eroded impact crater, but the circular structure of the buried Chicxulub crater has been determined largely from geophysical methods. Geophysical data for Chicxulub show substantial modifications to the vertical crustal structure. These vary depending on the scale of observation and resolution of data (Morgan et al. 1997), but can include large-scale (50-100 km) regional mantle penetrating faults, low angle faults (possibly associated with melting), displaced and centrally uplifted Moho, and local (~1-10 km) scale displaced and rotated fault blocks. We emphasise that all criteria established for large impacts are restricted to craters in continental crust, which have undergone brittle failure and not penetrated the crust. We have no comparable criteria for a large impact, which punctured oceanic crust and mantle, though this must have occurred. Furthermore, we do not know how the morphology of such a large oceanic impact crater might be further modified through the massive melting event and the transfer of these melts to the surface.

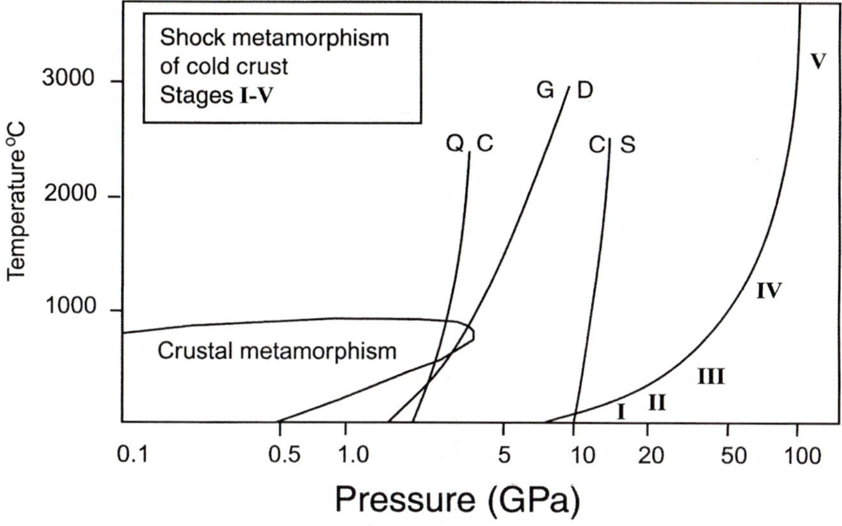

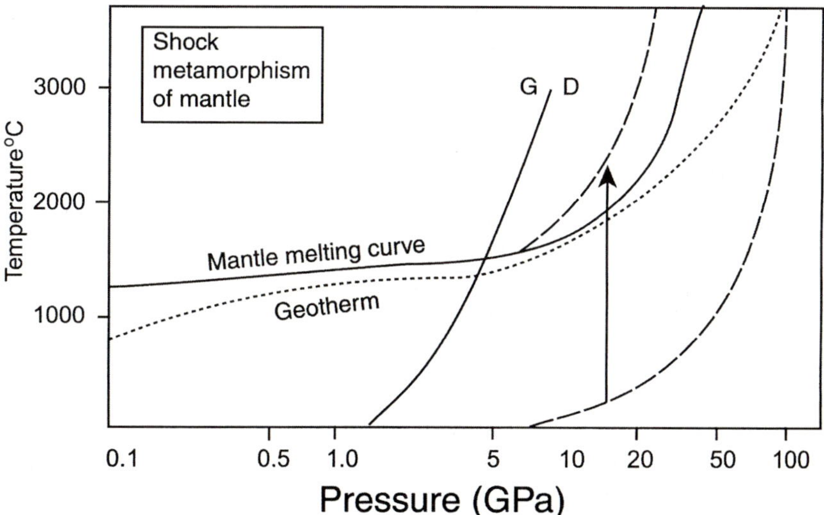

Fig. 5. Shock metamorphism of (a) continental crust with initial low ambient temperature and (b) hypothetical melting of lithosphere/mantle, due to ambient much higher temperature (geotherm). Mineral equilibrium phase transitions shown for G-D (graphite-diamond; Bovenkerk et al 1959) Q-C (quartz-coesite), and C-S (coesite-stishovite; Fei and Bertka 1999).

8
Impact-Plumes (I-Plumes)?

As for lunar melt extraction (Wilson and Head 2001), we hope in future to model the distribution and extraction of melts from beneath the crater floor. Our indicative model develops saucer-like sub-horizontal sill-like bodies at different depths. This reflects conventional impact melts within craters except that these decompression melts are far below the crater itself. We conjecture that melt extrusion would start with highly energetic eruption of low viscosity peridotitic melts, which would be bouyant compared to solid surrounding lithosphere. Interaction of these hot fluid melts with surface water would be likely to produce ultramafic and mafic pyroclastic rocks (cf. Siberian traps). Extraction of such large volumes of melt could lead to secondary mantle flow at ever decreasing rates due to bulk increasing viscosities with secondary melting, and associated metasomatism. These regions of zoned partially molten mantle represent a massive thermal perturbation resembling a conventional hotspot, and share a number of characteristics with mantle plumes. Such impact-plumes (or "I-plumes") could produce similar magmatic and geochemical signatures, but differ from traditional hot-spot plumes (or "H-plumes") in that; I-plumes neither require pre-magmatic thermal doming (see, e.g., Siberian traps) nor would they be related to a deep geophysical fingerprint. I-plumes may thus offer a possible alternative to H-plumes and are linked to shallow enrichment and depletion events restricted to the upper mantle, as an alternative to the widely perceived involvement of the D" layer at the core-mantle-boundary (e.g., Thompson and Gibson 2000).

9
Komatiites

The conclusion that high degrees of partial melting or even complete melting of mantle peridotite are possible following a large oceanic impact (Jones et al. 1999) strongly supports an old suggestion that komatiites (MgO > 18 wt%) can be generated by impacts (Green 1972); high-Mg lavas also occur in many LIPs including the Siberian traps. It avoids the problem of storage of high degrees of komatiitic melt and it does not constrain their petrogenesis to either wet or dry varieties. If this view is correct, then komatiite is unlikely to be a unique magma type but instead represents geochemical snapshots of mantle melting, or perhaps mixtures of multiple melting zones (subhorizontal layers in our models). Impact derived decompression melting may have been particularly effective during higher impact fluxes and periods of higher heat flow, as presumably during the early Archaean. Geologically young komatiites occur as spinifex-textured glassy flows of Mesozoic/Tertiary age from Gorgona Island (Gansser et al. 1979; Echeverria 1980; Kerr et al. 1997; Storey et al. 1991), and komatiites of Permian-Triassic age, have recently been described from northwestern Vietnam (Glotov et al. 2001). A feature of the Gorgona komatiites is their preservation of a large volume chaotic to stratified ultramafic breccia (23-27 wt% MgO), with glassy picritic blocks in a fine-grained matrix of plastically deformed high-Mg glassy globules (Echeverria and Aitken 1986). Conventional petrological and geochemical modelling requires a separate magmatic source for the komatiites, compared with associated basalts and picrites. The glassy breccias have been interpreted as evidence for violent submarine eruptions. We postulate instead, that the Gorgona komatiites might have resulted from decompression melting following an oceanic impact, and the ultrabasic breccias record violent interaction between variously melted peridotite and seawater. The classic Barberton komatiite sequence also indicates deep submarine eruption (Dann 2000) and is associated with enigmatic spherule beds with distinctive extraterrestrial Cr isotope ratios providing evidence of at least two major impacts at ~3.24 Ga from projectiles >20 km in diameter (Shukulyukov et al. 2000), suggesting that impacts might be reconsidered (Jones 2002). Lastly, very rapid extraction of komatiite melts formed by decompression partial melting of the deep mantle where diamond is stable, is perhaps the only way to preserve mantle diamonds in some komatiites (Capdevilla et al. 1999).

10
Candidates for Impact Volcanism

Our indicative model demonstrates the potential for large impact craters (~200 km) to trigger volcanism through decompression melting at any depth extending down to the low velocity zone (~200 km), with volumes of melt comparable to LIP's. The translation of released gravitational energy into melting depends on the geothermal gradient of the target region. Young oceanic lithosphere is most susceptible to this process (geotherm $>$~$17°C$/km), but in principle it could happen anywhere, including "cold" continental lithosphere (geotherm ~$13°C$/km), but with a proportionately larger impactor or higher velocity required. We have not yet determined the minimum size of event to initiate decompression melting, but we take an intuitive guide from the geological record. Since there are no known terrestrial impact craters greater than ~200 km diameter, we conjecture that this may be the lower size limit and larger craters in continental crust have auto-obliterated. Very little is known about oceanic impact craters, but these would require smaller impacts to trigger decompression melting, with the optimum target being an active ridge system with active volcanism before impact. Larger impacts produce more melt in a similar short time, with no upper volume limits; this is in contrast to mantle plumes where melting and melt delivery to the surface is a rate-controlled process related to mantle rheology. Here we present the case for two LIPs, one oceanic and one thin crusted-continental (or oceanic), which might represent impact-generated LIPs. Whether or not they are, remains to be tested.

10.1
Ontong Java Plateau

The Ontong Java Plateau is the largest and thickest oceanic plateau on Earth thought to have been formed by the coincidence of two plumes: a major mantle plume or superplume at ~120 Ma and a secondary plume at ~90 Ma (Phinney et al. 1999). It is not associated with major global mass extinctions (Coffin and Eldholm 1994; Wignall 2001). Geophysical data shows much greater and irregular crustal thickness (15 – 38 km) compared with normal oceanic crust (6-10 km) and a low velocity seismic "root" extending down to 300 km (Richardson et al. 2000). However the unexpectedly small subsidence history of the OJP lead Ito and Clift (1998) to rule out cooling of a large plume head; instead they suggested substantial magmatic underplating. Remnant surrounding seafloor magnetic anomalies show that the OJP formed in young oceanic crust perhaps only 10 Ma old, and may have formed very close to an active spreading ridge (Gladczenko et al. 1997). These fundamental indicators are sufficiently close to our model conditions (maximum melting in young oceanic lithosphere) that we suggest a large oceanic impact at around ~120 Ma, could have triggered this LIP; further details of this candidate for impact volcanism and the large scale effect of the impact on plate motions are presented elsewhere (Price 2001). In this case, the impact site is now represented by a massive layer of volcanic rock, which forms the oceanic plateau itself.

10.2
Siberian Traps

The Siberian Traps represents the single largest eruption of "continental" flood lavas. A somewhat larger impact would be required for our model to operate in continental crust. However, recent plate tectonic reconstructions constrained by seismic tomography indicate that Siberia may actually have been an oceanic environment with micro-continents and subduction zones (Van der Voo et al 1999). The lavas are dated at the end of the Permian (e.g., Campbell et al. 1992; see also Reichow et al. 2002), where a double extinction event may have occurred (Wignall 2001). Up to one third of the lower succession is represented by pyroclastic rocks, with individual tuff units covering up to 30,000 km^2; it was initially marine and developed in a massive subsiding basin that rules out a conventional mantle plume (Czamanske et al. 1998). Elkins-Tanton and Hager (2000) endorsed Sharma's view (1997) that the Siberian Traps cannot be the result of a traditional form of mantle plume. There is some independent global evidence that an impact occurred at the P-Tr boundary, although the evidence is by no means as convincing as for the K/T boundary. A weak Ir-anomaly together with possible shocked quartz were found both in Antarctica and Australia (Retallack et al. 1998). Chinese strata at Meishan placed the boundary at 251.4 ± 0.3 Ma and record rapid addition of isotopically light carbon over a time interval of 165,000 years, or less (Bowring et al. 1998), but problems with dating at this site have emerged (Mundil et al. 2001). Investigation of the marine faunal extinction including the same Meishan outcrops, lead Jin et al. (2000) to conclude that "a predicted true extinction level [occurred] near 251.3 Ma (94% of genera are included in a 0.1-Ma interval spacing). A more reasonable conclusion...is a sudden extinction at 251.4 Ma, followed by the gradual disappearance of a small number of surviving genera over the next 1 million years". An impact event is also supported by controversial evidence from extraterrestrial noble gases in fullerenes recovered from P-Tr boundary beds in China, Japan and Hungary (Becker et al. 2001), although the reliability of such techniques is seriously questioned (I. Gilmour, pers. comm., 2001; Farley and Mukhopadhyay 2001). Although the evidence for impact at the P/Tr boundary is much less clear than for bolide impact at the K-T boundary (Alvarez et al. 1980), there is a similar duality of signals between likely volcanic and impact sources. Therefore, it would seem important to test our hypothesis that the Siberian Traps could have been caused by decompression melting at the impact site, and that impact volcanism can uniquely explain the dual signals in the geological record. The geological record may be consistent with this idea, but we are not aware of any literature concerning the critical volcanic-sedimentary interface at the base of the Siberian traps. However, the onset of volcanism is everywhere an unconformity marked by tuffs uniformly above folded and variably missing palaeozoic strata (Czamanske et al. 1998). The thickest volcanic sequence is in the northern part (4,000 metres, Maymecha-Kotuy; 3,500 metres Norilsk) where massive Ni-sulphide mineralisation is related to mantle-dissecting faults (Hawkesworth et al. 1995). The large-scale occurrence of native nickel- iron (Oleynikov et al. 1985) in intrusive rocks related to the extrusive lavas, (including Pt-bearing nickel-rich iron; Ryabov and Anoshin 1999), is consistent with impact geochemical models that predict native iron and nickel iron (Gerasimov et al. 2001; Miura et al. 2001), and is reminiscent of native iron at the base of the flood lavas in west Greenland (Klöck et al. 1986). Also, the regional geology of the wider Siberian craton and bounding mountain fold belts (Baikal, Verkhoyansky, Taymyr) should be reconsidered in terms of the possible major plate tectonic effects of an impact, as confirmed by changing plate vectors at 250 Ma (Price 2001). The large-scale foundering of continental Siberian lithosphere at this time, recently proposed on the basis of geophysical data (Elkins-Tanton and Hager 2000) is consistent with our impact volcanism hypothesis. The recent recognition that the Siberian traps may have been double the volume than previously assumed, extending west as far as the Urals (Reichow et al. 2002), is easily accommodated in an impact volcanism model by relatively small changes in impactor parameters. The end-Permian event is complicated by the possible double epicentre implications required to produce the slightly older Emeishan flood-lava province in south China (Lo et al. 2002), which, if the dating is reliable (Mundil et al. 2001), were erupted a few million years earlier in a "marine" environment. This is not a problem for an impact volcanism explanation, simply requiring two impacts (Shoemaker-Levy 9 showed us that multiple impacts can occur; furthermore about 10% of known terrestrial craters >20 km are pairs, similar to the recent prediction (16%) that many near-Earth orbiting asteroids are double systems; Margot et al. 2002); however, it may require extraordinary pleading to explain two separate mantle superplumes. The Emeishan traps basal ash layers are characterized by concentrations of microspherules, whose origin is not fully understood (Yin et al. 1992) and earlier thought to have derived from the Siberian traps >2000 km away (Cambell et al. 1992). On the basis of exotic "impact metamorphosed" metallic Fe-Ni grains with up to 30% Ni (Kaiho et al. 2002) within the spherules (Miura et al. 2001) and an absence of shocked quartz, it has been suggested that an oceanic impact was the source of the Emeishan volcanism (Kaiho et al. 2001), but this work has been strongly criticised as being inconclusive (Koeberl et al. 2002). If subsequent investigations can demonstrate that the exotic grains are extraterrestrial (as for the K/T boundary) this would be the first direct evidence for impact at the base of the Emeishan traps, and would dramatically strengthen the claims of Kaiho et al. (2001) that the volcanism was triggered by an oceanic impact, as predicted by our model.

11
Discussion and Conclusions

Our indicative model shows that it is possible for the volume of decompressed mantle beneath a large ~200 km sized crater to greatly exceed the excavated volume of the impact crater itself, primarily due to reduction of lithostatic load. Under suitable conditions of geothermal gradient, this would lead to near instantaneous melting with volumes of the order of 10^6 km^3, similar to the characteristic volumes of LIP's. Optimum target conditions are represented by young oceanic lithosphere, close to or at an active ridge system and could be triggered by a smaller impact; the same process can operate in continental targets, perhaps requiring a somewhat larger impact depending on geothermal gradient and crust/lithosphere architecture. Our model ~200 km impact crater is formed by an initial transient crater, ~80-100 km deep, much deeper than the total crust, whether it is oceanic (~10 km) or continental (~30 km). The melting would take place under the entire crater, deep in the upper mantle where garnet is stable, and can extend down to the zone of stable diamond and the low velocity zone (~200 km). Initial melting may occur at various depths as sub-horizontal, saucer- or sill-like bodies, suggesting that mixing of melts from different depths (reservoirs) would be possible during the melt extraction process (volcanism). By comparison with conventional plume models, this would instantaneously trigger massive volcanism, with geochemical signatures dominated by a garnet-peridotite source mantle, and possible mixing of geochemical reservoirs.

The resultant thermal anomaly in the mantle could be long-lived, and the induced large-scale vertical and horizontal thermal gradients are expected to have a long-term effect on secondary mantle flow, leading to secondary mantle melting which may also be voluminous (see "Impact plumes" above). A secondary pulse of melting, from longer-term asthenospheric flow is currently being investigated by a group at MIT to reinvestigate the origin of lunar mare as post impact melts (Elkins-Tanton et al. 2002). Although this secondary melting is unlikely to approach similar volumes to the intial decompression melting, such adiabatic melting in convection currents nonetheless offers an attractive mechanism for sustaining volcanic activity at the impact site for up to 10 million years after the initial impact (Elkins-Tanton, pers. comm., 2002). We have demonstrated that the previously suggested but generally dismissed mechanism of pressure-release melting should indeed enable large impacts to generate excess volumes of mantle melt. The flux of large impactors expected is sufficient to explain many of the large igneous provinces seen on Earth, and might generally be considered where conventional plume explanations are untenable. As a result, this decompression melt may contribute more melt than conventional shock melting, and the cumulative melt volume may not scale linearly with crater dimension. We have presented arguments to support LIP candidates as impact volcanism-derived. Thus, the Ontong Java Plateau is a promising candidate for the auto-obliterated site of an oceanic impact crater, but has not yet been studied in sufficient detail, largely due to the inherent problems of studying submarine plateaus. We propose that the Siberian Traps, which are accessible and currently under considerable scrutiny, may be better explained by a large impact than by a conventional mantle plume. The closure of a former ocean between Siberia and Mongolia, as well as amalgamation with north and south China blocks may also have been occurring during Permian-Triassic times, and the impact target region may have been oceanic, with a mixture of micro-continents and subduction zones (Van der Voo et al 1999).; this would be much easier to fit with our model. If the end-Permian extinction requires two events separated geographically by 2-3000 km (Siberian traps, Emeishan traps China) this is no problem for impact models. A Siberian impact could explain, for example, the lack of thermal doming, their extreme osmium isotope geochemistry (Walker et al. 1997), and also the occurrence of cliftonite-bearing (cubic graphite) metallic nickel-iron in the intrusive traps (Oleynikov et al. 1985) as metamorphosed relics or products of meteoritic iron (as at Meteor Crater; Brett and Higgins 1967). Impact volcanism must have obscured the evidence for the original impact crater, and may at least partly explain why recognition of a global impact anomaly at the P/T boundary has been so difficult. The current day Moho topography beneath Siberia is variable but segmented, and has been interpreted as a series of mantle ridges and rifts (Kravchenko et al. 1997); seismic velocity structure shows a continuous substantial lateral velocity inversion (8.0 versus 8.4 km/s) at ~100 km depth underlying the entire Siberian platform (Mooney 1999). All of these are consistent with being relict impact features, albeit on a larger scale than commonly observed.

Our estimate of the amount of meteoritic material added to the Earth by large impacts since the end of the late heavy bombardment using cratering rate models is ~10^9 km^3. This represents only about 1% of the volume of the Earths crust, but could, for example, account for the entire PGE budget of the crust, and agrees with enrichment after core segregation supported by recent experimental models (Holzheid et al. 2000). Some of the largest impact craters in continental crust are associated with economic mineralisation, such as the nickel-rich massive sulphides at Sudbury (<200 km) (e.g., Molnar et al. 1999). The Witwatersrand gold deposits are concentrically zoned around the Vredefort structure and may be related to the acknowledged impact-driven hydrothermal activity (Gibson and Reimold 2000). Large impacts are expected to propagate significant hydrothermal activity, aided by intense rock fracturing and the thermal energy deposited in the crust (Kring 2000); they are potential mineral exploration targets, both terrestrial and extraterrestrial (Norman 1994).

There may be additional mantle signatures related to impact, such as the energy to drive large-scale mantle metasomatism and mineralisation. We have suggested that because of the combined effects of decompression and impact heating, complete melting of the mantle, and high degree partial melts are easily achievable, consistent with the production of ultrabasic melts like komatiites from large impacts. Komatiite Ni-PGE-sulphide ore systems typically have high Os concentrations, low Re/Os ratios, and near-chondritic Os isotope compositions, from which Lambert et al. (1998) concluded that large scale dynamic processes, including major lithospheric pathways, are critical to the development of these massive magmatic systems. The long-term effects of sustained melt extraction might result in rootless mantle hotspots, or impact plumes, which will require further modelling. The global consequences for plate tectonics throughout Earth history have recently been explored further by one of us (Price 2001).

We have concentrated on perhaps the most potent melting process in the Earth's mantle, specifically that triggered by decompression beneath a large impact crater, and agree with Boslough et al. (1986), who stated "the impact-produced flood basalt hypothesis is attractive because it is potentially testable on the basis of predictions of features that have not yet been discovered...unlike current plume models for flood basalts and hotspots". In conclusion, we assert that the concept of impact-induced volcanism has not been adequately examined and may offer a new framework for the interpretation of large-scale igneous and geological processes.

Acknowledgments

APJ and PdeC thank Philippe Claeys for supporting initial ideas, and together with Christian Koeberl and the ESF IMPACT programme for providing excellent discussion meetings. The paper has benefitted substantially from reviews provided by Boris Ivanov and Michael Rampino. We also thank colleagues at UCL for comments, discussions and different points of view.

References

Alvarez LW, Alvarez W, Asaro F, Michel HV (1980) Extraterrestrial causes for the Cretaceous-Tertiary extinction. Science 208: 718-720

Anderson D L (1998) The EDGES of the mantle. In: The Core-Mantle Boundary Region. American Geophysical Union, Washington, Geodynamics 28: 255-271

Becker L, Poreda RJ, Hunt AG, Bunch TE, Rampino M (2001) Impact event at the Permian-Triassic boundary: evidence from extraterrestrial noble gases in fullerenes. Science 291: 1530-1533

Bhandari N, Shukla PN, Gheveriya ZG, Sundaram SM (1995) Impact did not trigger Deccan volcanism: Evidence from Anjar K/T boundary intertrappean sediments. Geophysical Research Letters 22: 433-436

Boslough MB, Chael EP, Trucano TG, Crawford DA, Campbell DI (1986) Axial focussing of impact energy in the Earth's interior: a possible link to flood basalts and hotspots. In: Ryder G, Fastovsky D, Gartner S (eds) The Cretaceous-Tertiary Event and Other Catastrophes in Earth History. Geological Society of America Special Paper 307: 541-556

Bovenkerk HP, Bundy FP, Hall HT, Strong HM, Wentorf RH (1959) Preparation of diamond. Nature 184: 1094-1098

Bowring SA, Erwin DH, Jin JG, Martin MW, Davidek K, Wang W (1998) U/Pb zircon geochronology and tempo of the end-Permian mass extinction. Science 280: 1039-1045

Brett R, Higgins GT (1967) Cliftonite in meteorites: a proposed origin. Science 156: 819-820

Buchanan PC, Reimold WU (1998) Studies of the Rooiberg Group, Bushveld Complex, South Africa: no evidence for an impact origin. Earth and Planetary Science Letters 155: 149-165

Campbell IH, Griffiths RW (1990) Implications of mantle plume structure for the evolution of flood basalts. Earth and Planetary Science Letters 99: 79-83

Campbell IH, Czamanske GK, Fedorenko VA, Hill RI, Stepanov V (1992) Synchronism of the Siberian traps and the Permian-Triassic boundary. Science 258: 1760-1763

Canup RM, Asphaug E (2001) Origin of the Moon in a giant impact near the end of the Earth's formation. Nature 412: 708–712

Capdevilla R, Arndt N, Letendre J, Sauvage J-F (1999) Diamonds in volcaniclastic komatiite from French Guiana. Nature 399: 456-458

Cintala MJ, Grieve RAF (1994) The effects of differential scaling of impact melt and crater dimensions on lunar and terrestrial craters: Some brief examples. In: Dressler BO, Grieve RAF, Sharpton VL (eds) Large Meteorite Impacts and Planetary Evolution. Geological Society of America Special Paper 293: 51-59

Coffin MF, Eldholm O (1994) Large igneous provinces: crustal structure, dimensions, and external consequences. Reviews of Geophysics 32: 1-36

Courtillot V (1992) Mass extinctions in the last 300 million years: one impact and seven flood basalts? Israel Journal of Earth Sciences 43: 255-266

Czamanske GK, Gurevitch AB, Fedorenko V, Simonov O (1998) Demise of the Siberian plume: paleogeographic and paleotectonic reconstruction from the prevolcanic and volcanic record, north-central Siberia. International Geological Review 40: 95-115

Dann JC (2000) The 3.5 Ga Komati Formation, Barberton greenstone belt, South Africa, part I: new maps and magmatic architecture. South African Journal of Geology 103: 47-68

Echeverria LM (1980) Tertiary or Mesozoic komatiites from Gorgona island, Columbia: field relations and geochemistry. Contributions to Mineralogy and Petrology 73: 253-266

Echeverria LM, Aitken BG (1986) Pyroclastic rocks; another manifestation of ultramafic volcanism on Gorgona Island, Columbia. Contributions to Mineralogy and Petrology 92: 428-436

Elkins-Tanton LT, Hager BH (2000) Melt intrusion as a trigger for lithospheric foundering and the eruption of the Siberian flood basalts. Geophysical Research Letters 27: 3937-3940

Elkins-Tanton LT, Hager BH, Grove TL (2002) Magmatic effects of the lunar late heavy bombardment [abs.] Lunar and Planetary Science 32, abs. #1422, CD-ROM

Elston WE (1992) Does the Bushveld-Vredefort system (South Africa) record the largest known terrestrial impact catastrophe? [abs.] In: Abstracts submitted to the International Conference on Large Meteorite Impacts and Planetary Evolution, Lunar and Planetary Institute Contribution 790: 23-24

Farley KA, Mukhopadhyay S (2001) An extraterrestrial impact at the Permian-Triassic boundary? Science 293: 2343a

Fei Y, Bertka CM (1999) Phase transitions in the Earth's mantle and mantle mineralogy. In Fei Y, Bertka CM, Mysen BO (eds) Mantle petrology; Field observations and high pressure experimentation: a tribute to Francis R (Joe) Boyd. The Geochemical Society Special Publication 6: 189-207

Gansser A, Dietrich VJ, Cameron WE (1979) Paleogene komatiites from Gorgona island. Nature 278: 545-546

Gerasimov MV, Dikov YuP, Yakovlev OI, Wlotzka F (2001) Reduction of iron during an impact [abs.] In: Martinez-Ruiz F, Ortega-Huertas M, Palomo I (eds) Impact markers in the Stratigraphic record Abstract volume. Universidad de Granada, 31-32

Gibson RL, Reimold WU (2000) Deeply exhumed impact structures: a case study of the Vredefort structure, South Africa. In: Gilmour I, Koeberl C (eds) Impacts and the early Earth. Lecture Notes in Earth Sciences vol. 91, Springer, Berlin-Heidelberg, pp 249-277

Gladczenko TP, Coffin MF, Eldholm O (1997) Crustal structure of the Ontong Java Plateau: Modelling of new gravity and existing seismic data. Journal of Geophysical Research 102: 22711-22729

Glikson AY (1999) Oceanic mega-impacts and crustal evolution. Geology 27: 387-390

Glotov AI, Polyakov GV, Hoa TT, Balykin PA, Akimtsev VA, Krivenko AP, Tolstykh ND, Phuong NT, Thanh HH, Hung TQ, Petrova TE (2001) The Ban Phuc Ni-Cu-PGE deposit related to the phanerozoic komatiite-basalt association in the Song Da rift, nortwestern Vietnam. Canadian Mineralogist 39: 573-589

Grady MM, Hutchinson R, McCall GJ, Rothery DA (1998) Meteorites: Flux with Time and Impact Effects. Geological Society of London Special Publication 140, 278 pp

Green DH (1972) Archaean greenstone belts may include terrestrial equivalents of lunar maria? Earth and Planetary Science Letters 15: 263-270

Hamilton W (1970) Bushveld complex - product of impacts? In: Vissler DJL, Von Gruenewaldt G (eds) Symposium on the Bushveld Igneous Complex and other layered intrusions, Geological Society of South Africa Special Paper 1: 367-379

Hawkesworth CJ, Lightfoot PC, Fedorenko VA, Blake S, Naldrett AJ, Doherty W, Gorbachev NS (1995) Magma differentiation and mineralisation in the Siberian continental flood basalts. Lithos 34: 61-88

Hayhurst CJ, Clegg RA (1997) Cylindrically symmetric SPH simulations of hypervelocity impacts on thin plates. International Journal of Impact Engineering 20: 337-348

Hayhurst CJ, Ranson HJ, Gardner DJ, Birnbaum NK (1995) Modelling microparticle hypervelocity oblique impacts on thick targets. International Journal of Impact Engineering 15: 375-386

Holzheid A, Sylvester P, O'Neill HSC, Rubie DC, Palme H (2000) Evidence for a late chondritic veneer in the Earth's mantle from high-pressure partitioning of palladium and platinum Nature 406: 396-399

Ito G, Clift PD (1998) Subsidence and growth of Pacific Cretaceous plateaus. Earth and Planetary Science Letters 161: 85-100

Ivanov BA (1998) Large impact crater formation: Thermal softening and acoustic fluidization [abs.]. Meteoritics and Planetary Science 33: A76

Ivanov BA, Deutsch A (1999) Sudbury impact event: cratering mechanics and thermal history. In: Dressler B, Grieve RAF (eds) Large Meteorite Impacts and Planetary Evolution II, Special Paper of the Geological Society of America 339: 389-397

Ivanov BA, DeNiem D, Neukum G (1997) Implementation of dynamic strength models into 2D hydrocodes. applications for atmospheric breakup and impact cratering. International Journal of Impact Engineering 20: 411-430

Jin YG, Wang Y, Wang W, Shang QH, Cao CQ, Erwin DH (2000) Pattern of marine mass extinction near the Permian-Triassic boundary in South China. Science 289: 432-436

Jones AP (2002) Komatiites: new information on the type locality (Barberton), and some new ideas. Geology Today 18: 23-25

Jones AP, Smith JV, Dawson JB, Hansen EC (1983) Metamorphism, partial melting, and K-metasomatism of garnet-scapolite-kyanite granulite xenoliths from Lashaine, Tanzania. Journal of Geology 91: 143-165

Jones AP, Price GD, Claeys P (1999) A petrological model for oceanic impact melting and the origin of komatiite [abs]. In: Gersonde R, Deutsch A (eds) Oceanic impacts: mechanisms and environmental perturbations. Berichte zur Polarforschung 343, Alfred-Wegener-Institut für Polar- und Meeresforschung, Bremerhaven, pp 41-42

Jones AP, Price G D, DeCarli P, Price N, Hayhurst C (2001) Modelling impact decompression melting: a possible trigger for impact induced volcanism and mantle hotspots [abs.]. In: Martinez-Ruiz F, Ortega-Huertas M, Palomo I (eds) Abstracts, ESF Workshop on Impact Markers in the Stratigraphic Record. Universidad de Granada, pp 57-58

Kaiho K, Kajiwara Y, Nakano T, Miura Y, Kawahata H, Tazaki K, Ueshima M, Chen Z, Shi GR (2001) End-Permian catastrophe by a bolide impact: evidence of a gigantic release of sulfur from the mantle. Geology 29: 815-818

Kaiho K, Miura Y, Kedves M (2002) Spherules with Fe,Ni-bearing grains of Meishan Permian-Triassic boundary in China [abs.]. Lunar and Planetary Science 32, abs. #2052, CD-ROM

Kerr AC, Tarney J, Marriner GF, Alvaro N, Saunders AD (1997) The Caribbean-Colombian Cretaceous Igneous Province: The internal anatomy of an oceanic plateau. In: Mahoney JJ, Coffin MF (eds) Large Igneous Provinces: Continental, Oceanic And Planetary Flood Volcanism. American Geophysical Union, Washington, Geophysical Monograph 100: 123-143

Klöck H, Palme I, Tobschall HJ (1986) Trace elements in natural metallic iron from Disko Island, Greenland. Contributions to Mineralogy and Petrology 93: 273-282

Koeberl C, Gilmour I, Reimold WU, Claeys P, Ivanov B (2002) Comment on "End Permian catastrophe by bolide impact: evidence of a gigantic release of sulfur from the mantle". Geology 30: 855-856

Koulouris J, Janle P, Milkereit B, Werner S (1999) Significance of large ocean impact craters. In: Gersonde R, Deutsch A (eds) Oceanic impacts: mechanisms and environmental perturbations. Berichte zur Polarforschung 343, Alfred-Wegener-Institut für Polar- und Meeresforschung, Bremerhaven, pp 48-50

Kravchenko S, Schachotko LI, Rass IT (1997) Moho discontinuity relief and the distribution of kimberlites and carbonatites in the northern Siberian Platform. Global Tectonics and Metallogeny 6: 137-140

Kring DA (2000) Impact-induced hydrothermal activity and potential habitats for thermophilic and hypothermophilic life [abs.]. In: Catastrophic events and mass extinctions; impacts and beyond. Houston, Lunar and Planetary Institute Contribution no. 1053: 106-107

Lakomy R (1990) Implications for cratering mechanics from a study of the Footwall breccia of the Sudbury impact structure, Canada. Meteoritics 25: 195-207

Lambert DD, Foster JG, Frick LR, Ripley EM, Zienteck ML (1998) Geodynamics of magmatic Cu-Ni-PGE sulfide deposits: new insights from the Re-Os isotope system. Economic Geology and the Bulletin of the Society of Economic Geologists 93: 121-136

Langenhorst F, Deutsch A, Hornemann U (1998) On the shock behavior of calcite: Dynamic 85-GPa compression and multianvil decompression experiments [abs.]. Meteoritics and Planetary Science 33: A90

Lightfoot PC, Hawkesworth CJ, Olshefsky K, Green T, Doherty W, Keays RR Geochemistry of Tertiary tholeiites and picrites from Qeqertarssuaq (Disko Island) and Nuussuaq, West Greenland with implications for the mineral potential of comagmatic intrusions. Contributions to Mineralogy and Petrology 128: 139-163

Lo C-H, Chung S-L, Lee T-Y, Wu G (2002) Age of Emeishan flood magmatism and relations to Permian-Triassic boundary events. Earth and Planetary Science Letters 198: 449-458

Mahoney JJ, Coffin MF (eds) Large Igneous Provinces. American Geophysical Union, Geophysical Monograph 100, 438 pp

Margot JL, Nolan MC, Benner LAM, Ostro SJ, Jurgens RF, Giorgini JD, Slade MA, Campbell DB (2002) Binary asteroids in the near-Earth object population. Science 296: 1445-1448

McKenzie D, Bickle MJ (1988) The volume and composition of melt generated by extension of the lithosphere. Journal of Petrology 29: 625-679

Melosh HJ (1989) Impact Cratering - A Geological Process. Oxford Monographs in Geology and Geophysics 11: 245 pp

Melosh HJ (2000) Can impacts induce volcanic eruptions? [abs.]. In: Catastrophic events and mass extinctions; impacts and beyond. Houston, Lunar and Planetary Institute Contribution no. 1053: 141-142

Miura Y, Uedo Y, Kedves M (2001) Formation of Fe-Ni particles by impact process [abs.]. Lunar and Planetary Science 32, abs. #2149, CD-ROM

Molnar F, Watkinson DH, Everest JO (1999) Fluid-inclusion characteristics of hydrothermal Cu-Ni-PGE veins in granitic and metavolcanic rocks at the contact of the Little Stobie deposit, Sudbury, Canada. Chemical Geology 154: 279-301

Mooney WD (1999) Crustal and upper mantle seismic velocity structure of the former USSR. US Department of Defense. 21st Seismic Research Symposium: Technologies for Monitoring the Comprehensive Nuclear-Test-Ban Treaty, 162-171

Morgan J, Warner M, Brittan J, Buffler R, Camargo A, Christeson G, Denton P, Hildebrand A, Hobs R, Macintyre H, Kackenzie G, Maguire P, Marin L, Nakamura Y, Pilkington M, Sharpton V, Snyder D, Saurez G, Trajo A (1997) Size and morphology of the Chicxulub structure. Nature 390: 472-476

Mundil R, Metcalfe I, Ludwig KR, Renne PR, Oberli F, Nicoll RS (2001) Timing of the Permian-Triassic biotic crisis: implications from new zircon U/Pb age data (and their limitations). Earth and Planetary Science Letters 187: 131-145

Negi JG, Agrawal PK, Pandey OP, Singh AP (1993) A possible K-T boundary bolide impact site offshore Bombay and triggering of rapid Deccan volcanism. Physics of the Earth and Planetary Interiors 76: 189-197

Neumann GA, Zuber MT, Smith DE, Lemoine FG (1996) The lunar crust: Global signature and structure of major basins. Journal of Geophysical Research 101: 16,841-16,863

Norman MD (1994) Sudbury Igneous Complex: Impact melt or endogenous magma? Implications for lunar crustal evolution. In: Dressler BO, Grieve RAF, Sharpton VL (eds) Large Meteorite Impacts and Planetary Evolution. Geological Society of America Special Paper 293: 331-341

O'Keefe JD, Ahrens TJ (1993) Planetary cratering mechanics. Journal of Geophysical Research 98: 17011-17028

Oleynikov BV, Okrugin AV, Tomshin MD (1985) Native metals formation in basic rocks of the Siberian Platform, Yakutian Branch of the Soviet Academy of Sciences, Yakutsk, 188 pp (in Russian)

Phinney EJ, Mann P, Coffin MF, Shipley TH (1999) Sequence stratigraphy, structure, and tectonic history of the southwestern Ontong Java Plateau adjacent to the Solomon trench and Solomon Islands arc. Journal of Geophysical Research 104: 20449-20466

Pierazzo E, Vickery AM, Melosh HJ (1997) A re-evaluation of impact melt production. Icarus 127: 408-423

Poirier JP (2000) Physics of the Earth's Interior. Cambridge University Press, Cambridge, 312 pp

Pond RB, Glass CM (1970) Metallurgical observations and energy partitioning. In: Kinslow R (ed) High Velocity Impact Phenomena. Academic Press, New York 579 pp

Price NJ (2001) Major Impacts and Plate Tectonics. Routledge, London, 416 pp

Rampino MR (1987) Impact cratering and flood-basalt volcanism. Nature 327: 468-468

Rampino MR, Stothers RB (1988) Flood basalt volcanism during the last 250 million years. Science 241: 663-668

Reichow MK, Saunders AD, White RV, Pringle MS, Mukhamedov IA, Medvedev AI, Kirda NP (2002) ^{40}Ar/^{39}Ar Dates from the West Siberian Basin: Siberian flood basalt province doubled. Science 296: 1846-1849

Retallack GJ, Seyedolali A, Krull ES, Holser WT, Ambers CP (1998) Search for evidence of impact at the Permian-Triassic boundary in Antarctica and Australia. Geology 26: 979-982

Rhodes RC (1975) New evidence for impact origin of the Bushveld Complex, South Africa. Geology 3: 549-554

Richards MA, Duncan RA, Courtillot VE (1989) Flood basalts and hotspot tracks: plume heads and tails. Science 246: 103-107

Richardson WP, Okal EA, Van der Lee S (2000) Rayleigh wave tomography of the Ontong-Java plateau. Physics of Earth and Planetary Interiors 118: 29-51

Robin E, Lefevre I, Rocchia R (2000) Extraterrestrial spinel in marine sediments: an overview [abs.]. In: Catastrophic events and mass extinctions; impacts and beyond. Houston, Lunar and Planetary Institute Contribution no. 1053: 182

Roddy DJ, Schuster SH, Rosenblatt M, Grant LB, Hassig PJ, Kreyenhagen KN (1987) Computer simulations of large asteroid impacts into oceanic and continental sites-- preliminary results on atmospheric, cratering and ejecta dynamics. International Journal of Impact Engineering 5: 525-54

Rogers GC (1982) Oceanic plateaus as meteorite impact signatures. Nature 299: 341-342

Ryabov YY, Anoshin GN (1999) Platinum-iron metallization in intrusive traps of the Siberian Platform. Geologiya i Geofizika 40: 162-174 (in Russian)

Saunders AD, Storey M, Kent R, Norry M (1992) Consequences of plume-lithosphere interaction. In: Storey BC, Alabaster T, Pankhurst RJ (eds) Magmatism and the Causes of Continental Break-Up. Geological Society of London Special Publication 68: 41-60

Seyfert CK, Sirkin LA (1979) Earth History and Plate Tectonics. New York, Harper and Row, 96 pp

Sharma M (1997) Siberian traps. In: Mahoney JJ, Coffin MF (eds) Large Igneous Provinces: Continental, Oceanic and Planetary Flood Volcanism. American Geophysical Union, Washington. Geophysical Monograph 100: 273-295

Shoemaker EM, Wolfe RF, Shoemaker CS (1990) Asteroid and comet flux in the neighborhood of Earth. Sharpton VL, Ward PD (eds) Geological Society of America Special Paper 247: 155-170

Shukulyukov A, Kyte FT, Lugmair GW, Lowe DR, Byerly GR (2000) The oldest impact deposits on Earth – first confirmation of an extraterrestrial component. In: Gilmour I, Koeberl C (eds) Impacts and the early Earth. Lecture Notes in Earth Sciences 91, Springer, Heidelberg-Berlin, pp 99-105

Stöffler D, Deutsch A, Avermann M, Bischoff L, Brockmeyer P, Buhl D, Lakomy R, Müller-Mohr V (1994) The formation of the Sudbury structure, Canada: toward a unified impact model. In: Dressler BO, Grieve RAF, Sharpton VL (eds) Large Meteorite Impacts and Planetary Evolution. Geological Society of America Special Paper 293: 303-318

Storey M, Mahoney JJ, Kroenke LW, Saunders AD (1991) Are oceanic plateaus sites of komatiite formation? Geology 19: 376-379

Swegle JW (1990) Irreversible phase transitions and wave propogation in silicate materials. Journal of Applied Science 68: 1563-1579

Thompson RN, Gibson SA (2000) Transient high temperatures in mantle plume heads inferred from magnesian olivines in Phanerozoic picrites. Nature 407: 502-506

Turtle EP, Pierazzo E (2000) Impact heating during formation of the Vredefort structure [abs.]. In: Catastrophic events and mass extinctions; impacts and beyond. Houston, Lunar and Planetary Institute Contribution no. 1053: 233

Van der Voo, Spakman W, Bijwaard H (1999) Mesozoic subducted slabs under Siberia. Nature 397: 246-249

Walker RJ, Morgan JW, Beary ES, Smoliar MI, Czamanske GK, Horan MF (1997) Applications of the Pt-190-Os-186 isotope system to geochemistry and cosmochemistry. Geochimica et Cosmochimica Acta 61: 4799-4807

Weissman PR (1997) Long-period comets and the Oort Cloud. In: Remo JL (ed) Near-Earth Objects, the United Nations international conference. Annals of the New York Academy of Sciences, New York 822: 67-95

White R, McKenzie DP (1989) Magmatism of rift zones; the generation of volcanic continental margins and flood basalts. Journal of Geophysical Research 94: 7685-7729

Wignall P (2001) Large igneous provinces and mass extinctions. Earth Science Reviews 53: 1-33

Wilson L, Head JW (2001) Factors controlling the ascent of magmas on the moon [abs.]. Eos Transactions American Geophysical Union 82 (20) Supplement, pp 240-241

Wyllie PJ (1970) Magmas and volatile components. American Mineralogist 64: 469-500

Yin YG, Wang Y, Wang W, Sheng QH, Cao CQ, Erwin DH (2000) Pattern of mass extinction near the Permian-Triassic boundary in South China. Science 289: 432-436

Displacement of Target Material During Impact Cratering

Valery Shuvalov

Institute for Dynamics of Geospheres, Russian Academy of Sciences, Leninskiy prospekt 38-6, 119334, Moscow, Russia. (shuvalov@idg.chph.ras.ru)

Abstract. A model of impact cratering is designed to describe the excavation flow and expansion of ejecta through the atmosphere. A principal point of the model is the consideration of condensed ejecta as an aggregate of discrete particles. The motion of particles, and the exchange of heat and momentum with air and vapor are described by equations of multi-phase hydrodynamics. This allows the calculation of ejecta distribution at different distances from the crater. The presented model can treat all fragments as a bulk medium or as specific particles (i.e., melted, shock modified, ejected from specific depth, etc.). Here this model is applied to the study of the vertical impact of a spherical asteroid of 1 km radius on a granite target. The results of the numerical simulations demonstrate a separation of ejected condensed particles by size; the separation considerably influences the final distribution of the deposited ejecta.

1
Introduction

The investigation of the displacement of target material is an important problem in cratering mechanics, because the presence and distribution of impact ejecta (which are characterized by the occurrence of shock-modified rocks, and enrichment in meteoritic matter) provide the main evidence for impact events. The shock-metamorphic effects in target material occur mainly during the shock wave propagation through the target and depend on the maximum pressure experienced by the target material (Melosh 1989). Later the mechanical and chemical interactions between different parts of the ejecta and ambient air can cause additional modification. The vertical extent of the zone of target material displacement can be estimated from the depth of the transient cavity, and the horizontal scale is defined by ejecta spreading and deposition.

The purpose of this study is to develop a numerical model for the accurate description of the target material motion both due to the cratering flow and the extension of the ejecta through the atmosphere. In most previous models the motion of the ejected material (which was treated as continuous medium) was considered in the frame of hydrodynamic equations (e.g., O'Keefe et al. 2001). In other models the initial positions of the ejected particles were determined from hydrodynamical models and the particles were assumed to move through the atmosphere on ballistic trajectories, atmospheric drag taking into account (Maxwell 1977; Tauber 1978). Both approaches do not properly describe the interaction between solid ejecta and gas flow. The hydrodynamical approach implies the same velocity for all components at each point of space. However, in general, the velocity of solid particles differs from the gas velocity, and, particles of different sizes have different velocities (Boothroyd 1971). The ballistic approximation does not take into account the motion of ambient gas and the influence of solid particles on this motion. Furthermore, the motion of a cloud of solid fragments (ejecta curtain) cannot be described as the sum of the independent motions of all fragments (Boothroyd 1971).

The expansion of solid ejecta through the air was studied experimentally (Schultz 1992). It was shown that ejecta distributions might reflect entrainment of the ejecta in winds and turbulence generated in response to the outward moving ejecta curtain. Different styles of ejecta emplacement may reflect the degree of ejecta entrainment in the dynamics of atmospheric response, which, in turn, depends on crater and ejecta size. However, the direct extrapolation of experimental data to large scales and high impact velocities isuncertain and should be tested by numerical simulations.

In this paper a new approach is developed, which describes the extension of the ejecta curtain in the frame of multi-phase hydrodynamic equations (Valentine and Wohletz 1989). This new approach is believed to allow considering the atmosphere-ejecta interaction in more detail and to explain some specific features of the ejecta deposition. Of special importance is calculation of the distribution of different types of particles (i.e., shock modified, ejected from specific depth, etc.).

2
Numerical Model

The numerical model is based on the SOVA (SOlid, VApor) multi-material hydrocode (Shuvalov 1999), which is used to model all stages of the impact. The SOVA code is a Eulerian material response code with some Lagrangian features. It allows considering strong hydrodynamic flows with an accurate description of the boundaries between different materials (e.g., air, vapor, solid impactor material, etc.). The code is similar in conception to the CTH (Chapter THree) hydrocode (McGlaun et al. 1990), which is widely used in the USA.

In order to simulate the crater modification stage, a model of strength is included into the SOVA hydrocode. I used a rheology model based on the idea of "Acoustic Fluidization" (Melosh and Ivanov 1999) and the "Block Model" (Ivanov and Turtle 2001). This rheology model assumes that the rocks surrounding the crater structure are fractured and their behavior is similar to that of the classical "Bingham Media", where the stress is a linear function of strain rate. About one hundred thousand passive tracer particles are used to follow the motion of the target.

One of the main characteristics of this model is a mechanism for the transformation of solid and melted bulk ejecta into discrete particles (Shuvalov 2001). This transformation occurs when the bulk ejecta density falls below $0.5\rho_0$, where ρ_0 is the initial target rocks density. An equation of state of the target material and the phase equilibrium curve are used to determine the thermodynamic state of the ejected material. In order to transform solid ejecta into discrete particles it is necessary to determine the size distribution of the ejected particles. Following Melosh (1989), I use the following power-law relation

$$N(m)=Cm^{-b},$$

where $N(m)$ is the cumulative number of fragments of mass equal to or greater than mass m, the exponent b commonly ranges between 0.8 and 0.9, and C is a constant defined by the total ejected mass. The mass of the largest fragment m_m is determined from the statistical strength theory of Weibull (1951). In the case under consideration the basic equation of this theory can be written in the form:

$$(p/p_0)=(m_0/m_m)^{\alpha},$$

where p is the maximum pressure experienced by ejected mass, p_0 is the maximum pressure near the boundary of the excavated crater, m_0 is the mass of the largest intact block in the ejecta blanket, and the exponent equals 0.25. The values of p_0 and p are determined from numerical simulations, and m_0 is assumed to be determined by the relation

$$m_0=0.8M^{0.8},$$

where M is the total ejected mass (Melosh 1989). For melted ejecta the size of the largest fragment is assumed to be 3 cm. This model assumes that the mass of the largest fragment m_m is different for different parts of ejecta (e.g., excavated from different depth and distance from the crater center). In particular, the largest fragments are ejected from the periphery of the excavated crater (and fall near the rim) and from near the surface spall zone (and form secondary craters). It should be noted that any other ejecta size distributions derived from other theories or/and observations can be easily included into the model.

The following evolution of ejecta is described as a motion of discrete particles interacting with the gas. The motion of the particles and the heat and momentum exchange with air and vapor are described within the frame of equations of multi-phase hydrodynamics (Valentine and Wohletz 1989). To solve these equations I used the method of representative particles or markers (Teterev 1999). Each marker describes the motion of a great number of real fragments that have approximately the same sizes, velocities, trajectories, etc. The largest fragments are considered separately and are described by their own tracer particles.

All information about the ejected particles (velocity and angle of ejection, pre-impact location, maximum temperature and overpressure experienced during the impact, etc.) is conserved by markers. On the one hand, this allows to calculate the bulk ejecta distribution at different distances from the crater. On the other hand, the distribution of different types of particles (i.e., melted, shock modified, ejected from specific depth, etc.) can be displayed.

A specific implicit algorithm (Shuvalov 1999) provides a correct solution of the multi-phase hydrodynamic equations both for large fragments, which practically do not experience any atmospheric drag and follow ballistic trajectories, as well as for micro-particles, which move with local gas velocity. The method also provides the correct velocity of the settling of dust due to gravity (which depends on the size of the particles).

3
Numerical Simulations of the Impact of a 1-km-Radius Asteroid

The approach described in the previous section was applied to model the vertical impact (this allows using a 2D version of the code) of a 1-km-radius spherical asteroid on a granite target. The impact velocity was assumed to be 20 km/s. Impact energy equals 2×10^{21} J. The ANEOS (ANalytical Equation Of State) equation of state (Thompson and Lauson 1972) was used to describe the thermodynamical properties of both impactor and target materials. The thermodynamical characteristics of the air were described with the use of a tabulated equation of state (Kuznetsov 1965).

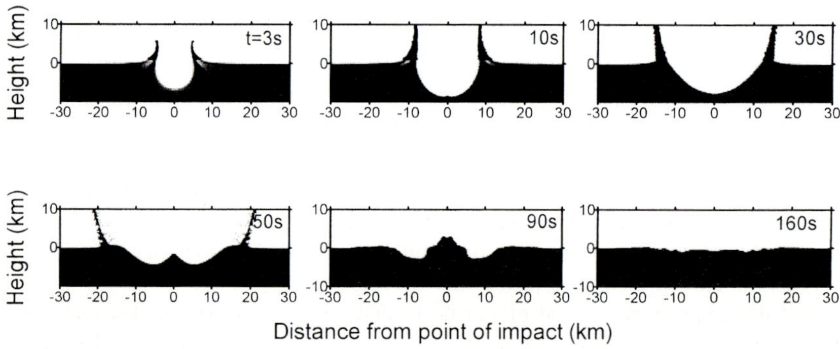

Fig. 1. Evolution of transient cavity and subsequent crater collapse after the vertical impact of a 1-km-radius spherical asteroid on a granite target. Impact velocity is 20 km/s.

The evolution of the transient crater is shown in Fig. 1. Ten seconds after the impact, the transient cavity reaches its maximum depth of about 8 km. However, the excavation continues for about twenty more seconds. At 30 s, the cavity achieves a horizontal extent of 11 km (radius). At the same time the depth of the crater begins to decrease due to the rising crater floor. Fifty seconds after the impact, a well-defined central uplift is formed. The radius of the visible crater increases during the late stage due to the slumping of crater walls and reaches a value of about 17 km. The final image in Fig.1 shows the crater at 160 s, which is close to the final crater shape.

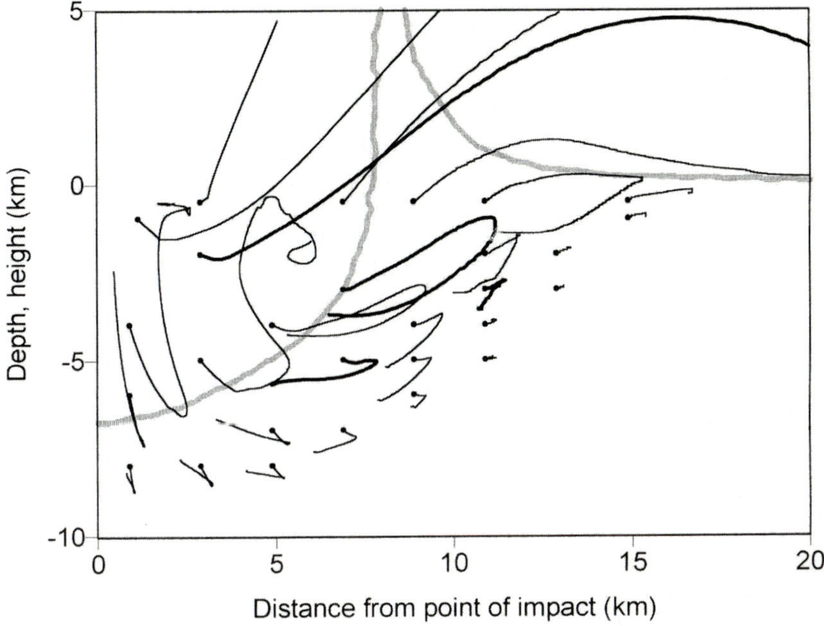

Fig. 2. Trajectories of selected Lagrangian particles in the target (initial positions are depicted by solid circles). The thick gray line displays transient cavity at the 10 s moment.

Figure 2 shows trajectories of selected Lagrangian points in the target during the impact process. The gray line shows the transient cavity at the moment when it reaches its maximum depth. Several typical trajectories can be distinguished. Some parts of target material escape from the crater and travel large distances. The initial positions of these particles define the excavated crater. Other parts of the excavated material remain within the area of the final crater and slump inwards during the process of crater modification (collapse). A large amount of target material is displaced for a significant distance, but remains within the crater and is not excavated. In general, these particles first move outward (following the shock wave), then revise their flow direction and even can be displaced to the target surface.

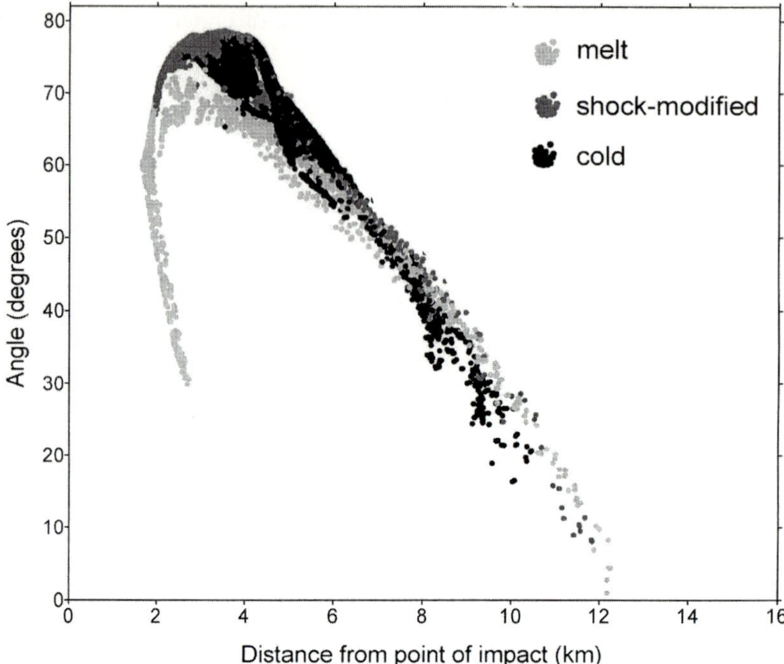

Fig. 3. Angle of ejection versus distance of ejection point from center of impact. This distance is approximately equal to the current (at the moment of ejection) radius of the transient cavity.

Figure 3 shows ejection angles versus a starting point of ejection with respect to the impact center. The ejection angle and the radial location (i.e., the radial distance from the crater center) of the point where the particle was ejected to atmosphere are shown for each marker. All ejecta can be roughly separated into two parts. The first, early, part of ejecta is formed at the initial stage of impact, before total impactor deceleration. This stage (during the first few seconds) is characterized by strong shock modification of the ejected material and a very wide range of ejection angles. The total mass of this early ejecta is not large (only several per cent of the total ejected mass); however, this ejecta has high velocities and reaches significant distances.

The second part of ejecta forms during the late stage of the cratering flow. The angle of ejection gradually diminishes from maximum values of 60 to 80 degrees at the beginning of the ejection to a few degrees at the end of the excavation.

The upper panel of Fig. 4 shows ejecta distributions and crater at the initial stage of impact (1.6 s). On the right side of the cross-section dark gray dots show selected particles of melted material; light gray dots show shock-modified material (i.e., material experiencing pressures of 15 to 40 GPa during the impact process); black dots show ejected fragments that do not experience pressures above 15 GPa. The leading fast ejecta consist of only vapor and melt, followed by shock-modified and cold fragments. Distribution of the early ejecta looks similar to an expanding cloud rather than an ejecta cone (or ejecta curtain).

On the left side of the cross-section different shading marks particles of different sizes. The smallest fragments, indicated by light gray dots, are subjected more significantly to atmospheric drag and lag behind heavier particles shown by dark gray dots. Large fragments with a radius exceeding 10 cm (shown by black dots) appear later when less shocked material is ejected.

The lower panel of Fig. 4 shows the temperature and density distributions, also at 1.6 s after the impact. At the beginning of the plume evolution, the temperature of shock-compressed air reaches 10,000 K and considerably exceeds the temperature of the expanding vapor. The temperature of small volumes of air surrounded by much more dense vapor reaches 20,000-30,000 K.

Figure 5 shows the same distributions 48 s after the impact. At this moment, the excavation is complete. Ejecta consisting of discrete particles form a permeable ejecta curtain. Expanding vapor penetrates through this curtain outward and carries away small (about 1 cm or less) ejecta particles. Large fragments are not subjected to the action of the expanding vapor and move ballistically. This leads to the radial separation of ejecta by particle sizes and changes in the trajectories of small particles.

A mixture of vapor and small melted particles moves through the curtain consisting of larger fragments, because at the moment of ejection the melt is concentrated near the inner boundary of the ejecta curtain. Small melted particles can be deposited on the surface of large fragments to form objects similar to melt-covered bombs, which were found near the Popigai crater and in some other places (Masaitis 1994).

The expanding vapor and condensed ejecta generate strong shock waves in the upper atmosphere. Shock-heated air expands upward. This expansion leads to adiabatic cooling of the air and the formation of a pall of dense air above the impact site. The temperature of the expanding vapor becomes larger than the temperature of the surrounding air because of energy release due to partial condensation. The temperature of the two-phase mixture cannot be lower than the temperature of the phase (vapor to solid) transition.

There are no markers in the central part of the expanding plume. However, this region consists of vapor and very small melted particles resulting from condensation. The size of these particles is estimated to be 1 to 100 μm (Nemtchinov et al. 1998). Such small particles have the same velocity as the surrounding gas and can be considered within the frame of usual hydrodynamics. These particles are important from the viewpoint of climatic effects (Toon et al. 1997), because the time of their settling is very long (months and even years). A small part of the plume, which has velocity exceeding 11 km/s, escapes from the Earth. If these particles are of interest they can also be described separately with the use of representative particles (as has been done for other parts of the ejecta).

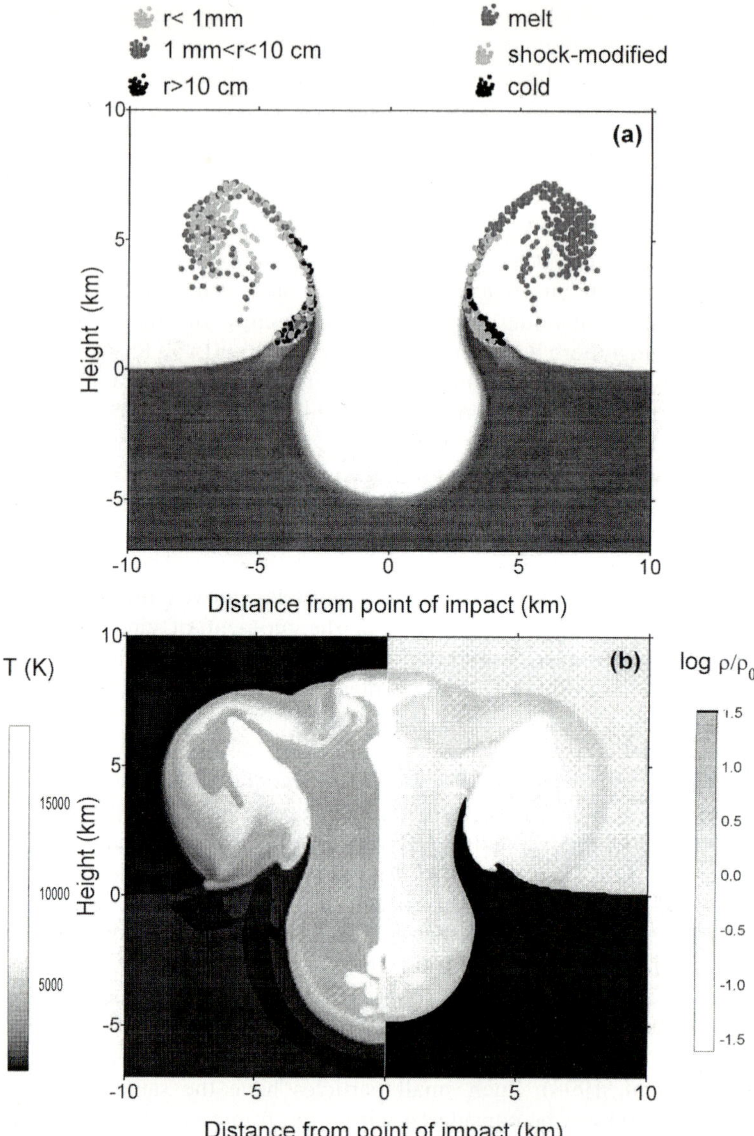

Fig. 4. The upper panel (a) shows the ejecta distributions and the crater shape 1.6 s after the impact. On the right side, dark gray dots display melted particles, light gray dots show shock-modified (but not melted) particles, and black dots mark cold ejecta material. On the left side, the particle size is shown. The lower panel (b) shows the spatial distribution of temperature (on the left) and relative density $\rho/\rho0$ (on the right), also at 1.6 s. $\rho0(h)$ is the equilibrium air density at altitude h.

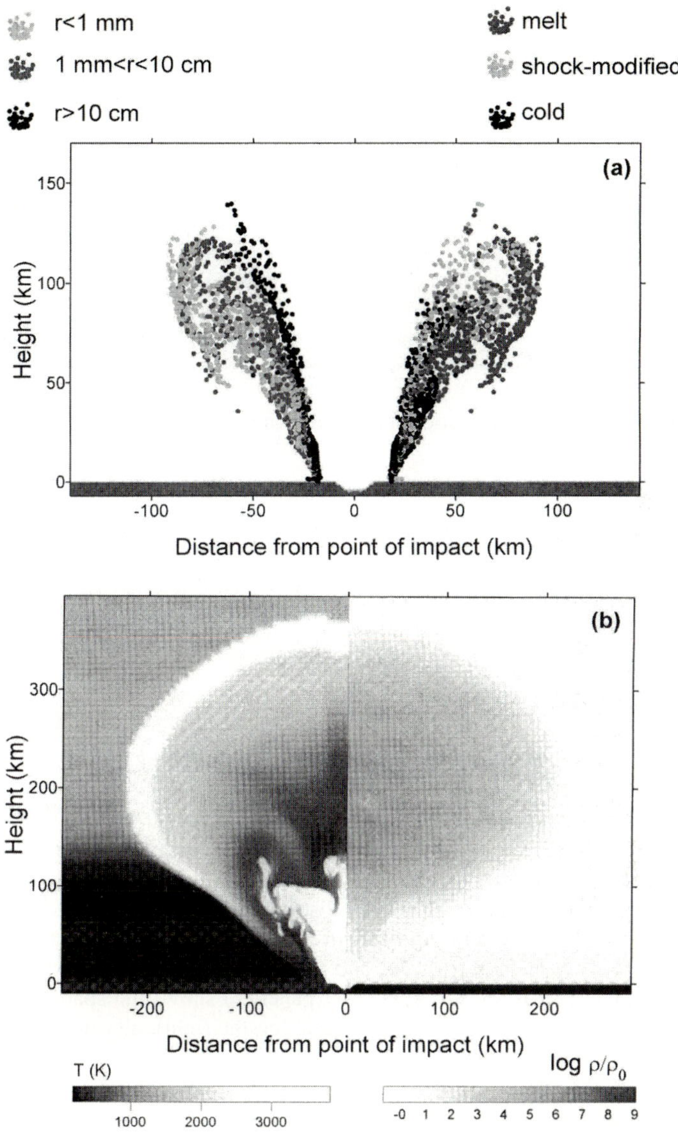

Fig. 5. The upper panel (a) shows the ejecta distributions and the crater shape 48 s after the impact. On the right side, dark gray dots display melted particles, light gray dots show shock-modified (but not melted) particles, and black dots mark cold ejecta material. On the left side, the particle size is shown. The lower panel (b) shows the spatial distribution of temperature (on the left) and relative density $\rho/\rho0$ (on the right), also at 48 s. $\rho0(h)$ is the equilibrium air density at altitude h.

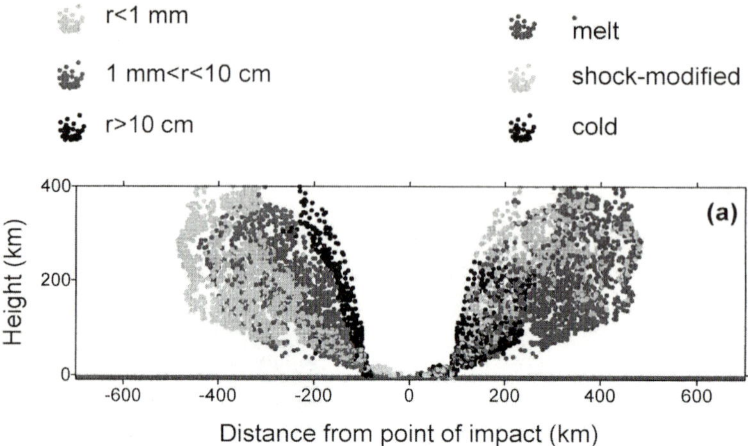

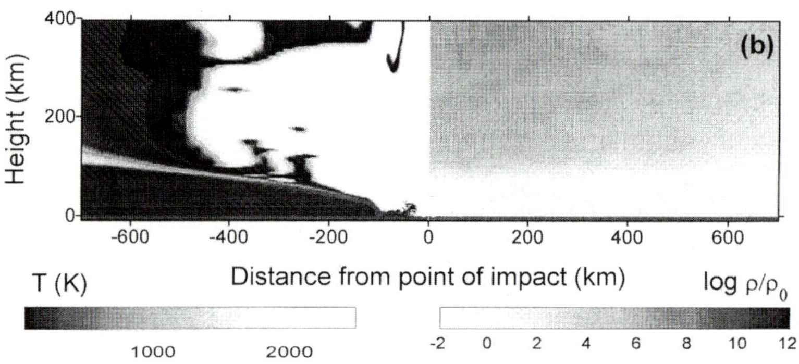

Fig. 6. The upper panel (a) shows the ejecta distributions and the crater shape 200 s after the impact. On the right side, dark gray dots display melted particles, light gray dots show shock-modified (but not melted) particles, and black dots mark cold ejecta material. On the left side, the particle size is shown. The lower panel (b) shows the spatial distribution of temperature (on the left) and relative density ρ/ρ_0 (on the right), also at 200 s. $\rho_0(h)$ is the equilibrium air density at altitude h.

Figure 6 shows the late evolution of the ejecta. Two hundred seconds after the impact, the ejecta cloud extends for a distance of 500 km from the crater center. At high altitudes (above 200 km) all ejecta moves ballistically. This is not the case in the dense atmospheric layers near the surface. Large fragments (shown by black dots) fall back ballistically, but small particles settle down slowly. The model suggests that a vertical separation of ejecta by particle sizes occurs.

Ten minutes after the impact, almost all large fragments with sizes exceeding 10 cm have fallen to the ground. Slowly settling fine particles may be transported for considerable time (hours) and distances (thousands of kilometers). Settling of the ejecta occurs in a strongly disturbed atmosphere, which experiences large-scale oscillations induced by the impact.

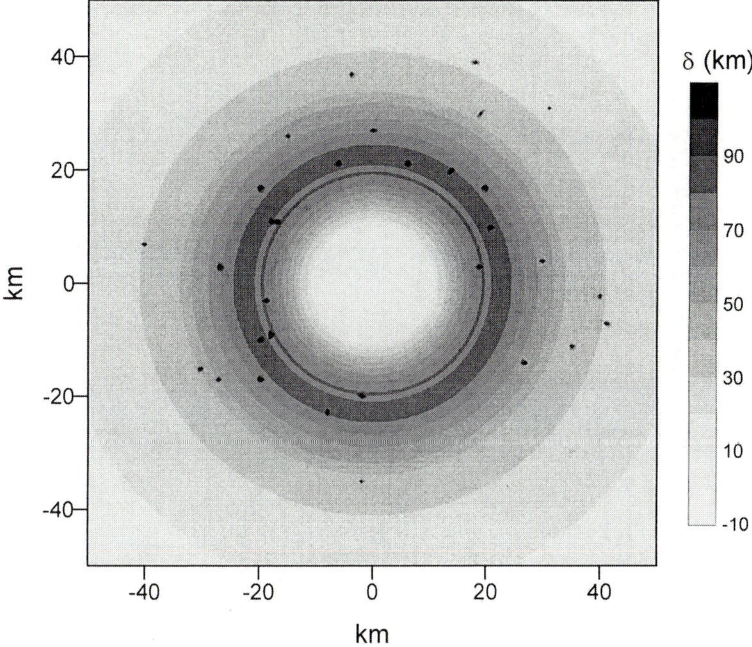

Fig. 7. The thickness δ of the ejecta blanket. Dark spots correspond to the places of secondary impacts of large fragments (probable formation of secondary craters).

Figure 7 shows the thickness of the ejecta blanket δ, which, at any surface S, is determined as the sum of the volumes of all particles, deposited on this surface, divided by the surface area. Dark spots correspond to the places of secondary impacts of large (a few hundred meters) fragments (i.e., locations of probable secondary craters).

Figure 8 shows a comparison between the thickness of the ejecta blanket obtained in the numerical simulations and the dependence derived from explosion experiments (McGetchin et al. 1973):

$$\delta = 0.04R(r/R)^{-3},$$

Here R is the transient cavity radius. A considerable part of the ejecta forming the rim moves as a dense continuous medium and does not rise above a few hundred meters. This part of the ejecta was not transformed into discrete particles and is not shown in Fig. 8. The results coincide with one another at distances (from the crater center) exceeding 40 km. In the vicinity of the crater the numerical results demonstrate some irregularities, wave-like deviations from the r^{-3}-law. The deviations result from the separation of different parts of ejecta in space.

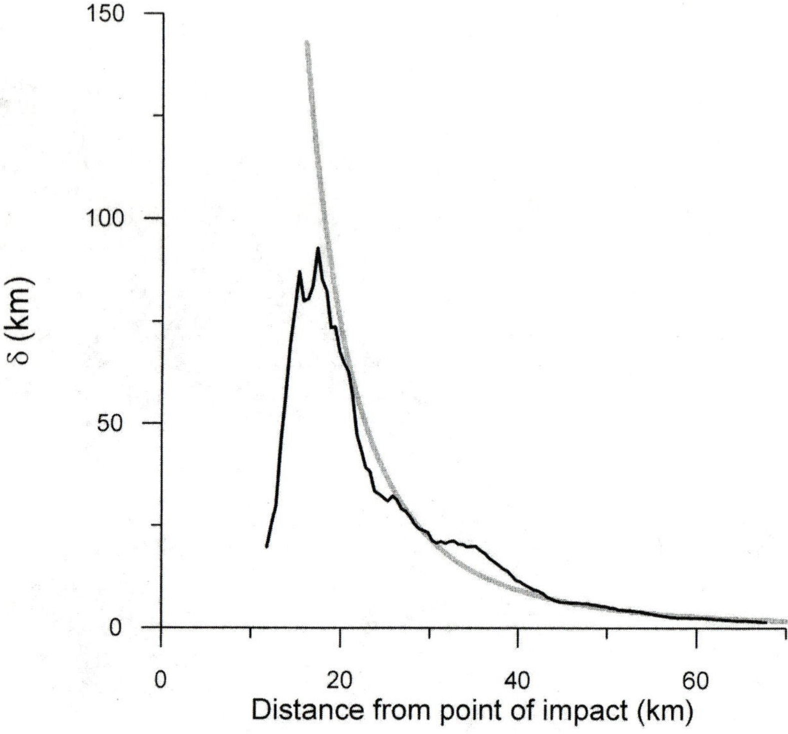

Fig. 8. The thickness δ of ejecta blanket versus distance from point of impact. The thick gray line displays the McGetchin et al. (1973) dependence, the thin black line displays the results of numerical simulations.

Figure 9 shows the time dependence of ejecta mass, lifted above the altitude h, for different values of h. Only 2.6×10^{10} tons of solids, or $2.3M_0$ (where M_0 is the mass of impactor), are ejected to altitudes above 80 km. This mass strongly decreases with time due to settling of the particles, and only $0.03M_0$ remains at high altitudes (above 80 km) ten minutes after the impact. The main source of air contamination at high altitudes is the condensation of the expanding vapor. Numerical simulations show that the mass of vapor ejected above 14 km reaches about $2M_0$. Approximately one half of the vapor is transformed into small (1 to 100 μm) droplets during the process of expansion (Nemtchinov et al. 1998). Such small particles settle very slowly in the atmosphere. Several minutes after the impact they can be considered as the main source of solids in the upper atmosphere.

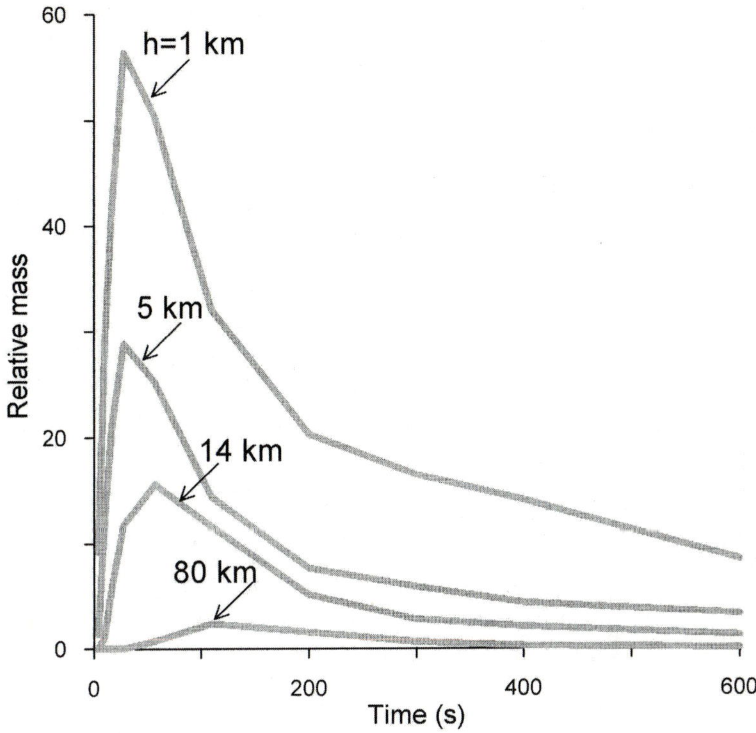

Fig. 9. The mass of ejecta rising above altitude h versus time.

The mass of solids ejected into the troposphere (above 14 km) reaches 1.7×10^{11} tons, or $16M_0$. The total excavated mass (defined as the mass rising above the crater rim and being transformed into discrete particles) is calculated to be about $100–150M_0$. In the case of a vertical impact the major fraction of this mass falls outside the transient cavity. A more exact estimate depends on the definition of excavated mass. This definition is not an obvious one, because there is no sharp boundary between excavated and displaced target material.

3
Summary

A model has been developed that allows to consider the displacement of the target material both due to the cratering flow and due to the expansion of ejecta through the atmosphere. This model treats ejecta as a large number of discrete particles interacting with the gas flow by momentum exchange.

An important effect that is not taken into account by purely hydrodynamic simulations (which treat ejecta as a continuous medium) is the distinction between the velocities of vapor and solid or melted particles of different sizes at the same point in space. This distinction causes the separation of different parts of ejecta in space. This separation, in turn, strongly influences the final spatial distribution of the deposited ejecta. In particular, this leads to wave-like irregularities in the thickness of the ejecta blanket. Similar irregularities are clearly seen in so-called rampart craters (Melosh 1989). This effect is usually explained by incorporation of liquid water in the ejecta. However, atmospheric modification of the ejecta curtain can be also important. The fine ejecta (about 1 cm or less) settle slowly (from a few minutes to several hours) and forms dense turbulent particulate flows along the surface. These flows in many respects behave as a liquid medium.

In the purely hydrodynamic simulations an ejecta curtain is impermeable for expanding vapor. The expansion of the vapor is restricted by the extent of the ejecta cone (i.e., ejecta curtain). In the model of discrete particles used in this study (similar to what happens in nature) the vapor passes from inside the curtain to the outside, in the gaps between condensed particles, and carries away small grains. This results in a layering of the curtain itself - the size of particles decreases in the outwards direction.

The present model allows to calculate the distributions of ejecta at different distances from the crater both for all fragments and for particles of a specific type (i.e., melted, shock modified, ejected from specific depth, etc.). In particular, the model can be used to study tektites and martian meteorites.

Acknowledgements

The author would like to thank Prof. C. Koeberl, Prof. J. Melosh, and Dr. T. Kenkmann for their valuable remarks; also to thank I. Trubetskaya for the assistance with preparing this paper.

References

Boothroyd RG (1971) Flowing gas-solids suspensions. Chapman and Hall Ltd., London, 289 pp

Ivanov BA, Turtle EP (2001) Modeling impact crater collapse acoustic fluidization implemented into a hydrocode [abs.]. Lunar and Planetary Science 32, abs #1284, CD-ROM

Kuznetsov NM (1965) Thermodynamic functions and shock adiabats for air at high temperatures. Mashinostroyenie, Moscow, 464 pp (in Russian)

Masaitis VL (1994) Impactites from Popigai crater. In: Dressler BO, Grieve RAF, Sharpton VL (eds) Large Meteorite Impacts and Planetary Evolution II: Geological Society of America Special Paper 293, pp 153-162

Maxwell DE (1977) Simple Z model of cratering, ejection and the overturned flap. In: Roddy DJ, Pepin RO, Merill RB (eds) Impact and Explosion Cratering. Pergamon Press, New York, pp 1003-1008

McGlaun JM, Thompson SL, Elrick MG (1990) CTH: a three-dimensional shock wave physics code. International Journal of Impact Engineering 10: 351-360

McGetchin TR, Settle M, Head JW (1973) Radial thickness variation in impact crater ejecta: Implications for lunar basin deposits. Earth and Planetary Sciences Letters 20: 226-236

Melosh HJ (1989) Impact Cratering: A Geologic Process. Oxford University Press, N.Y. & Clarendon Press, Oxford, 245 pp

Melosh HJ, Ivanov BA (1999) Impact crater collapse. Annual Review of Earth and Planetary Sciences 27: 385–425

Nemtchinov IV, Shuvalov VV, Artem'eva NA, Ivanov BA, Kosarev IB, Trubetskaya IA (1998) Light flashes caused by meteoroid impacts on the lunar surface. Solar System Research 32: 99-114

O'Keefe JD, Stewart ST, Ahrens TJ (2001) Chicxulub ejecta dynamics [abs.]. Lunar and Planetary Science 32, abs #2190, CD-ROM

Schultz PH (1992) Atmospheric effects on ejecta emplacement. Journal of Geophysical Research 97: 11,623–11,662

Shuvalov VV (1999) Multi-dimensional hydrodynamic code SOVA for interfacial flows: Application to thermal layer effect. Shock Waves 9: 381-390

Shuvalov VV (2001) Dust ejection induced by small meteoroids impacting Martian surface [abs.]. Lunar and Planetary Science 32, abs #1126, CD-ROM

Tauber ME (1978) An analytic study of impact ejecta trajectories in the atmospheres of Venus, Mars, and Earth. Icarus 33: 529-536

Teterev AV (1999) Cratering model of asteroid and comet impact on a planetary surface. International Journal of Impact Engineering 23: 921-932

Thompson SL, Lauson HS (1972) Improvements in the Chart D radiation-hydrodynamic CODE III: Revised analytic equations of state. Report SC-RR-71 0714. Sandia National Laboratory, Albuquerque, 119 pp

Toon OB, Turco RP, Covey C (1997) Environmental perturbations caused by the impacts of asteroids and comets. Reviews of Geophysics 35: 41-78

Valentine GA, Wohletz KH (1989) Numerical models of plinian eruption columns and pyroclastic flows. Journal of Geophysical Research 94: 1867-1887

Weibull W (1951) A statistical distribution function of wide applicability. Journal of Applied Mechanics 18: 140-147

Obscure-bedded Ejecta Facies from the Popigai Impact Structure, Siberia: Lithological Features and Mode of Origin

Victor L. Masaitis

Karpinsky Geological Institute, Sredny prospekt 74, St.-Petersburg, 199106, Russia. (vicmas@vsegei.sp.ru)

Abstract The lithology and distribution of proximal ejecta facies that formed during the late stages of formation of the Popigai impact crater, Siberia, are described. These deposits arc distinct from other rock facies of the crater fill due to obscure bedding (or unclear layering), poor or moderate sorting, and some other specific lithological features (character of debris, presence of accretionary lapilli etc). Three types of such ejecta facies occur in the upper parts of the crater fill; these are are composed of lithic microbreccias and suevites (A type), suevites with minor tagamites (B type), and suevites (C type). Similar rock facies, according to previously published data, may be distinguished in some other impact craters (e.g., Ries, Sudbury, and Logoisk). The preliminary interpretation of the mode of origin of these ejecta facies are air fall (A), pyroclastic-like flow (B), and base surge (C) deposits, which resemble some features of volcanic pyroclastic formations. It is possible that similar rock facies may be found in the crater fill of some other well-preserved impact craters.

1
Introduction

General impact lithologies, which occur both inside and outside of terrestrial impact craters, do not show any layering. However, some types of coarse stratification (or pseudostratification) of proximal and distal ejecta from these craters has been observed (Melosh 1989; Grieve and Pesonen 1992; French 1998; Montanari and Koeberl 2000). The individual geological bodies comprising crater fill, composed of impact lithic breccias and impactites (melted rocks), usually show a random inner texture and little or no differentiation in vertical sections. All principal facies patterns of these rocks formed due to very rapid deposition of ejected material composed of melted and crushed target rocks. Meanwhile, in some terrestrial impact structures, where the uppermost part of crater fill is preserved from erosion, proximal ejecta, which display unclear, indistinct layering ("obscure bedding") and poor sorting, are known (Mosebach 1964; Stöffler et al. 1977; Masaitis et al. 1980; Muir and Peredery 1984; Glasovskaya et al. 1991). These types of facies are not yet well studied, although their future exploration may provide important data on the mode of their origin during the late stages of accumulation of crater fill deposits. Furthermore, they may provide data on some processes in the impact plume.

The well-preserved large Popigai impact structure, Siberia, provides an opportunity for facies analysis of these types of ejected impact rocks, which can be observed in outcrops and drillcores. These rocks have been studied during the course of regular geological mapping and drilling over several decades. Their specific lithological features, e.g., obscure bedding, can easily be distinguished in large outcrops, but are more difficult to discern in drillcores due to their small size. These obscure-bedded impactites (mostly suevites) and lithic microbreccias (coptoclastites) were briefly described in some previous publications (Masaitis et al. 1980; Masaitis 1983; 1998), but were not described in detail and not distinguished as separate ejecta facies. It is probable that those rock facies occur more widely in the Popigai crater than is known at present.

The goals of this paper are to summarize the current knowledge of the data regarding these facies in the Popigai crater, to discuss some assumptions on their origin, and to compare them with similar facies that are found in some other impact craters.

2
Crater Fill of the Popigai Structure

The Popigai impact crater (diameter 100 km, age 35.7 Ma, Bottomley et al. 1997) is filled with a thick sequence of allogenic lithic breccias and melted rocks, i.e., tagamites (impact melt rocks) and suevites (melt-bearing polymict impact breccias). The target rocks are Archean and Lower Proterozoic gneisses and minor granites, which are overlain by a 1.5 km thick cover of Upper Proterozoic, Cambrian, and Permian sandstones, siltstones, slates, marls, dolomites, and limestones. Lower Triassic basaltic tuffs and dolerites sporadically occur to the north of the impact structure. The remnants of friable Cretaceous coaly sands, which were likely target rocks, also occur locally. Quaternary gravels, sands, and clays are widely distributed, especially in the eastern part of the crater area.

The crater inner structure is complex (Figure 1) and comprises a central depression, annular uplift (peak ring), annular trough, and a flat outer annular terrace, where disturbed sedimentary rocks occur. The disturbances caused by the impact (folds, thrusts, faults) gradually attenuate radially outwards (Masaitis 1998; Masaitis et al. 1980; 1998; 1999; Vishnevsky and Montanari 1999).

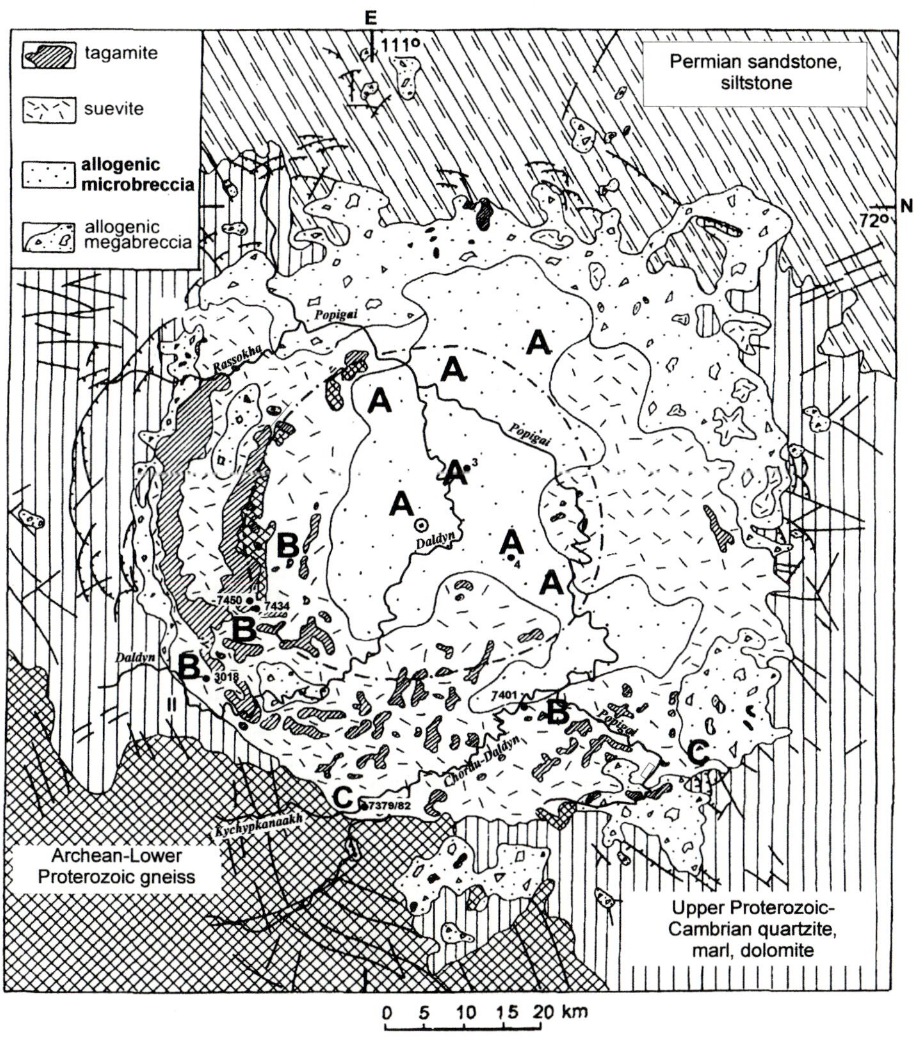

Fig. 1. Schematic geological map of the Popigai impact crater (after Masaitis et al. 1998). Quaternary deposits are not shown. Thick lines: faults and thrusts (undivided); dashed-dotted line: axis of the peak ring (annular uplift); dotted circle: crater center. Locations of certain studied sections of A, B, and C ejecta facies, which compose the upper member of the crater fill, are shown by corresponding letters.

The crater fill is the product of shock transformation, brecciation, and melting of primary sedimentary and crystalline target rocks, followed by ejection and deposition of this material. Allogenic impact lithologies, composed of impact lithic breccias and impactites (tagamites and suevites) have a total thickness of 1.5 to 2 km. [Note: The rock terminology used in this study is adopted from previous publications on the Popigai structure (e.g., Raikhlin et al. 1978; Masaitis 1983; Masaitis et al. 1998). In classifying Popigai rocks standard sedimentological and petrological approaches have been used (cf. Laznicka 1988; Le Maitre 1989; Mikhailov 1995). However, it should be considered that the nomenclature of impact-generated and impact-related rocks is not yet well established or uniform (cf., e.g., Stöffler and Grieve 1994; King and Petruny 2000; this volume)].

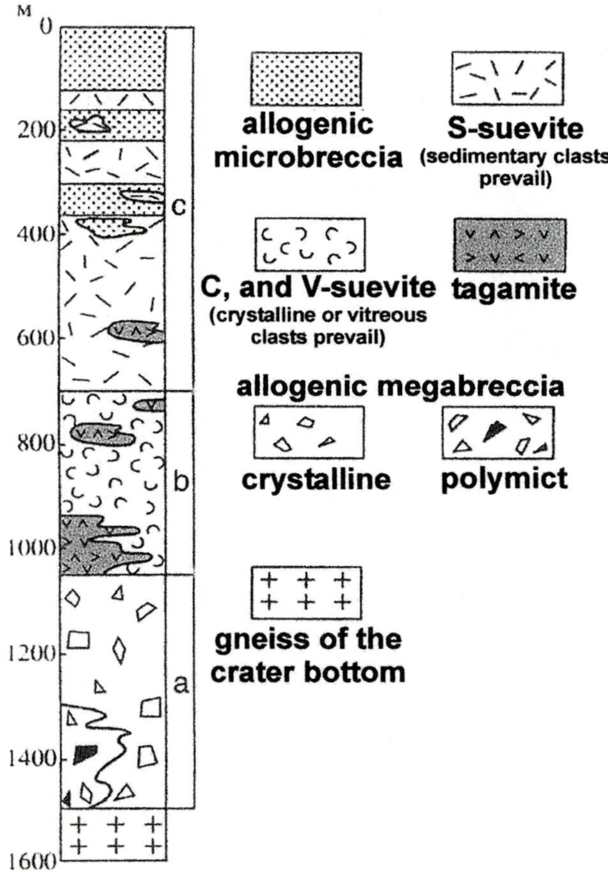

Fig. 2. Generalized section through the Popigai crater fill (modified after Raikhlin 1996). a - lower member, b - middle member, c – upper member.

The crater fill may be subdivided into two (Masaitis et al. 1998) or three (Raikhlin 1996) main units (Figure 2). In the latter work, the lower part of the crater fill is subdivided into two members, and this scheme is accepted in the present study. According to this subdivision the lower member consists of polymict and monomict (crystalline) blocky lithic breccias (megabreccias). The matrix of monomict crystalline megabreccia is usually made of suevites and tagamites. The middle member consists of intercalated lens-like and sheet-like bodies of tagamites and suevites, with the latter occurring mostly in the upper part of the sequence. The upper member comprises alternating suevites (enriched in sedimentary clastic material, but sometimes in impact glass too), and fine-grained lithic breccias or microbreccias, these breccias are found mostly in the uppermost part of the upper member. Small tagamite bodies may be present in its lowermost part. This general succession of the principal rock units of the crater fill is variable and in some areas some parts may be missing.

Some explanations on the subdivision of suevites that are widespread in the Popigai crater are necessary. Suevites are subdivided here according to four clast types in various quantitative ratios: 1) vitroclasts (fresh or altered impact glass particles); 2) lithoclasts (sedimentary and crystalline rock fragments); 3) crystalloclasts (mineral fragments derived from crystalline rocks); 4) granoclasts (rounded or subangular mineral grains, mostly derived from loose sandstones and sands). From prevalence of any of these clast types, the suevite varieties are named accordingly: "vitrogranoclastic", "granovitroclastic", "crystalloclastic" etc. Vitrogranoclastic (rare vitrolithoclastic) and granovitroclastic (rare lithovitroclastic) suevites are usually weakly lithified. Various rocks that comprise the upper member, and exhibit features of unclear layering and poor sorting may be subdivided into three rock facies (lithofacies), A, B, and C, which differ in lithological characteristics and position in a generalized vertical section. They are distinct from other suevite varieties that mostly form the lower and middle members and are characterized by irregular chaotic rock structures and by enrichment in unsorted glass lapilli, shards, bombs, and numerous fragments and blocks of target rocks. It should be emphasized that features characteristic of obscure bedding do not develop throughout the whole section of the upper member, but occur only sporadically within the rock facies A and B, and only the C type facies ccomprise exclusively obscure-bedded rocks. Ejecta deposits of type A facies

Type A facies deposits, composed of lithic microbreccia (coptoclastite) and suevite, occur in the uppermost part of the upper member and are distributed mostly within the central depression of the Popigai crater. These deposits were studied in drillcores, where their apparent thickness varies from 87.2 to 185.5 m, but there is evidence that they could be up to twice this thick. According to drilling data, there are no regular changes of thickness within the crater area. Similar deposits have been found in places within the annular trough (Figure 1).

The sequence, made of type A facies, overlies suevites of B-type facies, and locally occurs directly over the inner slope of the crystalline peak ring (i.e., the annular uplift) in the northwestern sector of the crater (i.e., the Majachika Upland). The deposits are represented by alternating, fine-grained, lithic breccias (microbreccias), and vitrogranoclastic suevites and rare granovitroclastic suevites (Figure 3, drillhole 3).

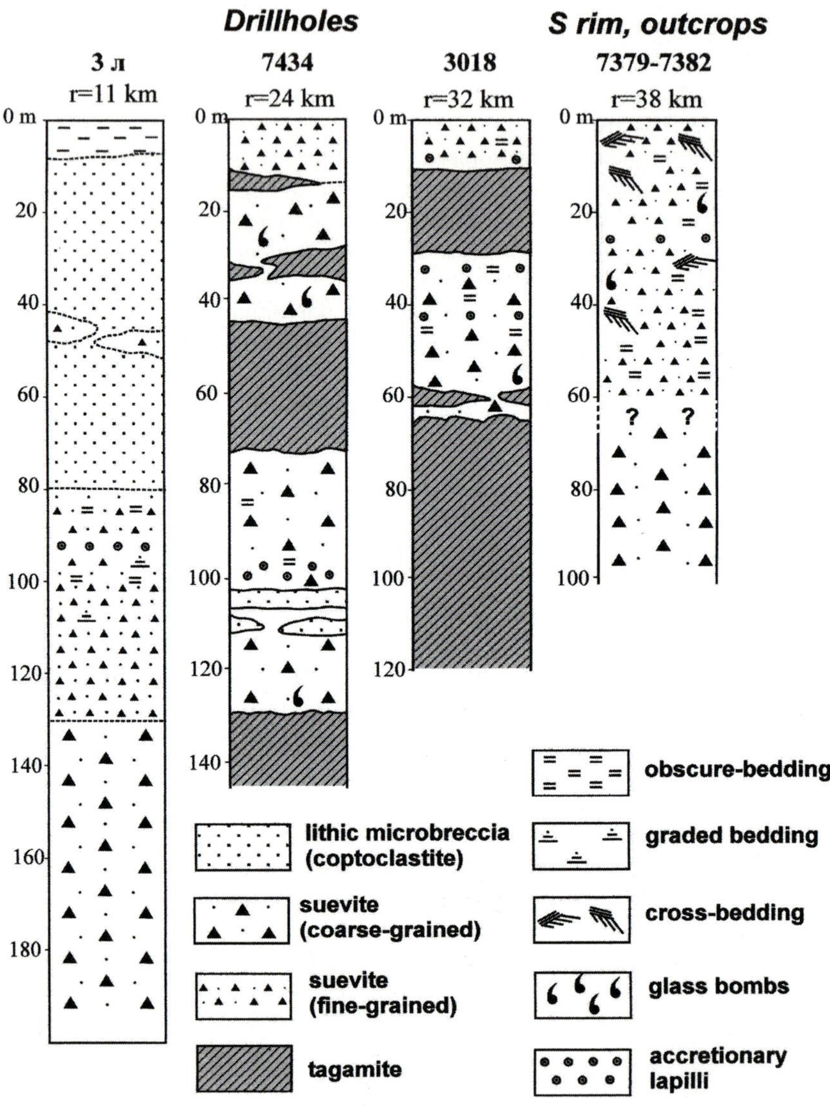

Fig. 3. Lithological columns through the upper member of the Popigai crater fill. Drillhole 3 (northeastern sector) illustrate type A facies, drillholes 7434 and 3018 (southwestern sector) – type B facies, generalized section of outcrops 7379-7382 (southern sector) – type C facies. r – distance from the crater center.

Fig. 4. Core sample of the lithic microbreccia. The groundmass is made up of partly rounded silty-sandy mineral grains (granoclasts) derived mostly from unconsolidated Cretaceous target rocks. Some larger lithic fragments are present. Drillhole 4, depth 73 m (central depression, southeastern sector).

Microbreccias are derived from comminuted Cretaceous unconsolidated coaly sandstones and sands, and they contain a minor admixture (no more than 1 vol.%) of rock fragments and glass bombs. The rock groundmass is made of silty-sandy angular or subrounded quartz and feldspar grains and grains of some other minerals (all of them are regarded as granoclasts), and small lithic particles. All these clasts are cemented by clayey matrix (Figure 4). The debris of pebble-sized and cobble-sized fragments is irregularly distributed. The fragments consist of sandstones, quartzites, claystones, carbonate rocks, brown coal, and petrified wood. Furthermore, there are some shocked and unshocked crystalline rocks, and an insignificant admixture of impact glass particles may be present. In some places cobble-sized rock fragments occur, forming lens-like bodies.

The glass content in vitrogranoclastic suevite is 10-20 vol.%, but reaches 50 vol.% in the granovitroclastic type. The size range of the constituent clasts is about 0.2-10 mm, and clasts about 1-2 mm in size dominate. In general, the clasts are unsorted, and are embedded within clayey interstitial matrix, which contains small clasts (less than 0.1 mm in size), minor zeolite, and carbonate impregnations. The suevites are matrix-supported and contain irregularly dispersed angular rock and glass fragments that are usually 1 to 20 cm across, but, in some cases, even larger.

The measured contents of main components of the described suevites, enriched in sedimentary clastic material, and located in the northwestern and southwestern crater sectors, are as follows (in vol.%): fractions >0.5 cm – glass 8-10; crystalline

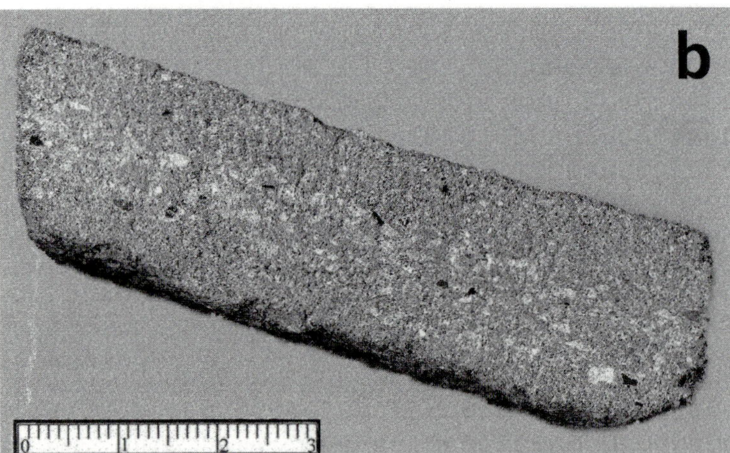

Fig. 5. Core samples of the vitrogranoclastic suevites, showing obscure bedding. Thin parallel beds dipping at the angle 15-20° contain abundant coarser clasts (core axes are vertical). Drillhole 3, depths 80 m (a) and 81.5 m (b) (central depression, northeastern sector).

lithoclasts 2-3; sedimentary lithoclasts 1; fraction 0.5 - 0.01 cm – glass 20-40; lithoclasts, granoclasts and crystalloclasts 25-40; and cementing groundmass (less than 0.01 cm) 20 (Masaitis et al. 1998). Sedimentary clasts dominate over crystalline clasts in vitrogranoclastic suevites (ratio is about 5:3); the granovitroclastic suevites are characterized by the opposite ratio of different clasts derived from disintegrated target rocks and are usually more coarse-grained.

The typical column of type A facies penetrated by drillcore 3, located in northeastern sector, 11 km from the crater center, is shown in the Figure 3. Microbreccias of the upper part of the section are monotonous in structure, but in drillhole 3 and in some others they show obscure bedding and unclear beds of fine-grained clayey material with a thickness of 1 to 10 cm. In hand specimen, the microbreccias show inclusions of small rock fragments. Suevite lenses, up to 1-m thick, also occur having sharp or gradational contacts. The lower part of the section is completely made of suevite. In several drillholes, these rocks display obscure bedding, which has a dip of about 10-15° (Figures 5a, b). The thickness of individual, thin, obscure beds does not exceed 1 to 5 cm. The total thickness of obscure-bedded varieties, which are irregularly distributed in drillhole sections, may reach 20 to 80 m. Accretionary lapilli occur in these rocks, and are of the rim-type with a diameter of up to 1 cm. Accretionary lapilli represent spherical bodies that are composed of very fine-grained clastic silicate material and glass particles, and are characterized by concentric zonation. Clastic aleurite- and pelite-sized material, which forms repeated zones or shells (rim-type and multirim-type lapilli); the size of the particles decreases outwards.

Gradational bedding was observed in these suevites. Specifically, the amount of glassy material increases down-section. The vitrogranoclastic fine-grained suevites are underlain by massive, unsorted coarse-grained granovitroclastic and vitroclastic suevites, which are more strongly lithified and enriched by glass bombs. It is possible that all these suevites belong to the B-type facies.

3
Ejecta Deposits of Type B Facies

The deposits of type B facies are distributed in the central portions, and, probably, also in lower portions, of the upper member. They are represented by suevites with minor tagamites, and have been observed in outcrops and drillcores in the central depression and in southwestern and southeastern sectors of the annular trough (Figure 1). Obscure-bedded suevites of type B facies were traced by drilling in local areas of the annular trough at distances ranging from 2 to 10 km. Their apparent thickness varies from 20 to 130 m, but may reach 300 m. Most probably these facies in contrast to type A facies, do not form a coherent layer, but occur mainly in isolated radial elongated bodies combined laterally with suevite that lacks any ordering in its internal structure.

The rocks are represented mostly by granovitroclastic, rare vitroclastic coarse-grained suevites, which contain to 10-15 vol.% of cobble-sized and pebble-sized rock fragments, glass shards, and glass bombs. The principal lithology of vitroclastic suevites is similar to that described above for the granovitroclastic variety, but the abundance of glass fragments may reach 75 vol. %. Some constituent glass particles show fluidal structure. The average clast size of these rocks is about 2 to 5 mm. Tagamite bodies, occurring together with suevites, have crystalline or hyaline matrix, including small clasts of gneiss and its constituent minerals. The matrix has a hemicrystalline or holocrystalline fine-grained or cryptocrystalline texture, in some cases a vesicular structure, and is made up of small crystals of pyroxene and plagioclase, residual glass, and secondary minerals (mostly montmorillonite, chlorite, hydromica, zeolites, and calcite). The content of clast inclusions (less than 0.5 cm across) varies, but usually is about 10-20 vol.%. Rare larger fragments (to 0.5-1 m) may also occur. All of these clasts were subjected to shock transformation and recrystallization.

The type B facies was penetrated by several drillholes in the southwestern sector of the Popigai annular trough. Two typical core sections are shown in Figure 3. Vitrogranoclastic, granovitroclastic, and more coarse-grained vitroclastic, partially sintered, suevites are interbedded here with lens-like and sheet-like tagamite bodies, and small lenses of lithic microbreccias occur as well. The intervals of the cores locally show obscure bedding, which varies in thickness from 10 to 55 m. In some cases such suevites occur twice within one section. The thickness of individual obscure beds, grading one into another, and slightly differ in clast size, abundance of rock fragments, and glass shards and bombs, is about 5 to 20 cm. In drillhole 3018, the dip of the obscure bedded layers reaches 45-55°. These suevites occur between tagamite bodies or overlie them. Typically, obscure-bedded suevites grade upward and downward into vitroclastic suevites, which lack any bedding and are enriched in glass bombs. At the same time, we should recall that obscure bedding in many instances cannot easily be recognized in core samples. In reality, this feature may be relatively common, but it has not been recognized.

Vitrogranoclastic suevites in some drillholes contain numerous accretionary lapilli of rim- and multirim type. Their average diameters are about 8-10 mm, but some reach 30 mm in size (Figure 6).

Similar suevites attributed to type B facies are exposed on the right bank of the Chordu-Daldyn River, 10 km upstream from its mouth (annular trough, southeastern sector). The rock sequence dipping to the northwest at an angle about 30-40° may be subdivided into three parts. The lower part of the massive granovitroclastic suevites contains numerous dolomite fragments up to 30 cm in size. The intermediate composite layer contains obscure-bedded granovitroclastic suevites. The upper part comprises vitroclastic suevites, which contain lithic fragments and numerous bombs of impact glass and glass-coated gneiss bombs that are up to 20 cm across (Figure 7).

Fig. 6. Core sample of the granovitroclastic suevite, bearing accretionary lapilli. Drillcore 5250, depth 65 m (annular trough, southwestern sector).

Lechatelierite bombs of the same size and particles were also found here. The intermediate layer, about 2 m thick (Figure 8), is composed of multiple beds and lenses 5-25 cm thick, which have indistinct borders and are composed of sorted granovitroclastic suevites. These suevites are distinct in their granulometry and content of lithic fragments and accretionary lapilli (with diameters of up to 20 mm, usually 8-10 mm). The lapilli are irregularly distributed in the host rocks, or form layers and lenses 10 to 30 cm thick, which are enriched in lapilli (10-30 vol.%). Broken lapilli are found as well (Figure 9a, b).

Similar obscure-bedded suevites, characterized by indistinct layers 2 to 10 cm thick, are exposed on the left bank of the Popigai River, 2 km upstream from the mouth of the Chordy-Daldyn River. They dip to southeast at an angle of 20°, thus this outcrop (located about 5 km from the former) may be regarded as the southeastern flank of a large asymmetric buried dune-like structure. It should be noted that its strike is parallel to the crater rim in this southeastern sector.

Fig. 7. Obscure-bedded thin sequence composing of the granovitroclastic suevite supported of large dolomite fragments (lower part of the outcrop), and vitroclastic suevite supported of glass bombs (upper part). The intermediate composite layer of granovitroclastic suevite (dark grey) dips left at angle of 30-40°. The height of the outcrop is 30 m. Outcrop 7404, Chordu-Daldyn River, downstream water (annular trough, southeastern sector).

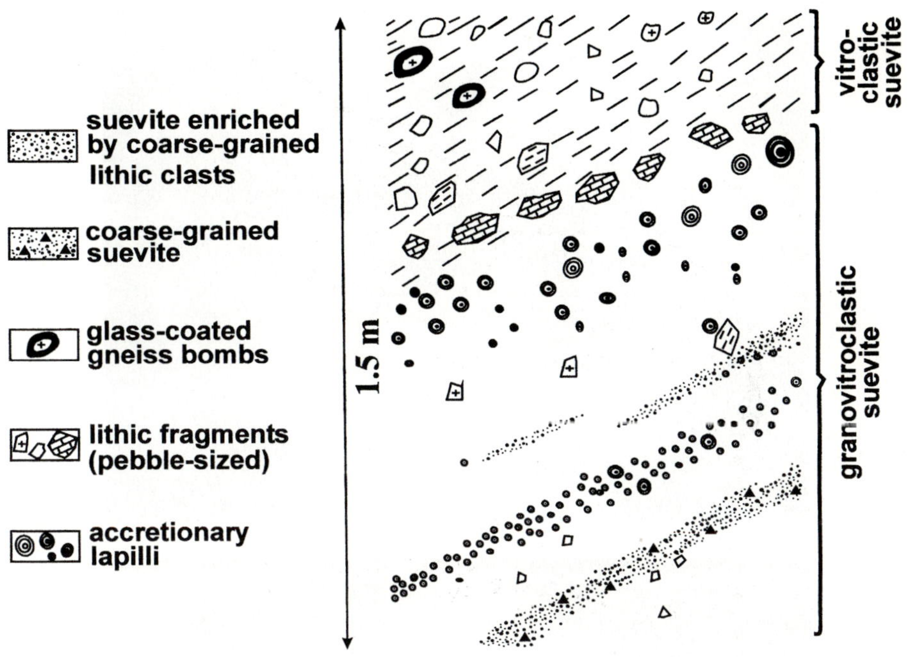

Fig. 8. Sketch section through the upper part of the intermediate composite layer shown on the Fig.7. 1-vitroclastic suevite, 2-granovitroclastic suevite. The individual obscure beds and lenses have gradational borders. Lithic fragments compose of crystalline rocks (crosses), claystones (dashes), dolomites (bricks) and sandstones (unshaded).

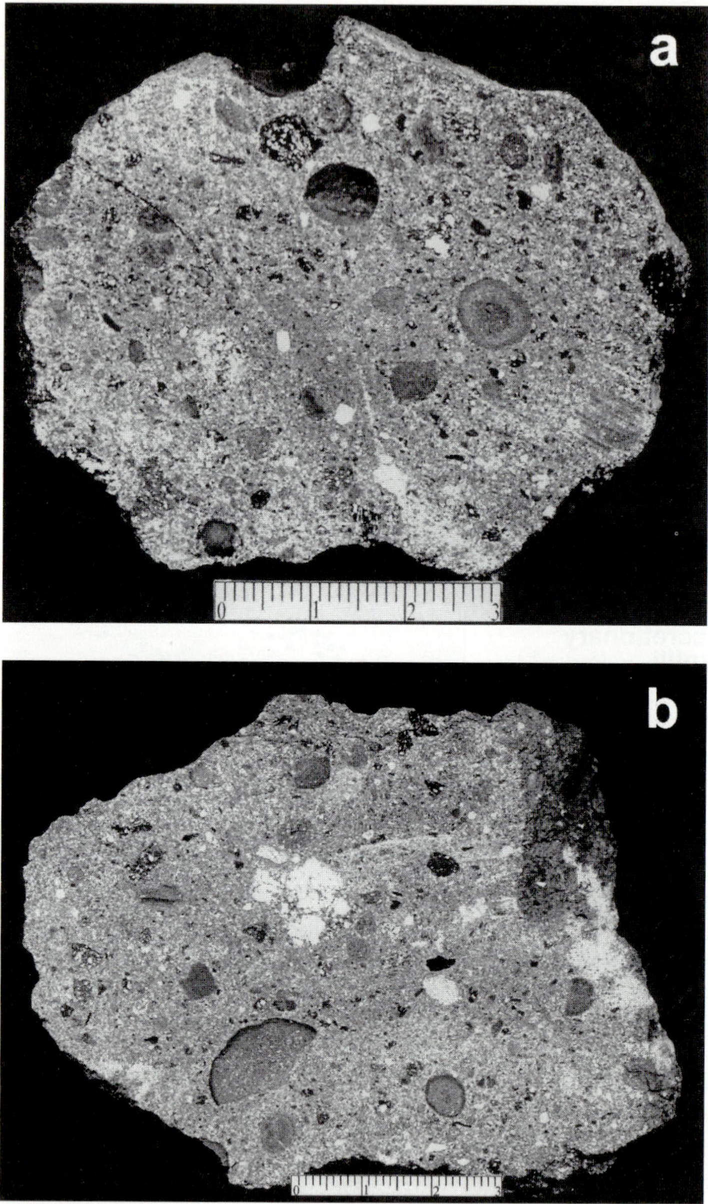

Fig. 9. Samples of the granovitroclastic suevite that composes the intermediate layer (see Fig.8). The suevite contains accretionary lapilli of rim- and multirim types (a), some of which are broken (b).

4
Ejecta Deposits of Type C Facies

Obscure-bedded suevites of type C facies were observed in the southern and southeastern sectors near the crater rim. The relatively thick sequence of these deposits exposed on the right bank of the Kychypkanaakh River (southern sector). The area in which these suevites occur is connected to the outer edge of annular trough, which is located 2.5 km to the south, and is composed of gneiss. Allogenic polymict megabreccia, which underlie suevites, is observed in small outcrops. The insignificant thickness of the megabreccia was probably caused by movement on faults striking parallel to the crater rim. Similar faults, which disturb the suevite sequence, are also suggested. The suevites are represented completely by obscure-bedded rocks, with a total thickness that reaches 60 to 100 m. The suevite comprises numerous indistinct beds or layers, each differing in their granulometry, admixture of large lithic fragments, and content of bombs (Figs. 10a-c). Their thickness may vary from 2 cm to 1 m. This sequence was traced in several elongated outcrops, 6-8 m high, located along the bed of the Kychypkynaakh River over a distance of about 2 km. The principal rocks of the sequence are vitrogranoclastic suevites. In some beds, they are very fine-grained and close to lithic microbreccia composition due to the low content of vitroclasts (about of 10-15 vol. %). The composition of the suevite is similar to that of type A facies, but in the described sequence, sorting is much better developed, and very fine-grained suevite varieties (average grain size less than 0.5 mm) are found. In some places, obscure cross-bedding exists (Figure 10a, b). Numerous dispersed irregularly accretionary lapilli (diameter approximately 0.5 cm) and small particles of fresh glass were observed in one of the fine-grained suevite layers. Some of the well-sorted rocks are a laminated suevite variety (Figure 10c). Sorting of the dispersed fragments has been observed locally, and their size decreases upwards. Coarse-grained suevite varieties lack any sorting. These varieties usually contain lenses that are parallel to the obscure bedding, and contain abundant pebble-sized rock fragments and glass bombs. The indistinct layer is about 1 m thick and is enriched in glass bombs and large cobble-sized fragments that are 20 to 30 cm across. This layer was traced within the northern part of the exposed suevites. In some cases gneiss bombs are embedded into fine-grained suevites and form bomb sags (Figure 11). These bombs are rounded and coated with glass, and their interiors are zoned, the result of thermal annealing within the impact fireball (Figure 12).

The interior structure of the suevite sequence displays several well-expressed, asymmetric dune-like forms, which are 4-6 m high and about of 20-40 m wide. These forms are elongated, with their northern slopes dipping at 40-60°, and their southern ones at 15-30°. Southern slopes are possible foreset beds, and northern may backset beds. Locally these dune-like formations overlap one another. In the central and southern parts of the outcrops, the indistinct layering shows dips of 60-70°. This probably caused by late-stage slumps, which occurred along the crater rim (Figure 13).

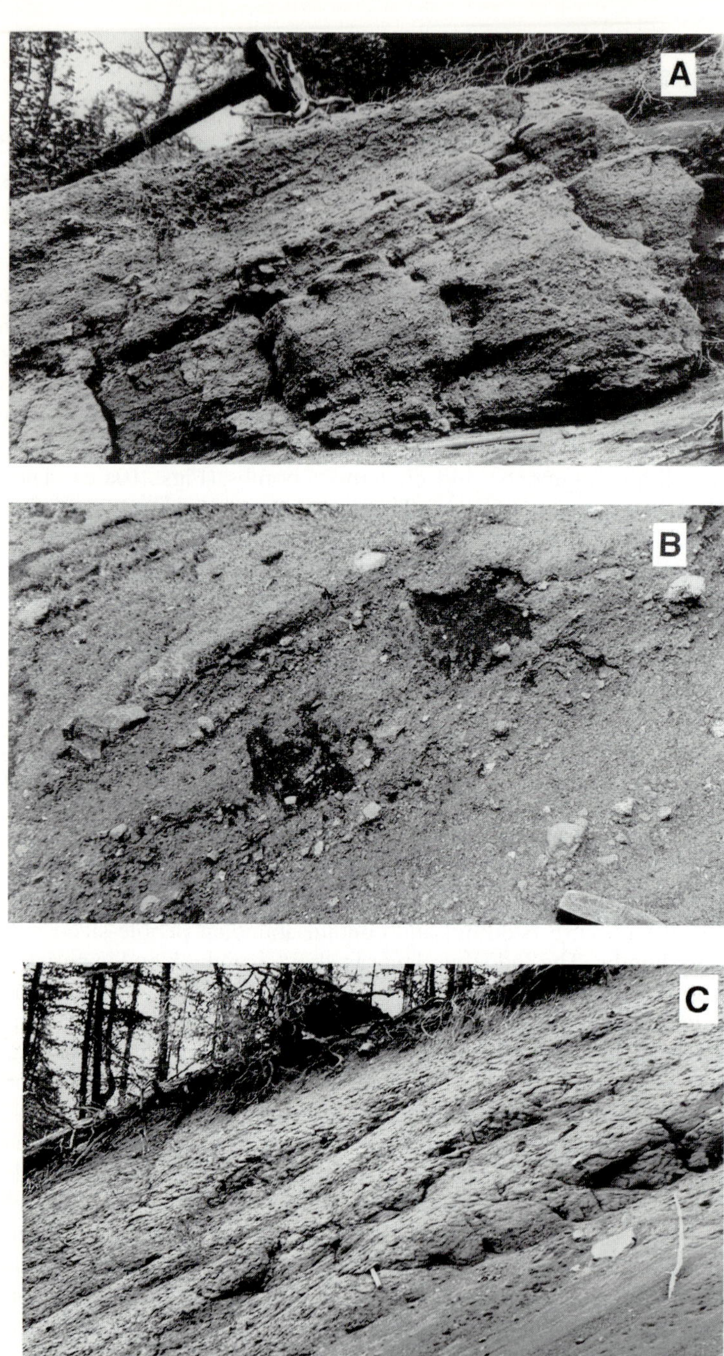

Fig. 10. Details of the outcrops (7379-7381) of the granovitroclastic obscure-bedded suevites, containing irregularly distributed pebble-sized lithic fragments. Weakly expressed cross-bedding is seen at center bottom (a) and upper left (b). Laminated suevite shows wavy layering, the underline bed composes of fine-grained well-sorted variety (c, right bottom). Obscure bedding dips at angles 20-40°. Hammers for scale. Kychypkanaakh River (southern sector, close to crater rim).

Fig. 11. Vitrogranoclastic suevite contains irregularly distributed pebble-sized lithic clasts, most of them sedimentary (outcrop 7382). Obscure bedding, which dips at angle 30-40°, is seen in upper left. Large rounded bomb of shocked gneiss coated with impact glass occupies bomb sag in the host suevite (lower left). Hammer for scale. Kychypkanaakh River (southern sector, close to crater rim).

Fig. 12. Coated with impact glass rounded gneiss bomb, its outer gneiss zone is thermally annealed. Kychypkanaakh River (southern sector, close to the crater rim).

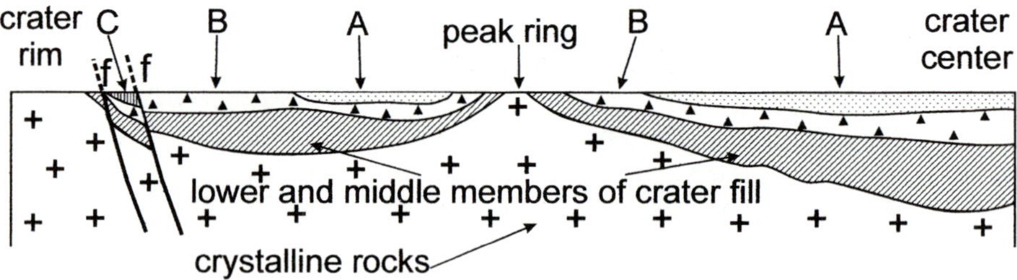

Fig. 13. Schematic section through the Popigai crater (not to scale), showing the interrelations of ejecta facies of the upper member of the crater fill: type A facies – air-fall deposits; type B facies – pyroclastic-like flow deposits; type C facies – base surge deposits. f – proposed faults.

5
Obscure-Bedded Ejecta Facies in some Other Craters

Sorted and obscure-bedded proximal and distal ejecta facies are known, respectively, inside and outside of several impact craters. Sorted suevites were first described from the Ries crater (Mosebach 1964; Förstner 1967; Graup 1981; Stöffler 1977; Jankowsky 1977; Newsom et al. 1990), and were subsequently found in the Sudbury impact structure (Peredery 1972; Muir and Peredery 1984; Bunch et al. 1999), and at the Logoisk crater (Glazovskaya et al. 1991). Sorted and graded distal ejecta facies, which are composed of lithic breccias and which formed mainly in subaqueous settings, are found close to the Chicxulub crater (Pope et al. 1999; Salge et al. 2000). Similar ejecta comprise the uppermost parts of the Alamo breccia (Nevada), the source crater of which is so far unknown (Warme and Kuehner 1997; 1998).

The proximal sorted and indistinct bedded ejecta lithologies from the above mentioned craters may be easily compared with equivalent rock facies of the Popigai crater described above. Sorted suevites in the Ries crater from the Denningen 1001 and FBN-73 boreholes may be attributed to facies similar to type A from the Popigai crater. The thickness of these sorted suevites ranges from 20 m to 66 m close to the crater center, and they are overlain by crater-lake sediments. The suevites show graded bedding and contain accretionary lapilli (Mosebach 1964; Förstner 1967; Graup 1981; Jankowsky 1977; Newsom et al. 1990). Another horizon of sorted suevite in the Ries crater, known as "middle sorted suevite" (Stöffler 1977; Newsom et al. 1990), is approximately 8 m thick and is overlain and partially underlain by suevites resembling type B facies from the Popigai crater.

Type A ejecta facies may be also distinguished in the uppermost part of the crater fill in the Logoisk impact structure, Belarus.The thickness of these deposits, which were penetrated by four drillholes in the central part of the structure, reaches 25 m. They underlain by coarse-grained lithic breccia, alternating with suevites. These facies are represented by microbreccias (coptoclastites) with graded bedding, which continue downward into fine-grained suevites that contain accretionary lapilli (Glazovskaya et al. 1991;, Masaitis 1999). It is possible that sorted lithic microbreccias and suevites in the uppermost part of the Kara crater fill, having similar lithological features (Selivanovskaya et al. 1990), may be also attributed to type A facies.

A thick sequence (200-500 m) of lower and middle parts of the Upper Member of the Black Onaping formation, Sudbury structure, Canada (Peredery 1972; Muir and Peredery 1984; Bunch et al. 1999), may be similar to the type B ejecta facies of the Popigai crater. Suevites-like rocks, which were deposited partially in an aquatic environment (Avermann 1994), show indistinct layering, insignificant sorting, and intercalations of fine- and coarse-grained varieties. These suevites contain small bodies of impact melt rocks similar to tagamite from the Popigai crater. Accretionary lapilli have been found in these suevites as well (Bunch et al. 1999).

6
Mode of Origin of Obscure-Bedded Ejecta Facies

The observations of specific lithological features of obscure- bedded and sorted proximal ejecta deposits of the Popigai impact crater (as well as of similar deposits from other craters) allow to deduce some mechanisms of their mode of formation during the late stages of impact cratering. Most of the observed lithological features are caused by lateral movements of the settled clastic material derived from the impact plume. The modes of transportation, accumulation, and burial of these deposits strongly depend on the composition and physical properties of jets, currents, and flows with respect to (1) ratios of ejected melt, solid clasts, and gas, (2) clast granulometry, (3) temperature, (4) density, (5) dynamic viscosity. Furthermore, velocity and trajectory of jets, the fluid flow regime, etc. are factors (Melosh 1989).

Type A ejecta may form as air-fall deposits during the last stage of the crater-fill process (Stöffler 1977; Melosh 1989; Newsom et al. 1990). They were produced by vertical settling of mainly fine-grained clastic material from the relatively cool, descending parts of the impact dust cloud. The admixture of large lithic clasts and bombs was caused by their steep ballistic trajectories and their fall into fine-grained material. Accretionary lapilli in these facies show that the dust cloud was saturated with water vapor, which condensed during cooling. Accretion of fine-grained clasts and formation of lapilli occur due to binding forces within moist, turbulent media. The most probable formation of accretionary lapilli is analogous to similar processes that occur in the course of volcanic deposition (Schumacher and Schminke 1991; 1995; Gilbert and Lane 1994). A slight tilting of the obscure bedding occurs because of insignificant horizontal movements of the deposited clasts, and the formation of a sand-wave , and/or deposition on an uneven surface of the crater fill. These air-fall facies encompass all other impact formations that are deposited slightly earlier, and were initially distributed over a wide area outside the crater. In many ways these facies resemble volcanic air-fall deposits (Fisher and Schminke 1984; Easton and Jones 1986; Cas and Wright 1987).

It is suggested that type B facies have been produced before deposition of facies A during the collapse of a giant vertical column of gas, solid particles, melt drops, and shards that were in places mixed by jets of impact melt. The descent and rapid radial spreading of hot flows carrying melt particles and lithic fragments could lead to the deposition of these materials, which were partially accompanied by more fine-grained clastic rocks transported in relatively colder base surges. Accretionary lapilli were deposited together with these moistened clastic rocks. Their occurrence within several horizons shows that this process involved multiple events. This observation, as well as combination of these deposits with unsorted and significantly coarser suevites (distinct in their ratio of fragments of different crystalline and sedimentary rocks), shows that several contemporaneous jets, currents, and flows must have had differences in their velocities, temperature, gas content, and clasts saturation. It is probable that deposition of type B facies can produce large dune-like deposits, which were rapidly buried below other ejecta.

Impactite flows that are enriched in hot clastic material, especially melt shards and particles in different state of solidification, may be similar to well-known facies of volcanic eruptions deposited by pyroclastic flows (Fisher and Schminke 1984; Easton and Johns 1986). These impactite flows ("tekoclastic flows", from Greek τηκω "teko" to melt) differ from volcanic pyroclastic ones not only by presence of shock metamorphic features in the composing rocks but also by mixing between jets composed of liquid impact melt, and flows composed of colder ejecta material.

Type C ejecta facies have so far only been found in the Popigai impact structure. The vertical position of these facies in the generalized section of the upper member is not well known. Most probably they occupy the uppermost part of the sequence, and were later subjected to subsidence close to the rim due to crater collapse during the modification stage (Figure 13). It cannot be ruled out that type C facies can laterally grade into type A, which also partially overlie them.

It is quite probable that type C facies were deposited in the outer parts of the crater, and also outside, from relatively cold, vapor saturated base surges moving radially and carrying ash and dust particles. The thickness of type C facies may vary in lateral directions, from several meters to hundreds of meters. During ascent, the impact plume, which produces the principal material of base surges, mixes with cold atmosphere, thus inverting the ejecta cloud inwards (Melosh 1989; Barnouin and Schultz 1996). As this hot material cools, melt particles are quenched and water vapor condenses. This explains the accretion of dust particles under wet conditions, as well as the origin of accretionary lapilli. The upper regime of the base surge then grades into the lower regime. During their deposition these surges produce dune-like deposits. These processes are accompanied by the fall of bombs and fragments, which are entrained in the impact fireball (Masaitis and Deutsch 1999) and are ballistically ejected along steep trajectories. It cannot be ruled out that deposition from base surges and subsidence movements along the outer limits of the annular trough were simultaneous, and that landslides of loose clastic material occurred at that time. Compared to volcanic rocks, the impact ejecta of facies C type are most similar to pyroclastic base surge deposits (Fisher and Schminke 1984; Easton and Johns 1986).

The general facies characteristics determined from the data obtained in the Popigai crater are given in Table 1, which allows to compare their principal features.

Despite the fact that these impact ejecta facies are in some aspects similar to volcanic analogues, there are considerable differences in the lithology and mode of origin of impact facies. First, petrographic characteristics are specific to impact lithologies (e.g., Deutsch and Langenhorst 1998; French 1998). Second, the impactites (part of which were superheated), were ejected at much higher velocities than any volcanic flows. The very rapidly expanding impact fireball caused inward-directed atmospheric winds and cooling of hot ejecta. Then the alternating deposition of hot and cold clastic materials occurred. Most of these processes produced lithological features of obscure-bedded impact ejecta that are not characteristic of pyroclastic formations, even those originating from terminal volcanic explosions.

Table 1. Obscure bedded ejecta facies comprise the upper member of the Popigai crater fill

Features	Type A	Type B	Type C
Principal lithologies	Lithic microbreccia (coptoclastite) and vitrogranoclastic suevite	Granovitroclastic and vitrogranoclastic suevites, minor tagamite	Vitrogranoclastic suevite
Average sizes of clasts	About 1 mm	2- 5 mm	About 0.5 mm
Sorting	Sorted, graded upwards	Unsorted, poorly sorted	Well sorted to poorly sorted
Specific lithological features	Rare debris 2-5 cm diameter, accretionary lapilli	Debris 30 cm diameter, glass shards, accretionary lapilli	Debris 1-5 cm diameter, isolated outsized clasts, bomb sags, accretionary lapilli,
Type of layering, inclination	Obscure bedding, inclination to 15°	Obscure bedding, inclination to 40°	Obscure bedding, plane-parallel and cross-bedding, inclination to 60°
Apparent thickness	to 185 m	20-130 m	60 m
Mode of occurrence	Coherent cover	Radial stretched bodies	Local radial tongues (?)
Mode of transportation	Air fall	Pyroclastic-like flow	Base surge

Proximal impact ejecta deposited during the late stages of formation of the Popigai crater, as well as in some other terrestrial craters, must be considered as objects for further studies. Future explorations may help not only with understanding some processes in the impact plume, but also to interpret remote sensing data on impact ejecta on the other planets, where late-stage ejecta deposits are much better preserved than on Earth.

7
Conclusions

1. The upper member of the Popigai impact crater fill includes ejecta deposits that are distinguished by the presence of obscure bedding and some other unusual lithological features. Three types of constituent rock facies, represented mostly by suevites, can be distinguished.

2. The three types, A, B, and C, can be characterized as follows. Type A facies occurs predominantly in the central depression of the crater and is composed of fine-grained, locally sorted suevites, which grade upward into lithic microbreccia (coptoclastites). Type B facies, which underlies type A ejecta, is found within the central depression and annular trough, and is composed mostly of coarse-grained glass-supported suevites and tagamites. Type C facies occur in isolated locations near the crater rim. These are represented by fine-grained and sorted suevites, sometimes cross-bedded, and include isolated layers and lenses of larger lithic fragments and glass bombs. All types of obscure-bedded facies contain accretionary lapilli, which can be considered a characteristic feature. In these facies minor admixtures of fragments and bombs ejected along steep ballistic trajectories occurs. Facies B and C form dune-like deposits.

3. The preliminary interpretation of the lithology and inner structure of the three rock facies described in this paper shows that they have been deposited from three distinct types of moving ejecta during the late stages of crater filling:
 - by air fall from a dust cloud (type A);
 - from pyroclastic-like flows caused by the collapse of an ejected column of melt droplets and solid rock fragments (type B); and
 - from base surges in areas distant from the crater center (type C).
These three ejecta facies are analogous to some types of volcanic deposits.

4. Some of the rock facies show that deposition occurred in multiple events due to alternating precipitation of hot and cool ejecta. Differences in composition and inner structure of adjacent thick layers, sheets, and lenses are caused by selective homogenization of brecciated and melted rock material within individual jets, currents, and flows. The presence of accretionary lapilli indicates that the impact event occurred within a wet target.

5. Obscure-bedded ejecta facies similar to those from the Popigai crater have been also found at, for example, the Ries, Sudbury, and Logoisk impact structures. Such deposits may also occur in some other large and middle-sized impact craters, in which the upper parts of the crater fill has not yet been eroded.

Acknowledgments

The author thanks his colleagues at the Karpinsky Geological Institute (St. Petersburg) M.S. Mashchak, A.I. Raikhlin, V.A. Maslov, and A.N. Danilin for help with the field observations and for further information on the rock facies. Also he is grateful to A.T. Maslov for preparing some of the figures. Detailed reviews by David King and Philippe Claeys were much appreciated. Their suggestions were very useful in improving the manuscript. The author is also grateful to Christian Koeberl for editorial assistance and his infinite patience during corrections.

References

Avermann M (1994) Origin of the polymict allochtonous breccias of the Onaping Formation, Sudbury structure, Ontario, Canada. In Dressler BO, Grieve RAF, Sharpton VL (eds). Large meteorite impacts and planetary evolution. Boulder, Colorado. Geological Society of America Special Paper 293: 265-274

Barnouin OS, Schultz PH (1996) Ejecta entrainment by impact-generated vortices: theory and experiments. Journal of Geophysical Research 101: 21.099-21.115

Bottomley R, Grieve R, York D, Masaitis V (1997) The age of the Popigai impact event and its relation to events at the Eocene/Oligocene boundary. Nature 388: 365-368

Bunch TE, Becker L, Des Marais D, Tharpe A, Schultz P, Wolbach W, Glavin DP, Brinton KL, Bada JL (1999) Carbonaceous matter in the rocks of the Sudbury basin, Ontario, Canada. In: Dressler BO, Sharpton VL (eds) Large Meteorite impact and planetary evolution. Geological Society of America Special Paper 339: 331-344

Cas RAF, Wright JV (1987) Volcanic successions. Modern and ancient. Allen & Unwin, London, 528 pp

Deutsch A, Langenhorst F (1998) Mineralogy of astroblems – terrestrial impact craters. In: Marfunin AS (ed) Advanced Mineralogy, 3. Springer-Verlag, Berlin-Heidelberg, pp 76-119

Easton RM, Johns GW (1986) Volcanology and mineral exploration: the application of physical volcanology and facies studies. Ontario Geological Survey, Miscellaneous Papers 129: 2-39

Fisher RV, Schminke H-U (1984) Pyroclastic rocks. Springer-Verlag, Berlin, 472 pp

Förstner U (1967) Petrographische Untersuchungen des Suevit aus den Bohrungen Denningen und Wörnitzostheim im Ries von Nördlingen. Contributions to Mineralogy and Petrology 15: 281-308

French BM (1998) Traces of catastrophe: a handbook of shock metamorphic effects in terrestrial meteorite impact structures. Lunar and Planetary Institute, Houston, LPI Contribution 954, 120 pp

Gilbert JS, Lane SJ (1994) The origin of accretionary lapilli. Bulletin of Volcanology 56: 398-411

Glazovskaya LI, Gromov EI, Parfenova OV, Il'kevich GI (1991) The Logoisk astrobleme (in Russian). Nauka Press, Moscow, 135 pp

Graup G (1981) Terrestrial chondrules, glass spherules and accretionary lapilli from the suevite, Ries crater, Germany. Earth and Planetary Science Letters 55: 407-418

Grieve RAF, Pesonen LJ (1992) The terrestrial impact cratering record. Tectonophysics 216: 1-30

Jankowsky B (1977) Die gradierte Einheit oberhalb des Suevits der Forschungsbohrung Nördlingen. Geologica Bavarica 75: 155-162

King Jr, DT, Petruny LW (2000) Astrostratigraphic units – some proposed nomenclature for terrestrial impact-derived and impact-related materials [abs]. Lunar and Planetary Science 31, abs. #2019, CD-ROM

Laznicka P (1988) Breccias and coarse fragmentites. Petrology, environments, associations, ores. Elsevier, Amsterdam, 832 pp

Le Maitre RW (ed) (1989) A classification of igneous rocks and glossary of terms.Blackwell Scientific Publications, Oxford London, 193 pp

Masaitis VL (ed) (1983) Textures and structures of explosion breccias and impactites (in Russian). Nedra Press, Leningrad, 159 pp

Masaitis VL (1998) Popigai crater: origin and distribution of diamond-bearing impactites. Meteoritics and Planetary Science 33: 349-359

Masaitis VL (1999) Impact structures of northeastern Eurasia: The territories of Russia and adjacent countries. Meteoritics and Planetary Science 34: 691-711

Masaitis VL, Deutsch A (1999) Popigai: gneiss bombs coated with impact melt – heating in the fireball? [abs]. Lunar and Planetary Science 30, abs.#1237, CD-ROM

Masaitis VL, Danilin AN, Mashchak MS, Raikhlin AI, Selivanovskaya TV, Shadenkov EM (1980) Geology of astroblemes (in Russian). Nedra Press, Leningrad, 231 pp

Masaitis VL, Mashchak MS, Raikhlin AI, Selivanovskaya TV, Shafranovsky GI (1998) Diamond-bearing impactites of the Popigai astrobleme (in Russian). VSEGEI Press, St-Petersburg, 138 pp

Masaitis VL, Naumov MV, Mashchak MS (1999) Anatomy of the Popigai impact crater, Russia. In: Dressler BO, Sharpton VL (eds) Large Meteorite Impact and Planetary Evolution. Geological Society of America Special Paper 339: 1-17

Melosh HJ (1989) Impact cratering. A geological process, Oxford University Press, NY, Clarendon Press, Oxford, 245 pp

Mikhailov NP (ed) (1994) Petrographic code. Magmatic and metamorphic rocks (in Russian). VSEGEI Press, St.-Petersburg, 127 pp

Montanari A, Koeberl C (2000) Impact stratigraphy. Lecture Notes in Earth Sciences vol. 93. Springer-Verlag, Berlin-Heidelberg, 364 pp

Mosebach R (1964) Das Nördlinger Ries, vulcanischer Explosionscrater oder Einschlagstelle eines Grossmeteoriten? Bericht der Oberhessischen Gesellshaft für Natur- und Heilkunde zu Giessen, Neue Folge, Naturwissenschaftliche Abteilung 33: 165-204

Muir TL, Peredery WV (1984) The Onaping Formation. In: Pye EG, Naldrett AJ, Giblin PI (eds). The geology and ore deposits of the Sudbury Structure: Ontario Geological Survey, Special Volume 1, pp 139-210

Newsom HE, Graup G, Iseri DA, Geissman JW, Keil K (1990) The formation of the Ries crater, West Germany: Evidence of atmospheric interaction during a larger cratering event. In: Sharpton VL and Ward PD (eds) Global Catastrophes in Earth History: an interdisciplinary conference on impacts, volcanism and mass mortality. Geological Society of America Special Paper 247: 195-206

Peredery WV (1972) Chemistry of fluidal glasses and melt bodies of the Onaping Formation. In: Gay-Bray JV (ed) New developments in Sudbury geology. Geological Association Canada Special Paper 10, pp 49-59

Pope KO, Ocampo AC, Fischer AG, Alvarez W, Fouke BW, Webster CL, Vega FJ, Smit J, Fritsche AE, Claeys P (1999) Chicxulub impact ejecta from Albion Island, Belize. Earth and Planetary Science Letters 170: 351-364

Raikhlin AI (1996) Suevites of the Popigai impact crater: internal structure and condition of formation. Solar System Research 1: 14-18

Raikhlin AI, Selivanovskaya TV, Masaitis VL (1980) Rocks of terrestrial impact craters: problems of classification [abs]. Lunar Planetary Science 11 : 911-913

Salge T, Tagle R, Claeys P (2000) Accretionary lapilli from the Cretaceous-Tertiary boundary site of Guayal, Mexico, preliminary insights into expansion plume formation [abs.]. Meteoritics and Planetary Science 35: A140-A141

Schumacher R, Schminke H-U (1991) Internal structure and occurrence of accretionary lapilli - a case study at Laacher See volcano. Bulletin of Volcanology 53: 612-634

Schumacher R, Schminke H-U (1995) Models for the origin of accretionary lapilli. Bulletin of Volcanology 56: 626-639

Selivanovskaya TV, Mashchak MS, Masaitis VL (1990) Impact breccias and impactites of the Kara and Ust-Kara astroblemes. In: Masaitis VL (ed) Impact Craters on Mesozoic-Cenozoic boundary (in Russian). Nauka Press, Leningrad, pp 55-96

Stöffler D (1977) Research drilling Nördlingen 1973: polymict breccias, crater basement, and cratering model of the Ries impact structure. Geologica Bavarica 75: 443-458

Stöffler D, Grieve RAF (1994) Classification and nomenclature of impact metamorphic rocks: a proposal to the IUGS Subcomission on the systematics of metamorphic rocks. In: Impact cratering and evolution of planet Earth (a Scientific Network of the ESF). Post-Östersund Newsletter, pp 9-15.

Stöffler D, Evald U, Ostertag R, Reimold WU (1977) Research drilling Nördlingen 1973 (Ries): composition and texture of polymict impact breccias. Geologica Bavarica 75: 155-162

Vishnevsky S, Montanari A (1999) Popigai impact structure (Arctic Siberia, Russia): geology, petrology, geochemistry, and geochronology of glass-bearing impactites. In: Dressler BO, Sharpton VL (eds) Large Meteorite Impacts and Planetary Evolution II. Geological Society of America Special Paper 339: 19-60

Warme JE, Kuehner HC (1997) Accretionary impact ejecta spherules in the Upper Devonian Alamo breccia, south-central Nevada [abs]. Geological Society of America, Abstracts with Programs 29 (7), p A-80

Warme JE, Kuehner HC (1998) Anatomy of an anomaly: the Devonian catastrophic Alamo impact breccias. International Geological Review 40: 189-216

Biostratigraphic Indications of the Age of the Boltysh Impact Crater, Ukraine

Anton Valter[1] and Ludmila Plotnikova[2]

[1]Institute of Applied Physics of National Academy of Science, Department N 50, Nauki Avenue 46, Kiev 39, 03650, Ukraine. (avalter@iop.kiev.ua)
[2]Institute of Geological Sciences of National Academy of Science, 55B Gonchara str., Kiev 54, 01601, Ukraine. (ignnanu@geolog.freenet.kiev.ua)

Abstract. The Boltysh meteorite crater (diameter $\approx$ 23 km) on the Ukrainian shield was formed in Precambrian ($\sim$ 2.1 Ga) granites and has a classical complex structure with a central uplift (average diameter $\sim$ 4 km). The annular trough around the uplift is filled with predominantly glassy impactites (impact melt rocks = tagamites) that are up to 230 m thick. The crater lake sediments overlie impactites and are represented by laminated siltstones, sandstones, claystones. These sediments have a maximum thickness of nearly 400 m in the center of the crater. Previous data about the geological age of the lowermost crater-fill sediments were contradictory (K_1-P_2), as weere the K-Ar whole rock ages of the tagamites (177-56 Ma).

The main subject of the present paper is the study of the ejecta of the Boltysh impact crater, relics of which exist outside of the crater proper. We studied the ejecta breccia from outcrops at distances of 25 – 30 km NW from the center of the crater. Two types of breccia exist: breccia composed of disintegrated basement granites (lower unit), and breccia with different rock fragments cemented by sand-like and more fine-grained material (upper unit). The granites and other basement rocks are predominant among the fragments in the upper unit, but a minor amount (near 1 vol%) of carboniferous sandstones, siltstones, cherts, as well as glassy impactites, also exist. Some mineral grains of the cement have a distinct shock metamorphic features, such as planar deformation features (PDFs) in quartz, deformation bands in feldspar, and kink-band in micas. In sedimentary dykes in the lower breccia unit, from an outcrop near the village of Lebedivka ($\sim$ 25 km NW of the crater center), small fragments of cherty rocks (up to 1 cm in size) with upper Maastrichtian foraminifera were found. From these data the time range of the crater formation in comparison with resent chronobistratigraphic scale was determined to be 66.8 - 65 Ma.

Near the village of Luzanivka ($\sim$ 30 km NW of the crater center) an abundant lower Paleocene fauna, including nannoplankton of the NP1-NP4 zones, occurs in clayey and carbonate sands that overlie the upper unit of distal ejecta mentioned before. Thus, the upper limit of the age of the deposition of the Boltysh ejecta can be set at 65 Ma. This age agrees with recent precise Ar-Ar radiometric ages. A more precise biostratigraphic age determination could be obtained from further biostratigraphic and litholological studies of the ejecta, in particular in the adjacent areas of the Dnieper-Donets depression and within the sedimentary filling of the older Rotmistrovka crater.

1
Introduction

An intriguing problem of modern geology is the question if large meteorite (asteroid, comet) impact events coincide with mass extinctions and geological boundaries. So far only one such association has been made with any degree of certainty, namely at the Cretaceous-Tertiary boundary, but suggestions exist for an extraterrestrial signature at other boundaries as well (for a recent review see, e.g., Montanari and Koeberl 2000).

The Boltysh crater age, commonly given as 88 ± 3 Ma, is attracting some interest because of recent fission track dates of its glassy impactites (Kashkarov et al. 1999), which confirmed a previous suggestion (Valter et al. 1984) that its age might be similar to that of the Cretaceous-Tertiary (KT) boundary. In addition, recent radiometric dating, using the Ar-Ar method by Kelley and Gurov (2002), resulted in an age of 65.2 ± 0.2 Ma; see also Gurov et al. (this volume). The aim of the present work is to refine the age of the Boltysh event by biostratigraphic methods.

2
Geological Setting and Previous Age Determination of the Boltysh Crater

The Boltysh crater is situated in the central part of the Ukrainian shield and was formed in Precambrian (2.1 Ga) granites and minor aplites, granulites, and gneisses (Fig. 1 a). The crater was studied by several different researchers, whose results are summarised by Masaitis (1999). The average crater diameter is $D_1 \approx 23$ km. This diameter refers to the crater rim (now eroded). The diameter of a low amplitude concentric annular structure around the crater is $D_2 \approx 30$ km. A value for $D_2/D_1 \approx 1.3$ is in satisfactory agreement with the ratio of the rim diameter to the crater diameter on a preimpact level for fresh craters (Melosh 1989). The classical complex structure of the crater with the central uplift (average diameter ~ 4 km) was buried under the Upper Paleogene, Neogene, and Quaternary sediments that attain a thickness of up to 200 m inside the crater depression. Outside the crater the thickness of these deposits is not more than 100 m. These deposits are underlain by well-preserved crater lake sediments, which are nearly 400 m thick in the central crater area and are represented by interbedded laminated siltstones,

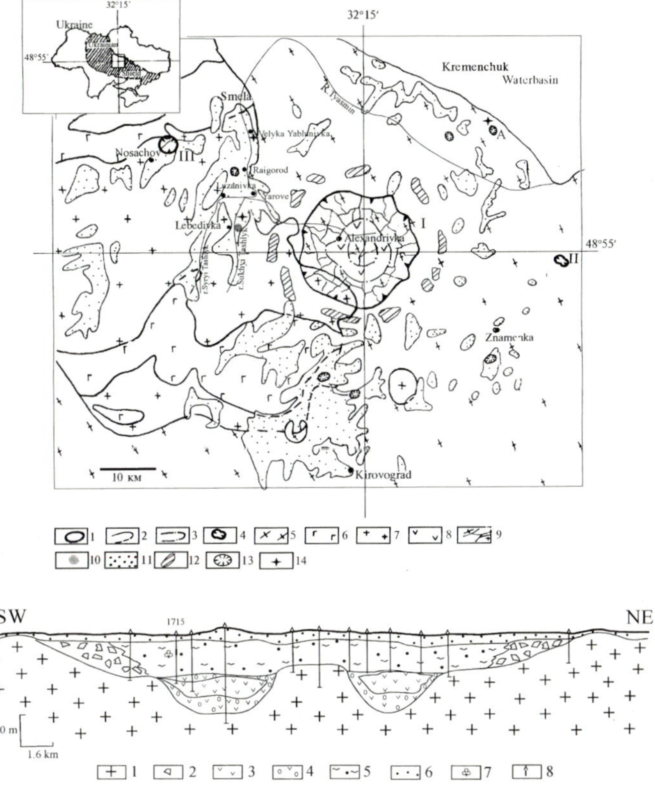

Fig.1. Geological setting of the Boltysh impact structure. a. Schematic Geology of the area of the Boltysh impact structure (from Valter et al. 1984, with minor corrections). 1- geological margins; 2- the same, underneath sedimentary cover; 3- the same, within Boltysh impact structure area; 4- impact structures: I – Boltysh; II – Zeleny Gay, III – Rotmistrovka; 5- Kirovograd granite; 6- gabbro; 7 - Rapakivi granite; 8- tagamites and suevites of the Boltysh crater; 9 - impact breccia within the Boltysh crater;10- the point of finding of the ejecta breccia fragments with the Upper Cretaceous fauna; 11- ejecta ouside the crater: sandy-gravel brecciated rocks, Paleocene 12- annular concentric uplift (~30-50 m) of basement rocks outside the Boltysh crater; 13 – suggested secondary craters, A – Adamovskiy crater; 14 - the points of sampling for granulometrical analysis.

 b. Geological section across the Boltysh impact structure (by Valter and Dobrianskiy (2001) with small corrections). 1- crystalline rocks; 2- breccia of the relics of the ejecta cover (i.e., the subject of the present investigation); basement rocks; 3- massive tagamites; 4- suevites and porous tagamites; 5- crater fill sediments (Paleocene); 6- sedimentary cover rocks (Eocene-Quaternary); 7- location of the Paleocene fauna in the 1715 borehole (see text); 8- bore-holes.

sandstones, and clays, and partly by carboniferous and clinoptilolite-bearing rocks with some oil shale layers.

Towards the crater margin, in the shallow outer part of the depression, fine-grained, partly chemogenic, lacustrine sediments are substituted by coarse-grained sands and allogenic lithic breccias with fragments and large blocks of basement crystalline rocks (Fig. 1 b). This not well studied unit is probably a mixture of proximal ejecta and the products of rim erosion. Now the friable rim is completely destroyed. The annular trough around the central uplift is filled predominantly with impact melt rocks (tagamites).

The geological age of the crater lake deposits, as well as the absolute age of tagamites, has been estimated and discussed in many previous publications (Bass et al. 1967; Vasiljev and Selin 1970; Yurk et al. 1975; Komarov and Rajchlin 1976; Valter et al. 1984; Bojko et al. 1985; Valter 2000; Kelly and Gurov 2002; Gurov et al. this volume).

The age of the crater formation may be dated biostratigraphically by the interval between the target rocks and the first crater sediments. As we did not have the necessary core material of crater sediments, we attempted to determine a crater age by specifying the stratigraphic position of the ejecta that cover fossils. Thus, in this chapter we consider only the coarse-grained sediments of ejecta outside the crater (Fig. 1a). These sediments mostly overlie directly the crystalline rocks of the Precambrian basement and are integrated into the Raygorod suite of the local age scale (Ryabchun 1970; Ryabchun and Gubkina 1972; Brjansky et al. 1978). These deposits were identified as ejecta from the Boltysh crater due to the occurrence of mineral grains with shock metamorphic effects, such as lanar deformation features (PDFs) in quartz, kink-bands and PDFs in biotite, as well as the presence of glassy impactite fragments (Valter and Ryabenko 1977; Gurov and Valter 1977).

During geological mapping in the area it was determined (Ryabchun 1970; Bryansky et al. 1978) that breccias of the Raygorod suite are not very common around the crater (Fig. 1 a). The ejecta are totally missing from a 4- to 5-km-wide zone directly outside the crater rim. They do occur in lower areas on the surface of crystalline rocks outside the crater, where they form separate "spots" from one to several kilometers in extent. They are also more widely distributed in depressions corresponding to river paleovalleys, which partially coincide with the present-day drainage system (Ryabchun 1970; Ryabchun and Gubkina 1972; Bryansky et al. 1978; Valter et al. 1984; Gurov and Khmel'nitskyi 1996). As a whole these depressions form a somewhat East-West elongated ring that is 50-60 km in its axial diameter and has a variable width (from about 5 to 15 km), surrounding the crater depression. Fragments of a second ring, about 70 km in axial diameter and 10 km wide, seem to occur in the west and in the south (Fig. 1a). The thickness of the ejecta blanket deposits outside the crater varies from zero to more than 80 m in local circular or slightly elliptical depressions up to 1 km, which were identified within the ejecta area and were previously suggested (Valter et al. 1984) to be secondary craters (Fig.1a). Within the same area the above-mentioned Rotmistrovka and Zeleny Gay impact craters are located. As was shown from analyses of new geophysical and drilling data (Valter et al. 1997, and unpublished data), it appears that Zeleny Gay is a double structure ($D_I \approx 0.8$ and $D_{II} \approx 0.7$ km). The clayey sediments of the crater lake contain pollen of *Pinur s/g Diploxylon, Pinus s/g Hapoxylon, Carpinus, Myrica, Castanea, Quercus, Rhus, Santalaceae, Ericaceae, Chenofodiaceae, Artemisia*, as determined by R.N. Rotman. This complex is undoubtedly Tertiary in age, possibly Paleocene (Valter et al. 1997). The crater sediments overlap the lithic breccia, with fragments of granites and gneisses, as well as with individual mineral grains in which the shock metamorphic features, such as PDFs in quartz deformation bands in microcline and kink-bands in micas are common.

It was shown by Valter et al. (1984) that the breccia outside the Boltysh crater is the result of the fragmentation of basement rocks by true ejecta and mixing with fall-back ejecta. The breccia is best and most completely exposed to the west of the crater, in the valleys of the Sukhyi Tashlyk, Syryi Tashlyk, and Tyassmin rivers (Fig.1a). A double layering of the breccia deposits was established by Valter and Ryabenko (1977) and Gurov and Valter (1977) in exposures near the village of Lebedivka (Fig. 1a). The lower layer is composed of basement rocks crushed up to gruss material. Its visible thickness is up to 8 m. Pockets and clastic dykes that are up to several tens of centimeters thick are observed on its surface. These are filled by more fine-grained material from the overlying layer. Outcrops of this layer were also observed to the north down the valley of the Sukhyi Tashlyk near the village of Yarove at a distance of almost 12 km from the crater rim (Fig 1a).

Fig 2. Clastic dyke with fragments of Upper Maastrichtian rocks in brecciated granites (the scale is shown by the white pen, which is 15 cm long).

The breccia of the upper layer in exposures near the village of Lebedivka is composed of a greenish gray, extremely inequigranular gravel-sandy-clayey rock, which contains abundant basement rock fragments as well as clayey pseudomorphs up to 5 cm across, probably from strongly metamorphosed rocks and recrystallized glasses. For the psephite-psammite rock component the absence of roundness and the sorting of the grains is cpecific (Fig. 3). Precambrian crystalline rocks are predominant among the fragments in the breccia. These are mainly granitic and gneissic rocks, as well as granulites, pegmatites, and aplites. Up to 5 vol.% of the comparatively small fragments (<10 cm in size) are sedimentary rocks: limestones, sandstones, cherty rocks, marls.

Close to the bottom of the layer large granite and gneiss blocks of up to 8 m in size are abundant. In some breccia outcrops in exposures near the village of Lebedivka indications of very coarse gradational bedding can be observed. With the whole thickness of exposed rocks being about 5 m, the lower parts of the layer (2-2.5 m) are characterized by 30-40 vol.% of fragments that are 25 cm and more in diameter, and by a very high content of centimeter-sized fragments in the matrix, whereas the upper part (up to 3 m thick) is characterized by breccia with a comparatively fine-grained matrix (Fig. 3), and with an abundance of large fragments of less than 5 vol. %. Such coarse gradational bedding is also observed, but in less complete sections, in outcrops near the village of Yarove and in a quarry near the village of Velyka Yablunivka (Fig. 1a). From our study of some of the outcrops and from core descriptions of the drill cores that intersected the filling of probable secondary craters, we suggest that sedimentary cycles with similar coarse gradational bedding were reproduced repeatedly while the deep depressions were being filled.

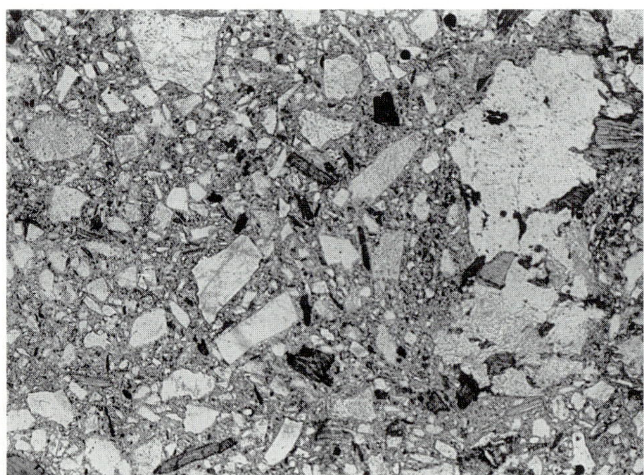

Fig. 3. The microstructure of the matrix of the upper breccia layer from an outcrop near the village of Lebedivka. Thin section under the polarizing microscope, 1 Nicol prism. The width of the image is 2.5 mm. One can see sharply angular fragments of granite (the largest fragment), partly altered grains of feldspar (gray) and quartz (light); biotite grains are dark gray or black.

Earlier the upper limit of the age of the breccia layers was determined from the observations that the breccia contain rock fragments with Cenomanian fauna

(Makarenko 1970), and that they overlie sandstone of probably Cenomanian age near the village of Nosachov (Ryabchun 1970); see Fig. 1a. An abundant occurence of Paleocene fauna was found by Ryabchun (1970) near the village of Luzanivka (Fig. 1a) in clayey and carbonate sands overlying the breccia strata. These deposits were named the Luzanivka layers and subdivided later by separation of the lower Makartite suite (Moroz and Soloviak-Krukovski 1993). This suite was dated to be of Early Paleocene age based on various micro- and macrofauna fossils (Makarenko 1970; Moroz and Soloviak-Krukovski 1993). This age can be determined accurately from the occurrence of nannoplankton and plankton foraminifera. The lower part of the Makartite suite corresponds to nannoplankton zone NP1 of the Martini scale and by plankton foraminifera to the lower part of *Acarinina inconstans*, the *Globoconusa daubjergensis* zone (roz and Soloviak-Krukovskiy 1993).

3
Methods and Results of Studies

The search for microfauna fossils and their identification in the fragments of sedimentary rocks from the impact breccias of the upper horizon and in the sedimentary dykes of the lower breccia horizon were done by conventional methods.The samples were taken from the location of the best exposure of the crater ejecta, near the village of Lebedivka, which is approximately 12 km from the crater rim. The exact point of sampling is shown in Fig. 1a.

The two types of samples were collected. The first one was represented by four fragments of sandy marls from the above-mentioned upper part of the breccia unit. The fragments were up to ten centimeters in size. The second type of samples was taken from the same exposure from the central zone of a sedimentary dyke (Fig. 2) in the lower part of the breccia unit. The mass of each sample to be analyzed was about 200 g. The samples were placed into water and carefully disaggregated. After 24 hours of settling some drops of the suspension of the fine clayey fractions were transferred with a pipette to a slide and, after drying, examined under the optical microscope with a high magnification, or, for nannoplankton, with an electron microscope (by S.A. Lulieva). Sandy fractions were studied for foraminifera by one of the authors (Plotnikova).

In two analyzed fragments from the upper layers of the breccia from outcrops near the village of Lebedivka, the foramenifera *Stensioeina emscherica Baryschn.*, *Anomalina infrasantonica Balakhm*, as well as the complex of nannoplankton CC15 (Sissingh 1977) zone *Lucianohabdulus cayexi*, were determined. This confirms a late Coniacian age of the rocks from which these fragments were derived.

We assume that any later sediments, if they existed at the moment of the impact event, could have been washed off during the catastrophic displacements of the loose ejecta and were involved in the filling of sedimetary dykes within the crushed crystalline basement rocks described above. The studied dyke (Fig. 2) is composed of inequigranular grained gravel-sandy-clayey sediments. In places its thickness is up to 0.5 m; it is filled with sandy material, which contain grains of granitic minerals (quartz, feldspar, biotite). Some grains show shock metamorphic features, such as PDFs in quartz, deformation bands in feldspars, and kink-bands in biotite. From the sandy fraction of the samples from the central zone of the dyke (NN 85-4; 85-4A and 85-4B) small fragments of cherty rocks were studied for the presence of foraminifera. The following Maastrichtian foraminifera were found: *Neoflabellina cf. reticulata* (Rss.), *Stensio iona cf. pommerana Brotz.*, *Brotzenella cf. praeacuta* (Vass.), *Cibicidoides cf. aktulagayensis* (Vass.), *C. cf. bembix* (Marss.), *Cibicides cf. voltzianus* (Orb.), *Bolivinoides cf. peterssoni Brotz.*, *Bolivina cf. incrassata* (Rss.), *Reussella cf.minuta* (Marss.), and *Globotruncana sp.*

For comparison with previous data of Valter and Efimenko (1981) (curve 6 in Fig. 4), the grain size distribution and the petrological and mineralogical compositions of the fragments and grains in the breccia were analyzed in two new samples by granulometry. These samples were taken from the quarry near the village of Velika Yablunivka (curve 6' in Fig. 4) and from the filling of secondary Adamovskiy crater of Dnieper- Tyasmin watershed (curve 6'' in Fig. 4). For locations of these sampling points, see Fig. 1a.

Each sample consisted of several sub-samples of slightly cemented breccia. The breccia were carefully crushed with a rubber pestle, and then the fraction larger than 10 mm was sieved off. The rock varieties and their contents were determined by standard petrographic techniques. Individual large fragments were picked out of the samples and were discarded. The upper grain size limits were taken to be 2 cm for a sample of 1 kg (Fig. 3, curve 6) (Valter and Efimenko 1981), 5 cm for a sample with an initial mass of 4 kg (curve 6'') and 7 cm for samples with an initial mass of 7.5 kg (curve 6'). After the removal of the large fragments the weight of the samples was reduced by quartering. Then the fraction less than 10 mm was additionally destroyed by boiling in water and the careful disaggregation was repeated. The content of the coarse-grained fraction (>0.1 mm) was determined by sieving. The amount of the fine fractions (<0.1 mm) was determined after reducing the samples to a weight of 50 g by sedimentation analysis. The results are shown in Table 1.

The endemic character of faunal fossils in crater lacustrine sediments (mollusks, fishes) caused a debate regarding the geological age of the fossil-bearing rocks, whether the age is Paleocene or Cretaceous (Bass et al. 1967). The results of fossil flora and palynological determinations seemed to be more reliable, and were taken to indicate a Paleocene age of the Boltysh crater sediments (Bass et al. 1967). In particular, F.A. Stanislavsky (1968) found – in sapropelite shale layers in the core of well 1715, at a depth of 100-130 m from the top of the lacustrine crater-fill sediments (Fig. 1b) – fossils that are undoubtedly of Paleocene age - *Hakea exulata Heer, Dryandroides antiqua Wat.*, especially *Dryophyllum furcinerve Schmalh., D.curticellense (Wat.)*, as well as *omptonia* of the same type as in Thanetian sandstones of the Paris Basin.

One of the reasons to discuss the age of the Boltysh crater were considerations that it might have formed coeval with the small Rotmistrovka crater (D ≈ 2.2 km) (Fig. 1 a). These suggestions (Bass et al. 1967; Vasiljev and Selin 1970) were based on the facial similarity of lacustrine sediments in both craters, including fossils of the same groups of flora and fauna (*Filices, Algae, Ostracoda, Pisces, Crustacea*). But a comparison of these lacustrine sediments in the Rotmistrovka crater with those at Boltysh indicate that they are older than Cenomanian and Aptian deposits that are characterized from fauna (Plotnikova and Jakushin 2002) and flora (Stanislavsky 1968). K-Ar dating of glassy impactites of the Rotmystrovka crater indicated ages of 130±10 Ma, which is in agreement with the paleontological data (Valter et al. 1984).

Table 1. Granulometric and compositional data of the upper part of the ejecta breccia from different locations

Dimensions of grains (mm)	Mass (wt%)		
	Lebedivka, 50/3+50/4 (Valter and Efimenko 1981)	Velika Yablunivka, 1-82E+1-82W	Adamovskiy crater
>10	1.44	7.00	5.68
7 - 10	0.30	1.55	2.93
5 - 7	0.71	1.85	3.42
3 - 5	1.80	3.30	7.32
2 - 3	2.88	2.50	7.58
1 - 2	5.84	6.30	10.50
0.5 - 1	9.00	3.75	5.04
0.25 – 0.5	22.31	9.60	10.55
0.1 – 0.25	12.27	6.30	5.11
0.01 – 0.1	17.46	20.35	15.01
0.001 – 0.01	24.06	36.10	26.53
<0.001	1.93	1.4	0.33
Modal compositions of rock fragments >10 mm	Granites: 95 Gneisses: 3 Quartzite: 2	Granites: 85.7 Gneisses: 11 Quartzite: 1.7 Clayey shale: 1.6	Granites: 89.50 Gneisses: 9.35 Sandstones: 0.70 Quartzite: 0.45

Biostratigraphic dating of the Boltysh event is of particular interest not only because it may be more precise than some isotopic dating, but also because it gives direct information on the geological position of this event relative to stratigraphic boundaries (cf. Montanari and Koeberl 2000).

For the Boltysh impactites, a number of K-Ar age determinations exist. Their results yielded a wide age range, from 177 to 56 Ma (Bass et al. 1967; Yurk et al. 1975; Komarov and Rajchlin 1976; Valter et al. 1984; Bojko et al. 1985; Valter 2000). The most widely cited age was the one by Bojko et al. (1985) at 88 ± 3 Ma.

The higher ages were explained (Yurk et al. 1975; Valter et al. 1984; Valter 2000) as an overestimation due to the presence of basement rock fragments within the suevites and tagamites. In addition, newly formed phases, such as impact diamond (Verchovski et al. 1991), together with glass (Kelley and Gurov 2002), may capture radiogenic argon. For somewhat weathered samples lower ages may be the result of partial loss of argon. In any case, Ar-Ar results obtained for fresh samples using modern instrumentation that require only small amounts of sample for analysis (e.g., Gurov et al. 2001 and this volume; Kelley and Gurov 2002) are much more reliable than those obtained earlier by the K-Ar method from larger samples, and indicate an age of 65.17 ± 0.64 Ma.

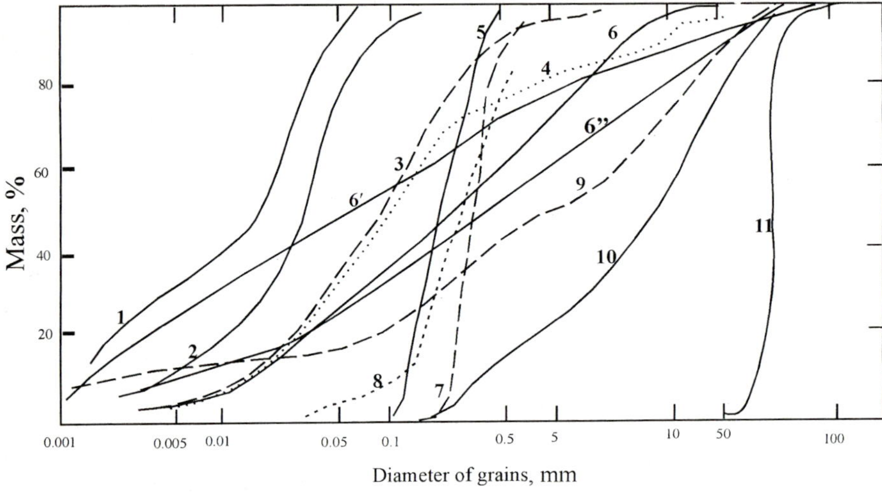

Fig. 4. Logarithmic cumulative curve of the granulometric composition of the Boltysh crater ejecta: curves 6, 6', 6'' (see text). For comparison: 1 – glacial lacustrine aleurite; 2 – loess; 3 – lunar soil; 4 – till; 5 – dune sand; 7 – beach sand; 8 – river sand; 9 – mud flow deposits; 10 – river gravel; 11 – beach gravel (for references, see Valter and Efimenko 1981).

4
Discussion

The results of the granulometric analysis of the ejecta show that size sorting is practically absent. From their granulometric features the ejecta are close to the characteristics of mud flow deposits, but are somewhat different from the distribution of particles in undisturbed ejecta from explosions (e.g., precipitates from nuclear explosions). Also some very coarse gradational bedding of the studied breccia series was observed in outcrops near the village of Lebedivka.

Taking into account both breccia abundance and fragment composition, we conclude that the filling of the river valleys and other depressions happened as a result of unloading of catastrophical, rapid, and possibly repeated temporary flows. The appearance of such flows could have been similar to those observed at artificial explosion craters (Roberts 1968). Such processes must have lead to gigantic landslides, producing the powerful mudflows on land and dynamically similar turbulent water flows in the sea.

Within the area where ejecta have been preserved, their redeposition obviously occurred under terrestrial conditions, but near the river Tyasmyn-Dnieper, depression fillings may have been deposited in a shallow sea. The foraminiferal microfauna (Fig. 5) is abundant here in breccias at the possible secondary crater Adamovskiy (Fig. 1a), but the granulometric composition of the breccia cement (matrix) (Figures 3 and 5) is practically the same as elsewhere. In a sample from this area, Kraeva and Plotnicova (Valter 2000; Valter 2001) identified well-preserved Paleocene benthic foraminifera: *Globulina amygdaloides Reuss, Guttulina ex gr. ipatovcevi Vass., Gyroidina octocamerata Cushm. et Hanna, Eponides toulmini Brotz., E. et gr. toulmini Brotz., Alabamina wilcoxensis Toulm., Lamarkina naheolensis Cushm. et Hanna., Cibicidoides praeventrutumidus Mast Ceratobulimina aft. tuberculata Brotz., Anomalina simplex Brotzs., Anomalinouides danicus Brotz.*

Fig. 5. The microstructure of the matrix of the upper breccia layer from the probable secondary crater Adamovskiy. Thin section of drill core sample from 77 m depth, polarizing microscope, 1 Nicol prism. Width of image is 2.5 mm. As in Fig. 3 one can see sharply angular fragments of partly altered grains of feldspars (gray) and some smaller quartz grains (light); biotite grains are dark gray or black. Near the center of the image there is a fossil of *Lamarkina cf. Naheolnensis Cushm et Hanha*.

The main result of the present work is, in our opinion, the discovery of small rock fragments, from the sedimentary dyke at the bottom of the breccia, which contain Maastichtian foraminifera. Most of these forams appear in the Upper Campanian or in the Lower Maastrichtian and pass into the Upper Maastrichtian, but do not continue into the Paleocene. *Brotzenella praeacuta (Vass.)* and *Roussella minuta (Marss.)* appear for the first time in the Upper Maastrichtian and continue into the Paleocene.

Moroz and Soloviak-Krukovskiy (1993) reported the discovery of rare samples of the zonal species *Nefrolithus frequens Gorka* in the Raygorod breccia cement. We assume that this microfauna entered the sandy cement as a result of weathering of clayey fragments.

A comparison with the cross-section of Upper Cretaceous deposits of the Dnieper-Donets depression (DDD) (Lipnik and Lulyeva 1981) helps to establish a correlation between the foraminifera in the impact ejecta and the lower subzone complex – Reusella minuta (Marss) of Hanzawaia ekblomizone, which are the basis of the Upper Maastrichtian subdivision in the DDD. The Nefrolithus frequens nanoplankton zone in the DDD also corresponds to the Upper Maastrichtian. Thus, the forms identified in the impact ejecta determine the age of the deposits to be Late Maastrichtian.

The data on absolute time scale of Cretaceous marker-species abundances (Berggren and Norris 1997; Thierry et al. 1998) allow us to date the described complex of Maastrichtian foraminifera from the youngest breccia fragments of post-crater ejecta to the time interval from 66.8 Ma (mass development of *Neoflobellina frequens*) to 65 Ma (disappearance of these forms at the K/T boundary). This age range corresponds to the time of formation of the Boltysh crater. The lower age limit of the crater formation is determined by the Earliest Paleocene age of well stratified Luzanovka series rocks, which overlie the ejecta. These rocks are characterized by the macro- and microfauna of the corresponding age, as well as by calcareous nanoplankton of the earliest (lower) zone of the Paleocene NP1, dated at 65 Ma. Thus, the new biostratigraphic data give the most reliable age of the Boltysh crater formation to be within the age range 66.8-65.0 Ma. This value is very close to, or even synchronous with, the K/T boundary. Our data agree with new Ar-Ar age determinations reported by Gurov and Kelley (2002). A summary of the stratigraphy in the area is given in Table 2.

A more definite answer to the question regarding the relation of the Boltysh ejecta to the K/T boundary could be obtained in the future from careful studies of cross-sections of the adjacent parts of the DDD, using drill core studies.

Table 2. Stratigraphic sequence of Upper Cretaceous and Paleocene sediments of the Boltysh crater region

Series	Sub-series	Stage	Sub-stage	Boltysh ejecta blanket
Paleocene	Lower	Danian		Lower Makartite suite of Luzanovka layers (Moroz and Soloviak-Krukovskiy 1993)sandy, clay limestones; thickness up to 4 m. contain Pulsiphonina prima, Cibicides leclus, Globoconusa daubergensis.Calcareous nannoplankton zones from NP1 to NP4
Upper Cretaceous		Maastrichtian	Upper	∧∧ Marls, limestone. Fragments in Raigorod breccia (Makarenko, 1970; Plotnikova and Jakushin 2002; this work): Neoflabellina cf. reticulata (Rss.), Stensioeiona cf. pommerana Brotz., Brotzenella cf. praeacuta (Vass.), Cibicidoides cf. aktulagayensis (Vass.), C. cf. bembix (Marss.), Cibicides cf. voltzianus (Orb.), Bolivinoides cf. peterssoni Brotz., Bolivina cf. incrassata (Rss.), Reussella cf. minuta (Marss.), Globotruncana sp.and Neoflobellina frequens Gorka
		Coniacian	Upper	≈≈≈ Sandy marls. Fragments in Raygorod breccia (Makarenko 1970; this work) with Stensioeina emscherica Baryschn., Anomalina infrasantonica Balakhm; nannoplankton - complex CC15 - Lucianohabdulus cayexi
		Cenomanian	Upper	≈≈≈ Sandstones, limestones. Fragments in Raygorod breccia (Makarenko 1970) with Lingulogaelinella globosa (Brotz.), Anomalina (parzevae Vass.), nannoplancton - Anomalina globosa Brotz., A. Fluensa Ploth., Cibicides jarzevae Vass., Gumbelitria cenomana Kell., Bolivinita eovigiriniformis Kell., Globigerina cf. caspia Vass. Nannoplankton - Clinorhabdus eximius Stov.,C .turriseiffeli (Delf.), Tranolithus variatus (Caratini), T. gabalus Stov., Parhabdolithus embergeri (Noel.), P. condylosus (Stov.), Lithastrinus floralis Str., L. planus (Stov.), Coccolitus actinosus Stov., C.ex. gr. pelagicus (Wallich), C. ircumradiatus Stov., Cyclolithus granosus Stov., Deflandrius intereisus (Defl.), Zygolithus diplogrammuus Delf., Stephanolithion crenulatum Stov., Chyphragmalithus achylosus Stov.
Lower Proterozoic				≈≈≈ Granites with minor granulites and gneisses

∧∧∧∧∧∧∧∧∧∧∧∧∧∧∧∧∧∧∧∧∧ - discontinuity, possible hiatus;

≈≈≈≈≈≈≈≈≈≈≈≈≈≈≈≈≈ - hiatus

Acknowledgements

We thank V.P. Brjansky (Geological Survey of Ukraine) for cooperation with the study of the Boltysh ejecta and Dr. S.A. Lulieva (Institute of Geological Sciences National Academy of Science) for the nannoplankton determinations, as well as the Ministry of Education and Sciences of Ukraine for partial support through grant N2M/253-99. We also thank the reviewers, Dr. K. Kirsimäe and Dr. E. Molina, and the editor, Dr. C. Koeberl, for their valuable efforts to improve this article.

References

Bass YuB, Galka A I, Grabovsky VI (1967) The Boltysh combusible shale. Geologia i ohrana nedr (USSR) 9: 9-15 (in Russian)

Berggren WA, Norris RD (1997) Biostratigraphy, phylogeny and systematics of Paleogene trochospiral planctic foramenifera. Micropaleonology, 43 (Suppl. 1): 17 -116

Bojko AK, Valter AA,Vishnyak MM (1985) About the age of Boltysh crater. Geologicheskiy Zhurnal (Ukraine) 45 (4): 86-90 (in Russian)

Bryansky VP, Zlobenko VG, Rjabtchun VK (1978) The breccia rocks of Paleogene in area of the Boltysh depression. Geologicheskii Zhurnal 2: 135-138 (in Russian)

Gurov EP, Kheml'nitskii AF (1996) Dissemination and Preservation of Ejecta from Impact Structures: the Boltysh and Acraman Craters. Solar System Research 30: 19-24

Gurov EP, Valter AA (1977) Ejecta of the Boltyshian meteoritic crater in the Ukrainian Shield. Geologicheskiy Zhurnal 27 (6): 79-84 (in Russian)

Gurov EP, Kelley S, Babina NV (2001) Ejecta of the Boltysh impact crater in the Ukrainian Shield [abs.]. 6th ESF - IMPACT Workshop Abstract Book, Granada, Spain, pp 45-47

Gurov EP, Kelley SP, Koeberl C (2002) Ejecta of the Boltysh impact crater in the Ukrainian shield. This volume

Kashkarov LL, Nazarov MA, Lorenz KA, Kalinina MA, Kononkova NN (1999) Fission-track dating of the Boltysh impact structure. Solar System Research 33: 291-298

Kelley SB, Gurov E (2002) Boltysh, another end-Cretaceous impact. Meteoritics and Planetary Science 37: 1031-1043

Komarov AN, Rajhlin AJ (1976) The comparative study of impactite age by the fission-track and K-Ar methods. Transactions (Doklady) of the Soviet Union Academy of Sciences, Earth Science Sections 228: 673-676 (in Russian)

Lipnik ES, Lulyeva SA (1981) Zones of bentos foraminifera and lime nanoplankton in Campanian and Maastrichtian deposits of the Dnieper-Donets depression. Preprint of the Institute of Geological Sciences of Academy of Sciences of Ukraine, Kyiv, 83-21: 57 pp (in Russian)

Makarenko DYe (1970) Early Paleocene mollusks of the Northern Ukraine. Naukova Dumka Press, Kyiv, Ukraine, 127 pp (in Russian)

Masaitis VL (1999) Impact structures of northeastern Eurasia: The territories of Russia and adjacent countries. Meteoritics and Planetary Science 34: 691-711

Melosh HJ (1989) Impact cratering: A Geological Process. Oxford Monographs in Geology and Geophysics 11. Oxford University Press, New York, 245 pp

Montanari A, Koeberl C (2000) Impact Stratigraphy: The Italian Record. Lecture Notes in Earth Sciences, Vol. 93, Springer Verlag, Heidelberg, 364 pp

Moroz SA, Soloviak-Krukovskiy YV (1993) European Paleocene stratoregion of Luzanivka. Lviv University Collection of Articles of Paleontology 29: 65-72 (in Ukrainian)

Plotnikova LF, Yakushin LN (2002) New data on stratigraphy of the Cretaceous sediments of the Ukrainian Shield. In: Teslenko YuV (ed) The biota evolution as a base of stratigraphy. Institute of Geological Sciences of National Academy of Sciences of Ukraine. Special Publication, Logos, Kyiv, Ukraine, pp 62-63 (in Russian)

Ryabchun VK (1970) Paleocene deposits of the North-Eastern part of the Ukrainian shield. Collection of Scientific Works of Kyiv University 5: 29-34, Kyiv University edition (in Russian)

Ryabchun VK, Gubkina TB 1 (1972) On question on Lower Paleocene of the Central and North-Western part of the Ukrainian Shield (middle-Dnieper area). Lviv University Collection of Articles of Paleontology 1 (9): 73-77 (in Russian)

Roberts WA (1968) Shock crater ejecta characteristics. In: French BM, Short NM (eds) Shock Metamorphism in Natural Materials. Mono Book Corp., Baltimore, pp 101-114

Sissingh W (1977) Biostratigraphy of Cretaceous calcareous nanoplankton. Geologie en Mijnbouw 56: 37-56

Stanislavsky FA (1968) The age and stratigraphy of sapropellites from Boltysh depression Geologicheskiy Zhurnal (Ukraine), 28 (2): 105-110 (in Russian)

Thierry J, Farley MB, Thierry Jacquin, De Graciansky P-C, Vail PR (1998) Cretaceous chronobiology. In: De Graciansky P-C, Hardenbol J, Thierry J, Vail PR (eds) Mezozoic and Cenozoic Sequence Chronostratigraphic Framework of European Basin. SEPM Special Publication 60

Valter AA (2000) Ejecta deposits surrounding crater and geological age of Boltysh astrobleme. In: Churjumov K.I. (ed) Proceedings of the 1st International Conference on Modern Problems of Comets, Asteroids, Meteors, Meteorites, Astroblems and Craters. Vinnitsa, Ukraine: 330-337 (in Russian)

Valter AA (2001) The throwout deposits of Boltysh crater as the probable local K/T impact marker on the Ukrainian Shield [abs.] 6th ESF - IMPACT Workshop Abstract Book, Granada, Spain, pp 135-136

Valter AA, Dobryanskiy YuP (2001) Cooling regimes of the stratum tagamites and their influence on impact diamond intactness. Mineralogical Journal (Ukraine) 23 (4): 56-66 (in Russian)

Valter AA, Efimenko VV (1981) The granulometric and mineral composition of Boltysh meteorite crater throwout on Ukrainian shield. Geologichesky Journal 41 (2): 29-37 (in Russian)

Valter AA, Ryabenko VA (1977) Explosion craters of the Ukrainian Shield. Naukova Dumka Press, Kyiv: 154 pp (in Russian)

Valter AA, Brjansky VP, Lazarenko EE (1984) The peculiarities of the astroblemes genesis of the Ukrainian Shield. In: Markov MS, Bazilevskiy AT, Katsura IK, Mukhin LM, Sukhanov AP (eds) 27th International Geological Congress 19 Comparative Planetology: 89-96. VNU Science Press, Utrecht, The Netherlands

Valter AA, Mosejchuk MA, Kalashnik AA (1997) The new interpretation of the structure of Zeleny Gay Astrobleme on the Ukrainian Shield [abs.]. Abstracts submitted to the 26th Microsymposium on Comparative Planetology, Moscow, October 13-17, 1997, pp 127-128

Vasiljev IV, Selin YuI (1970). New data on paleontological characterization of productive rock mass of Boltysh combustible shale deposits. Transactions of Academy of Sciences of Ukrainian SSR, series B (12) : 1059-1061 (in Russian)

Verchovsky AB, Valter AA, Shukolyukov YuA (1991) Noble gases in shock-produced diamond from Popigai meteorite crater [abs.]. European Geophysical Society, XVI General Assembly, Wiesbaden 1991, Abstract SE9-7: C57

Yurk YuYu, Erjomenko GK, Polkanov YuA (1975) Boltysh depression - fossil meteorite crater. Soviet Geology 2: 141-144 (in Russian)

Ejecta of the Boltysh Impact Crater in the Ukrainian Shield

Eugene P. Gurov[1], Simon P. Kelley[2], and Christian Koeberl[3*]

[1]Institute of Geology, National Academy of Sciences of the Ukraine, 55-b Gontchar Str., 01054 Kiev, Ukraine.
[2]Department of Earth Sciences, Open University, Milton Keynes MK7 6AA, United Kingdom.
[3]Institute of Geochemistry, University of Vienna, Althanstrasse 14, A-1090 Vienna, Austria.
[*](christian.koeberl@univie.ac.at)

Abstract. The Boltysh crater is an about 24-km-diameter complex impact structure, which is situated in the central part of the Ukrainian Shield. The crater is surrounded by an ejecta blanket represented by a polymict breccia layer that is preserved over an area of ~6500 km^2. The later fall-back or suevite ejecta are preserved only in the crater overlying an impact-melt sheet. The ejecta blanket outside the crater is extensively eroded and varies in thickness from tens of meters at a distance of about 2-3 crater radii to 1-4 meters at 4-5 crater radii from the center of the structure. The farthest recorded distance of the ejecta is ca. 66 km or 5.5 crater radii to the W-N-W of the Boltysh structure. Fall-back ejecta from the Boltysh impact event include a suevite breccia layer, which now occurs only within the crater rim, where it overlies an impact melt sheet and the top of the central uplift in the crater. The thickness of the suevite layer varies from a few meters to up to 97 m in the eastern part of the crater.

The Boltysh ejecta were deposited on top of Precambrian crystalline basement of the Ukrainian Shield over almost the entire extent of the ejecta blanket. Only within the nearby, about 140-Myr-old Rotmistrovka impact structure, 45 km NW of the Boltysh structure, do the Boltysh ejecta overlie older sediments, which comprise the post-impact sediments of that crater. There, up to 18 m of Boltysh ejecta overlie marls and chalk of Cenomanian-Turonian age, causing brecciation of the underlying chalk. Outside the crater rim the Boltysh ejecta layer is overlain by Middle Eocene sediments, but deposition of these sediments was preceded by extensive erosion of the Boltysh breccias.

The post-impact sediments in the Boltysh structure are up to 550 m thick in the deepest parts of the crater. The lower series are sandstones and siltstones of closed freshwater lake origin, without any visible floral and faunal remains. The overlying series of shales and oil shales contain abundant remains of the Paleocene and Eocene paleoflora. These sediments are, in turn, covered by deposits of early to mid Eocene rocks and by more recent sediments of Neogene-

Quaternary age. Thus, the stratigraphic age of the Boltysh impact crater, and its ejecta, is constrained to be between the Cenomanian-Turonian and the Paleocene.

The most commonly quoted age for the Boltysh age is 88 ± 3 Ma years, based on whole rock K-Ar ages of impact-melt rocks. However, our new data, derived by laser stepped heating and spot $^{40}Ar/^{39}Ar$ dating of impact melt rocks, yields an age of 65.17 ± 0.64 Ma. This age is in agreement with an earlier fission track age, and some recent biostratigraphic studies, indicating that the Boltysh crater was formed simultaneously with, or within a few hundred thousand years of, the Cretaceous-Tertiary boundary age Chicxulub impact structure.

1
Introduction

The Boltysh impact structure is centered at N 48°54' and E 32°15' in the basin of the Tyasmin river, the right tributary of the Dnieper river. The town of Alexandrovka is located near the center of the crater (Fig. 1, 2). The depression in the surface of the crystalline basement of the Ukrainian shield corresponding to the Boltysh impact structure was first recognized by L.G. Tkachuk in 1930-1932 (unpublished reports in the National Geological Archive in Kiev). The structure is covered by Quaternary sediments, and the circular form of the basin and its diameter of about 24 to 25 km were determined by later studies, using drilling and geophysical investigations. Numerous holes were drilled with the structure to explore deposits of oil shales with commercial reserves of more 3 billions tons (Bass et al. 1967). Although Bass and co-authors suggested a volcanic origin for the Boltysh structure, describing the rocks as "volcanic-like", an impact origin of the structure was suggested by Golubev in 1969 (reported in Golubev et al. 1974), and was confirmed by Masaitis (1973, 1974), who was the first to find shock metamorphic effects in rocks from the Boltysh crater.

The characteristics of the Boltysh impact structure and its ejecta, as presented in this paper, are based mainly on work by Gurov and Gurova (1985, 1991), with the addition of other work (Bass et al. 1967; Masaitis et al. 1980; Valter et al. 1982; Valter and Ryabenko 1977; Yurk et al. 1975). The main observations reported in this work are derived from the study of cores of the holes 11475 - 1148 m (these values refer to the depth to which the respective holes were drilled), 17 – 677 m, 18 – 527 m, 19 – 560.1 m, 20 – 517.6 m, 21 – 599.3 m, 29 – 118.8 m and 50 – 736 m (Figs. 3, 4), which were drilled in the 1970s and 1980s. Samples of impact-melt rocks for 40Ar/39Ar dating were selected from the cores of hole 50, which was drilled by a project of E.P. Gurov in 1984 for demonstration purposes at excursion 098 of Session XXVII of the International Geological Congress.

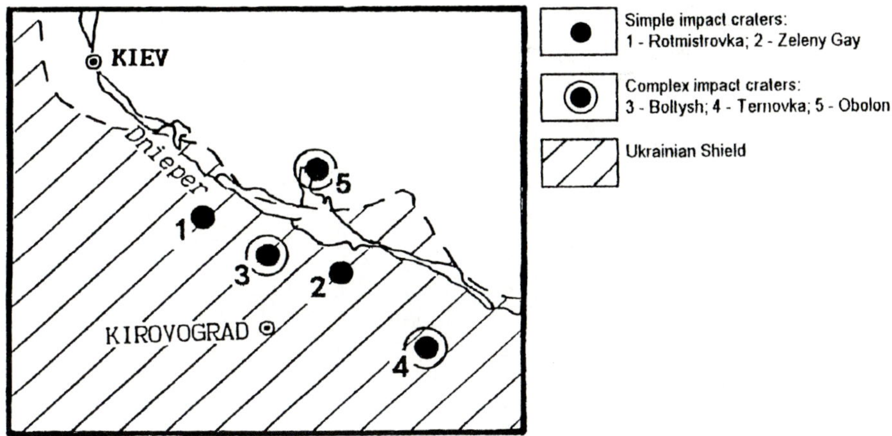

Fig. 1. Impact structures in the central part and northeastern part of the Ukrainian Shield.

The extended breccia layer overlies the crystalline basement of the Ukrainian Shield around the Boltysh structure and is exposed in the basins of the Tyasmin and Ingul rivers (Fig. 2). The breccias are poorly sorted and weakly consolidated rocks that are composed of crystalline rocks. The breccia layer is covered by Cenozoic sediments, and their outcrops occur only in the limited area in the Tyasmin river basin. The breccias were first described by G.G. Andreichik, V.P. Bryansky, G.M. Karpov, V.K. Ryabchun, V.G. Zlobenko and others during geological surveys in this region in the 1950s to 1970s (unpublished reports in the National Geological Archive in Kiev; Gurov and Valter 1977; Bryansky et al. 1978). The breccias were initially described as tuffs and tuffites of a paleovolcano. The ejecta layer was also interpreted as sedimentary rocks connected to extensive tectonism of the region (V.K. Ryabchun, unpublished data). Moroz and Sovyak-Krukovsky (1993) also interpreted some of the breccias as olistostromes.

The link between the breccia layer and a meteorite impact was made by discovery of shock metamorphic effects (PDFs in quartz and feldspars, kink bands and PDFs in biotite) in these rocks from the outcrops in the Tyasmin river basin (Bryansky et al. 1978; Gurov and Valter 1977; Valter and Ryabenko 1977); this followed the identification of the Boltysh structure itself as an impact crater (Masaitis 1973, 1974). The spatial distribution and the thickness of the breccia Masaitis 1973, 1974). The spatial distribution and the thickness of the breccia layer were determined from unpublished descriptions of numerous shallow holes

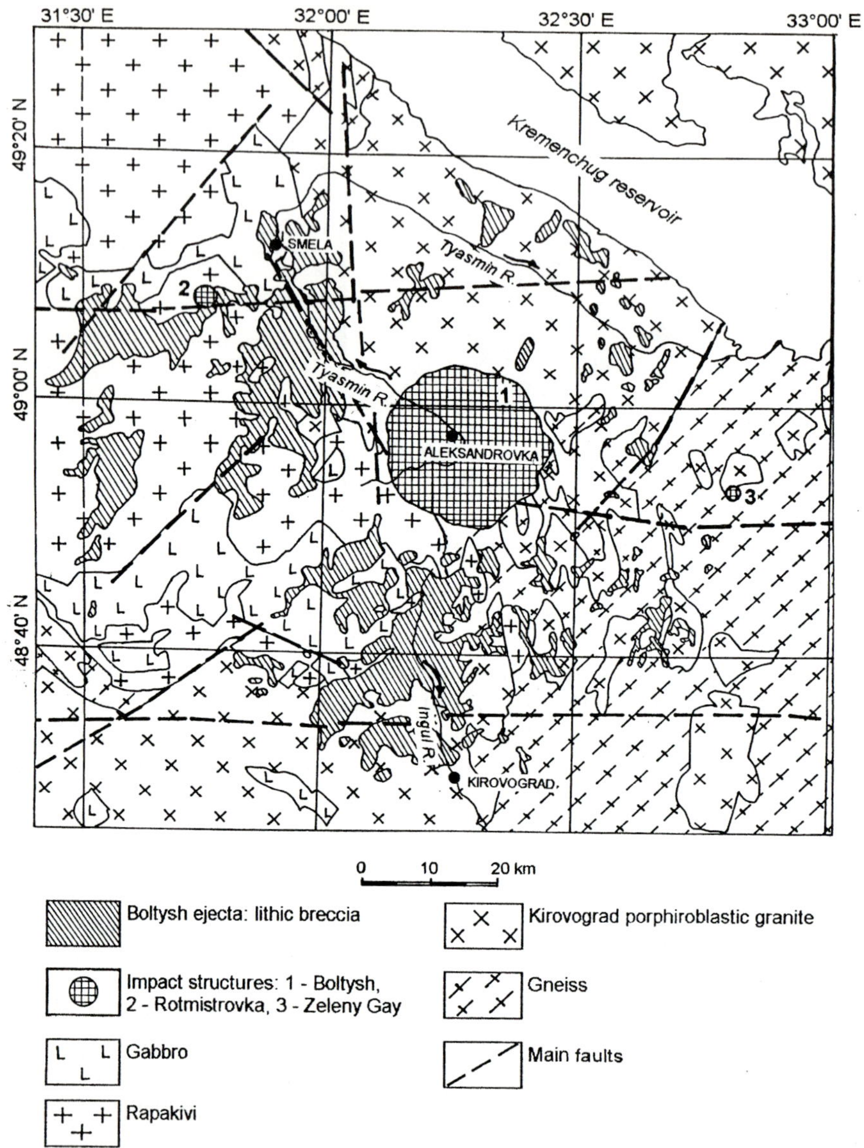

Fig. 2. Schematic map of the Boltysh impact structure and its ejecta (after Bryansky et al. 1978; Gurov and Khmelnitsky 1996). Regional geology after Shcherbak (1983) and Zaritsky (1992). The Cenozoic sediments that cover the area have been omitted.

(that were drilled in this area for geological surveys led by G.G.Andreichic, V.P.Bryansky, and other geologists mentioned above. In addition, the modes of occurrence of the Boltysh ejecta within the Rotmistrovka crater have been studied, using cores of some of the drillholes that penetrated the post-impact sediments and overlying breccias in the Rotmistrovka structure (Gurov and Babina 2000; Gurov and Gurova 1991).

2
Structure of the Boltysh Impact Crater

Boltysh is a complex impact structure about 24 km in diameter; the present-day depth of impact-affected and derived rocks is about 1 km (Gurov and Gurova 1985, 1991; Masaitis et al. 1980; Valter and Ryabenko 1977). It is presently covered with Quaternary sediments that are up to 30 m thick, is poorly expressed in terms of surface morphology, and, therefore, it is not visible on remote sensing images (aerial or satellite photography). According to V.I. Grabovsky (unpublished data), the area of the Boltysh impact crater is marked by a negative gravity anomaly of –19 mGal, about 22-23 km in diameter.

The crater was formed in Precambrian crystalline rocks of the Ukrainian Shield (Fig. 2) that are represented in this area by the so-called Kirovograd porphyroblastic granites (age ca. 1550 Ma) and by ca. 1850 – 2220 Ma biotite gneisses (Shcherbak et al. 1978) (Fig. 2). The volumetric ratio of these two rock types in the Boltysh target area is about 5:1 (Gurov and Gurova 1991; Masaitis et al. 1980).

The principal elements of the crater structure are the central uplift, deep inner basin, shallow peripheral depression, and uplifted original rim (Figs. 3b, 4a-c). The central uplift of the Boltysh structure is 4 km in diameter at the top and shows up to about 600 m stratigraphic uplift. It has a plateau-like surface. The uplift comprises blocky breccia of shock metamorphosed granites and gneisses, which are extensively altered as a result of their interaction with the heated water of the post-impact crater lake (Gurov 1996).

The deep inner part of the crater, 12 km in diameter, encloses the central uplift. The rim of the inner crater is expressed as an inflection of the crater floor. The stratigraphy of the inner crater of the Boltysh crater is best displayed in drillhole 11475, which penetrated the structure and terminated in the weakly fractured basement. The true depth of the inner crater in hole 11475 is 1065 m from the present-day surface and ~550 m from the floor of the surrounding peripheral depression. The basement of the inner crater is composed of weakly fractured granites that occur in the interval from 1065 m to 1148 m. Unconsolidated granitic rock flour with rare granite clasts occurs in the interval from 920 to 1065 m.

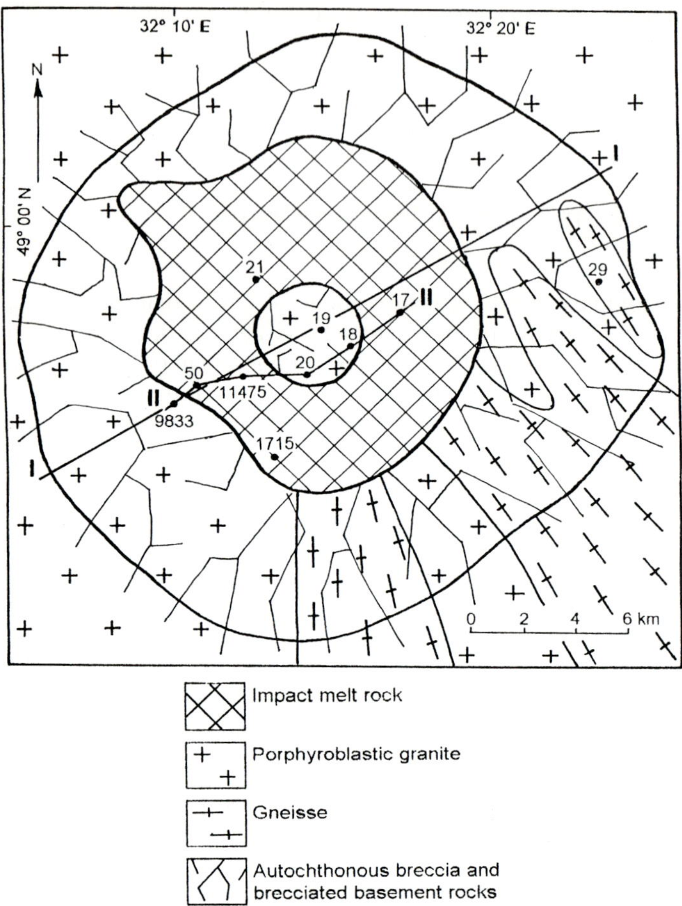

Fig. 3. Schematic map of the Boltysh impact structure. The annular impact-melt sheet occupies the inner crater around the central uplift. The post-impact sediments are omitted. The locations of some drill cores and the positions of cross-sections are indicated.

A series of lithic breccias and suevites overlies the rock flour and occurs between 920 and 792 m. The impact melt rocks occur in the interval from 792 to 573 m. The melt rocks form an annular sheet about 12 km in diameter and up to 220 m thick. The plateau-like top of the central uplift is elevated by about 80 m above the top of the melt sheet. The surface of the melt is subhorizontal, within

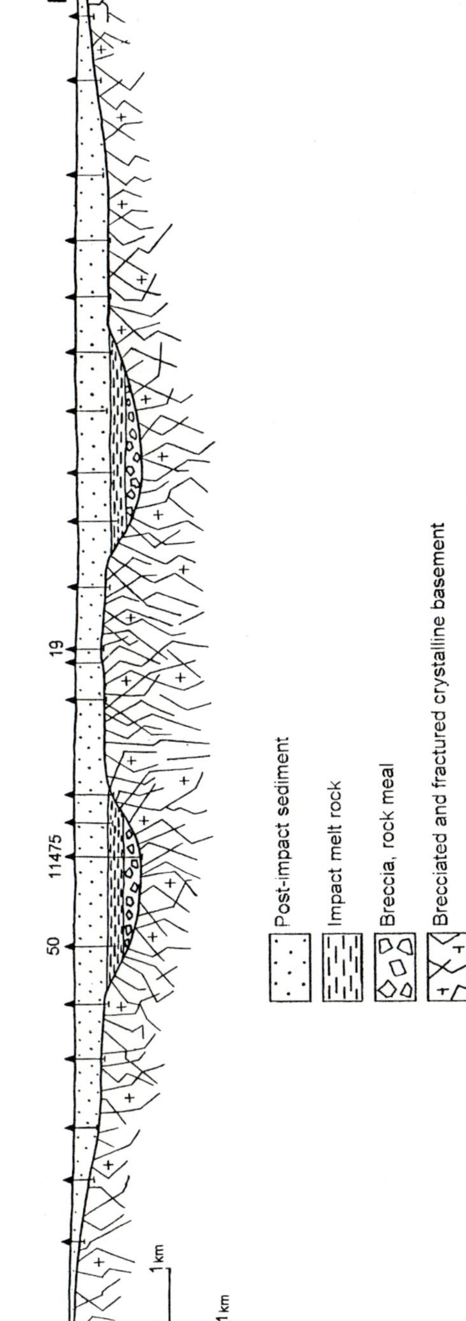

Fig. 4. (a) Schematic southwest to northeast cross-section through the Boltysh impact structure (I-I), as derived from drill core stratigraphy.

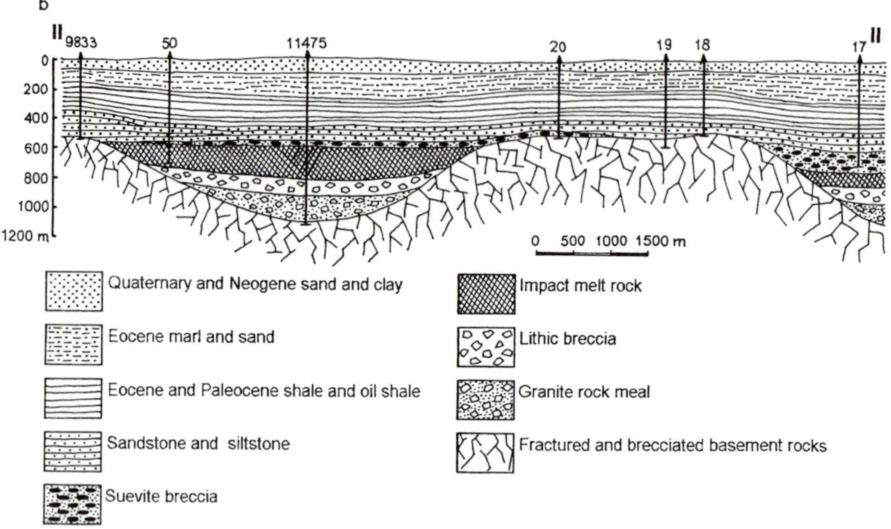

Fig. 4. (b) Detailed cross-section through the central part of the Boltysh structure (II-II). Location of cross-sections are indicated in Fig. 3.

around 10 m across its whole extent, and covers an area of ~85 km^2. It has been suggested that mobile high-temperature impact melt occupied the deepest part of the impact structure around the central uplift (Gurov and Gurova 1991). Patches of fall-back suevite, up to about 30 m thick, overlie the impact-melt sheet.

The peripheral part of the Boltysh structure is a shallow annular depression around the inner crater. Its depth is about 550 m at the limits of the inner crater and decreases gradually to the crater rim (Fig. 4). The basement to the peripheral depression is formed of brecciated and fractured crystalline rocks. The crater is surrounded by an uplifted rim, composed of weakly fractured granites, exposed in its north-western sector in the valley of the Tyasmin river.

The Boltysh crater is filled with post-impact sedimentary rocks that attain a thickness of about 470 m above the central uplift and up to about 550 m over the deepest part of the crater (Fig.4c). The lower series of the post-impact sediments, about 120 m thick, is represented by sandstones, siltstones, and sands with thin interlayers of breccias mainly in its basal horizons. The rocks of that series do not contain any determinable paleofloral imprints. The overlying sediments are a series of shales and oil shales that are about 250-300 m thick. Abundant paleofloral and paleofaunal remnants of the Late Paleocene to Early Eocene age occur in these rocks (Stanislavsky 1968), indicating that they are deposits of a closed freshwater basin. Siltstones, marls, and sandstones of Middle Eocene age, with a total thickness of up to 100 m, were deposited at the Middle Eocene transgression to the northern slopes of the Ukrainian Shield. The Neogene and Quaternary sands and clays are on top of the sediment filling of the Boltysh structure

3
Geology and Petrography of the Boltysh Crater Ejecta

The Boltysh impact structure is surrounded by patches of ejecta that are found, partly preserved, over an area of around 6500 km^2 surrounding the crater (Fig. 2). The present day ejecta outcrops represent the remnants of an initially extensive layer that covered the central part of the Ukrainian Shield after the impact. The state of preservation of ejecta seems to depend on the initial thickness of the breccias and surface morphology in the area at the time of the impact. The predominant occurrence of the breccias in the paleovalleys and depressions of the crystalline basement was distinguished by I.M. Etingof, V.N. Ryabchun, and others (unpublished data) and Bryansky et al. (1978). The ejecta are a complex system of individual patches and fields of the lithic breccias generally covering areas from a few square kilometers up to 100 km^2.

The ejecta were extensively eroded in an annular zone nearest to the crater edge, which is about 6-10 km wide and corresponds to the basement of the extensively eroded original crater rim. Relics of the rim are exposed in the valley of the Tyasmin river at the NW edge of the Boltysh crater. Numerous patches and fields of ejecta appear on the outer slopes of the original rim at distances of about 18-20 km from the center of the structure. An incomplete ejecta layer occurs to a distance of about 38-40 km to the N, NE and E of the crater center, 44-50 km to the SE, S and SW, and to 60-66 km (5.0-5.5 crater radii) to the W and WNW from the center of the crater. Two of the largest patches of ejecta occur in the Ingul river basin to the S of the crater and in the Tyasmin river basin to the WNW of the crater center. The most extensive erosion of the ejecta took place to the N and NE from the crater, on the north-eastern slope of the Ukrainian Shield, where only small patches of breccia are preserved (Fig. 2).

The thickness of breccias is up to 8 – 11 m in surface exposures and reaches up to tens of meters in some drillholes at distances of two to three crater radii from the center. The thickness variations of the preserved ejecta with the radial distance from the crater center was calculated for concentric annular zones around the crater, each 0.5 crater radii wide (Table 1). The weighted mean ejecta thickness was determined as the total thickness of breccias in all of the drillholes within each zone relative to the number of holes that penetrate the ejecta in this zone. Drillholes without breccias were not considered, because preservation of ejecta depends from the paleorelief of the region, and breccias would not have been preserved within uplifted areas (Gurov and Khmelnitsky 1996).

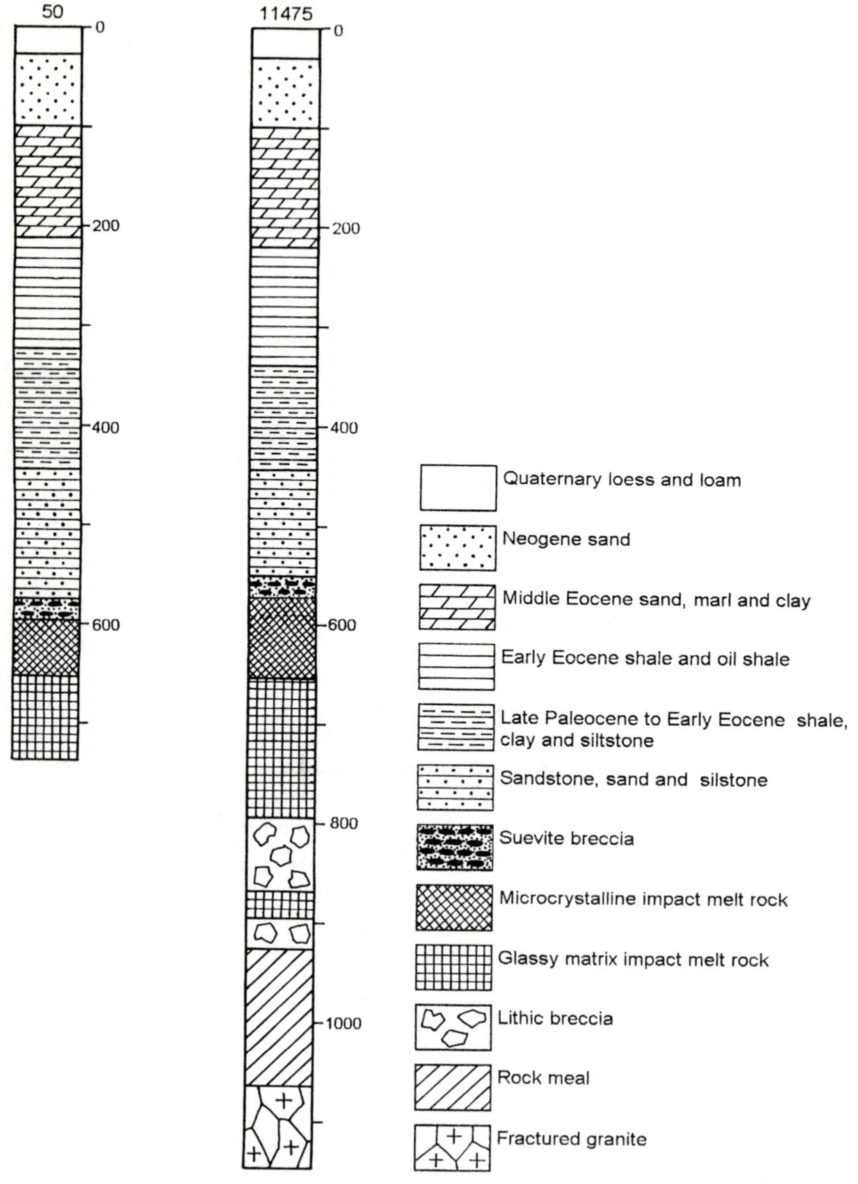

Fig. 5. Stratigraphic columns of drill cores 50 (drilled to 736 m) and 11475 (drilled to 1148 m). Core 11475 was drilled to the SW of the central uplift in the deepest part of the inner crater; whereas core 50 was drilled near its SW edge.

Table 1. Calculated initial ejecta thickness (in meters) of the Boltysh crater (I) and weighed mean thickness of preserved Boltysh ejecta (II) in relationship to their distances from the crater center (crater radii).

	Distance from the crater center									
	1.0-1.5	1.5-2.0	2.0-2.5	2.5-3.0	3.0-3.5	3.5-4.0	4.0-4.5	4.5-5.0	5.0-5.5	5.5-6.0
I*	307	112	52	28	17	11	7	5	4	3
II	26	25	15	15	8	14	9	2	<1	<1

*Calculations of ejecta thickness were made for the distance from the crater center to middle of each zone, using equations of McGetchin et al. (1973) and Stöffler et al. (1975).

Fig. 6. Outcrop of monomict breccia in the Tyasmin river valley. Granite clasts are cemented by fine-grained matrix.

Initial ejecta thicknesses at different distances from the crater center were calculated using equations of McGetchin et al. (1973) and Stöffler et al. (1975). The calculations were made for radial distances from the crater center to the center of each zone, for which the mean thickness of the preserved ejecta was calculated. The initial thickness of ejecta of an impact structure 24 km in diameter was ~600 m at the crater rim, dropping to about 10 m at a distance of 47 km (3.9 crater radii) and to ~1 m or less at a distance of about 90 km (7.5 crater radii) from the center of the structure. The projected initial area covered by the Boltysh ejecta to a depth of greater than 1 m is ~25,000 km^2. This method of using ejecta layer thickness to derive crater parameters has been confirmed by estimating the crater diameter of the Acraman structure (South Australia) from the thickness of its ejecta preserved in the Adelaide geosyncline at a distance of 300 km from the crater center (Gurov 1993; Gurov and Khmelnitsky 1996). Estimates of the crater diameter ranged from 85 to 150 km (Williams 1986; Williams et al. 1996), but Gurov (1993) derived a value of 30-40 km, which is very close to the value of 30 km quoted by Shoemaker and Shoemaker (1996). It can also be noted that Boltysh is of the same size as the much younger Ries impact crater in southern Germany, for which ejecta in the form of moldavite tektites have been found at distances of up to about 400 km from the crater. A comparison between ejecta distribution of these craters requires taking into account a variety of characteristics, such as target rock types and stratigraphy, impact angle and velocity, and preservation state.

Fig. 7. Outcrop of polymict breccia in the basin of the Tyasmin river. The largest angular clasts, up to 30 cm in diameter, are unshocked and weakly shocked granites.

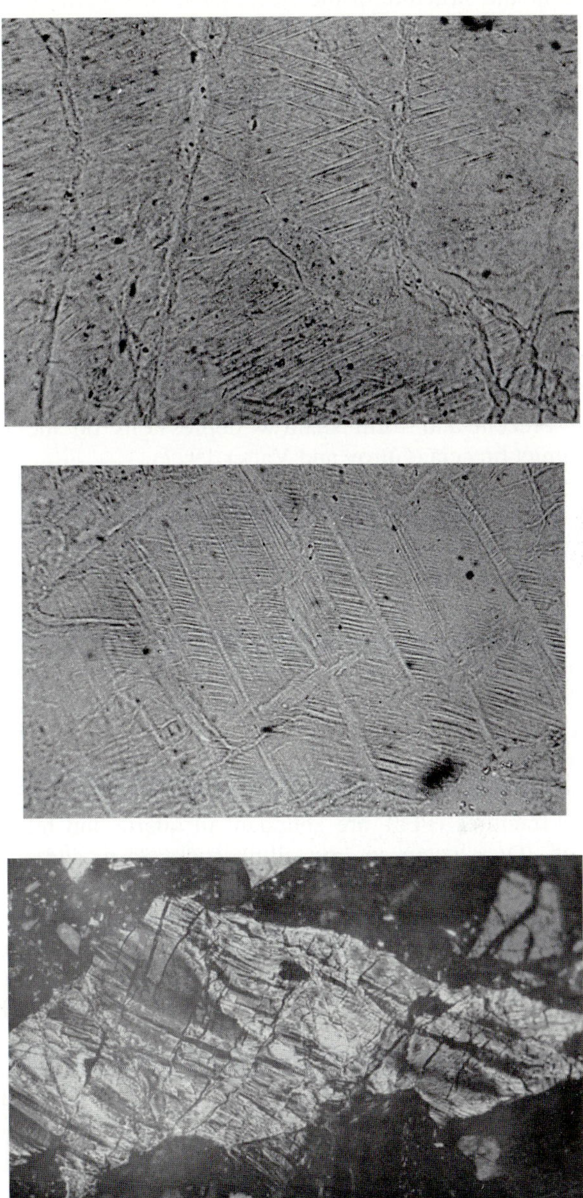

Fig. 8. Microphotographs showing the shock metamorphic effects observed in Boltysh ejecta. (a) Shocked quartz with PDFs in granite clast from lithic breccia (sample B-507a, 0.60 mm wide, parallel polars). (b) Shocked potassium feldspar with PDFs in granite clast from lithic breccia (sample B-507a, 0.55 mm wide, parallel polars). (c) Kink-banding and fracturing in biotite from matrix of breccia (sample B-507, 1.06 mm wide, crossed polars).

The composition and structure of the ejecta have been investigated by study of their outcrops in the Tyasmin river basin. Two main types of breccias occur in this area: a monomict breccia that is overlain by a polymict breccia. The first type, the monomict breccia, is composed of clasts and lumps of crystalline rocks consolidated by fine-grained material of the same rocks (Fig. 6). Brecciation and cataclasis are abundant in these rocks, but shatter cones and microscopic shock metamorphic effects have not been observed. It is suggested that monomict breccias may have formed during the passage of a shock wave through the rocks and during (secondary) impacts of large blocks of ejected material and that they represent autochthonous material.

The second type of ejecta is a polymict breccia composed of rock and mineral clasts set in a fine-grained matrix (Fig. 7). The contact of the polymict breccia with the underlying monomict breccia is exposed in some of the outcrops in the Tyasmin river basin. Granite blocks up to 10 m in size occur in the basal horizon of the polymict breccia and form shallow depressions on the surface of the underlying monomict breccia (Gurov and Valter 1977).

Clasts of unshocked and weakly shocked granites and gneisses predominate in the polymict breccia, whereas strongly shocked clasts of crystalline rocks are subordinate. The ratio of the relative contents of granite and gneiss clasts in breccias is 5:1, similar to the distribution of basement rocks in the area. The rare, extensively weathered glass clasts and particles have been altered to the friable clay-rich masses that partly preserve the initial fluidal structures of the glass. Rare clasts of white chalk, marl, and glauconitic sandstone occur in some breccia outcrops located about 22 km to the NW of the crater center. The matrix of the breccias is weakly consolidated material composed of fine-grained rock flour and some larger mineral clasts.

Shock metamorphic effects, in the form of planar deformation features (PDFs) and rarely planar fractures (PFs), are abundant in quartz and feldspars from the polymict breccias. They occur in crystalline rock clasts and in larger single mineral grains in the matrix (Fig. 8a and b). Up to three intersecting PDF orientations occur in quartz, with the $\{10\bar{1}3\}$ orientation dominating. The presence of the high-pressure quartz polymorph coesite in the breccia was confirmed by X-ray diffractometry of separated mineral fractions extracted from the matrix of the polymict breccia (for details of the method see Gurov et al. 1980). Shocked biotite contains kink bands and rare planar deformation features (Fig. 8c).

Fall-back ejecta or suevites (glass-bearing polymict breccia) of the Boltysh impact structure form a patchy layer on the surface of the impact-melt sheet within the crater. The formation of these rocks from material that originated from within the crater, was ejected, and fell back is demonstrated by the occurrence of aerodynamically shaped glass bodies. Suevites predominantly occur in the central part of the crater, but their distribution in the rest of the structure is not well studied. The thickness of the suevite over the central uplift varies from ~1 m in its eastern part (hole 18) to 28.5 m in the south in drill hole 20. The thickness of the suevites is 22 m in drillhole 50 and 12 m in hole 11475 in the SW part of the crater. Complex contacts and interlayering of suevites and massive impact-melt rocks is observed in drill hole 11475, where two suevite layers, 7 m thick (interval

566-573 m) and 5 m thick (interval 576-581 m), are separated by impact melt 3 m in thickness. The maximum thickness of suevites is 97 m in hole 17, 3.2 km to the NE from the crater center.

The fall-back material most likely formed "islands" of suevite breccia on the surface of the impact melt. The density of the massive impact-melt rocks of the crater is ~3.50 g cm^{-3} , and the density of fall-back suevites is ~3.35 g cm^{-3}. Some fractions of the suevitic material were partly digested in the impact melt, the top part of which is enriched in clasts. Also, the lower part of the suevite layer is partly melted and converted into mainly isotropic or very fine-grained material, in which the original suevite texture, with numerous glass clasts, is still weakly visible on the polished surface of cores and in thin sections (for example, hole 50, depth 593-595 m). Partial melting of the suevite matrix seems to have taken place at distances of 1-3 m from the contact with impact melt.

The fall-back suevites are composed of glassy clasts, clasts of the crystalline rocks and minerals, and a fine-grained matrix. The glassy clasts and particles are the main components of suevites, varying from 4-5 to 40-50%. A high content of glass (>50 vol%) occurs in suevite from cores 17 and 50. Glass particles often form irregular flattened shapes, whereas aerodynamically-shaped glass bodies are rare in the suevites. The color of the suevite glasses is generally gray to dark gray, although rare rose- and purple-colored glasses occur in the upper section of the suevite layer. The glasses preserve fluidal structures, although they are now extensively devitrified and composed of plagioclase, potassium feldspar, and quartz. Some of the rose-colored devitrified glasses contain hematite, as confirmed by X-ray diffractometry. The rock clasts in suevites are generally granites and, rarely, gneisses. No clasts of sedimentary rocks have been observed in fall-back suevites. The clasts range in size from millimeters up to 3 m. Shatter cones occur in some of the largest granitic clasts (core 20 – 471 m, 504 m) and PDFs are abundant in quartz and feldspar. The predominant orientations of PDFs in quartz from suevites are $\{10\bar{1}3\}$, $\{10\bar{1}2\}$, and $\{10\bar{1}4\}$.

The fall-back suevites are overlain by post-impact sediments. The basal horizons of sedimentary rocks in the central part of the crater are mainly coarse-grained and immature sandstones with thin interlayers of breccias and siltstones. The suevites closest to the sediments (interval from 573 to 575 m of hole 50) are extensively weathered and converted into weakly consolidated masses that still preserve their initial texture.

4

Stratigraphic Position of the Ejecta and the Age of the Boltysh Impact Structure

The Boltysh crater and its ejecta are located in the axial part of the Ukrainian Shield, an area that was not subjected to transgressions from the Late Paleozoic through the Mesozoic (Bondarchuk 1960). Therefore, the ejecta from the Boltysh impact were deposited directly on the surface of the Precambrian crystalline rocks in the whole area of their distribution. However, an ocean covered the north-eastern slopes of the shield, including part of the Tyasmin river basin, during a short period of the Cenomanian-Turonian transgression (Bondarchuk 1960). Cenomanian and Turonian sediments were deposited in this area and have been preserved within the Rotmistrovka impact structure. The post-impact sediments of the Rotmistrovka crater are overlain by a 18-m-thick breccia layer identified as ejecta of the Boltysh structure (Gurov and Gurova, 1991; Gurov and Babina, 2000).

The Rotmistrovka impact crater (centered at 49° 08' N, 31° 44' E) is located about 45 km to the NW of the Boltysh impact structure; it was formed in Rapakivi

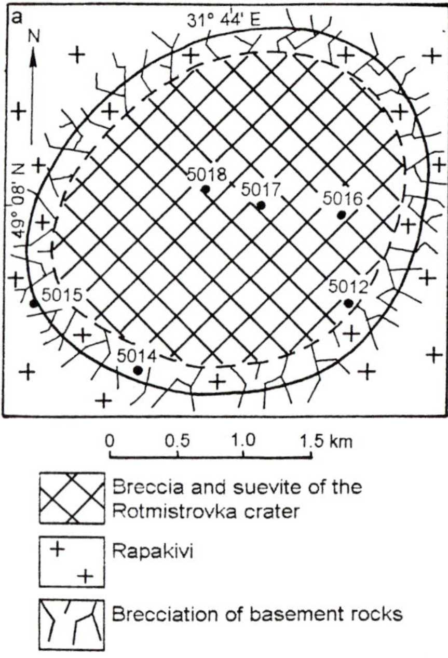

Fig. 9. (a) Location of some drill cores (discussed in the text) in the Rotmistrovka impact structure. Boundary of breccia and suevite occurrence is indicated by the hatched line. Post-impact sediments and more recent deposits are omitted.

granites of the Korsun-Novomirgorodsky Massif of Proterozoic age (Fig. 2). The Rotmistrovka structure is a simple crater it has a present-day diameter of about 3 km and its depth to the crater basement is about 400 m (Fig. 9a and b). The impactites include lithic breccias and suevites with a thickness of up to 70 m in the central part of the crater (Masaitis et al. 1980; Valter and Ryabenko 1977).

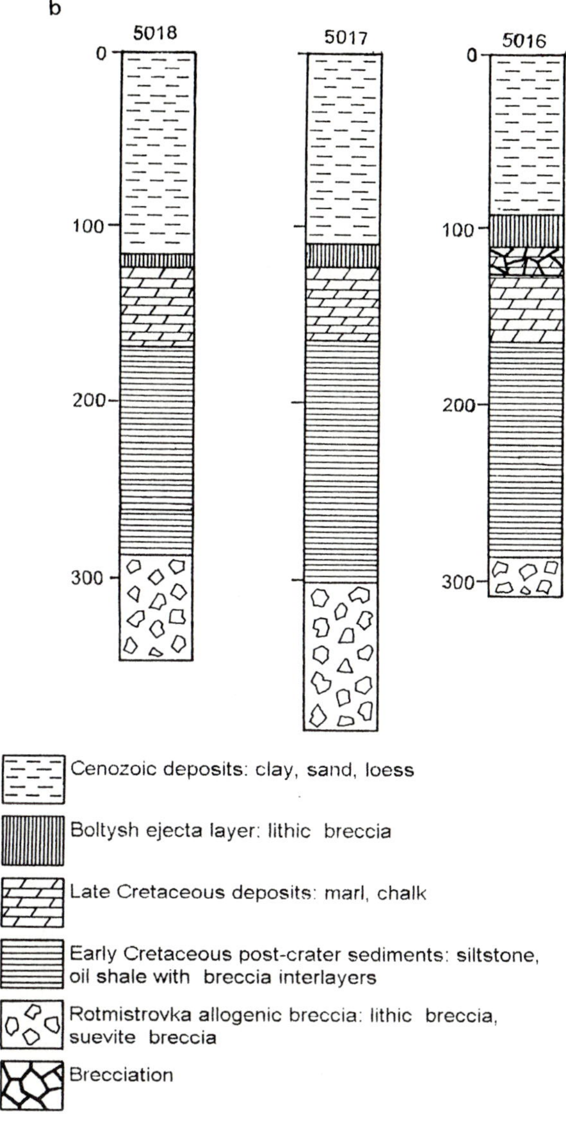

Cenozoic deposits: clay, sand, loess

Boltysh ejecta layer: lithic breccia

Late Cretaceous deposits: marl, chalk

Early Cretaceous post-crater sediments: siltstone, oil shale with breccia interlayers

Rotmistrovka allogenic breccia: lithic breccia, suevite breccia

Brecciation

Fig. 9. (b) Stratigraphic columns of drill cores 5016 (drilled to 309 m), 5017 (drilled to 387 m) and 5018 (drilled to 346.5 m). Cores 5017 and 5018 were drilled near the center of the crater, while core 5016 was drilled in its eastern part. The stratigraphic columns show overlapping of post-impact sediments of the Rotmistrovka crater by the Boltysh ejecta. Brecciation of Cenomanian - Turonian sediments under their contact with ejecta occurs in core 5016.

The post-impact sediments of the Rotmistrovka impact structure are a 120-m-thick series of sandstones, siltstones, and oil shales with rare interlayers of sedimentary breccia (Fig. 9b). The paleoflora in these strata are of Aptian-Albian age (Stanislavsky 1968). The upper series of the post-impact sediments are marl and chalk with Cenomanian and Turonian fauna (Stratigraphy of the Ukrainian SSR 8 1971). Although it has been suggested that the Rotmistrovka and Boltysh impact structures formed simultaneously (Valter and Ryabenko 1977), the stratigraphic age of the Rotmistrovka crater lies between Late Jurassic – Early Cretaceous (Masaitis et al. 1980) and Early Cretaceous, ca. 140 Ma (Grieve et al. 1995). A 15- to 18-m-thick layer of Boltysh ejecta occurs on top of the Cenomanian and Turonian deposits within the Rotmistrovka crater (Fig. 9b), further emphasizing the age difference. Brecciation of the upper horizon of marl and chalk was observed in drillhole 5016 in the interval from 110 to 125 m. Clasts in the Boltysh breccia within the Rotmistrovka crater are shock-metamorphosed granites and gneisses; in some cases mixing with the underlying sediments onto which the breccia fell is indicated.

Clasts of marl and chalk, likely of Late Cretaceous age, occur in the Boltysh ejecta in some breccia outcrops in the Tyasmin river basin outside the Rotmistrovka crater (Gurov and Valter 1977; Valter and Ryabenko 1977). Thus, the lower age limit for the Boltysh structure and its ejecta is Cenomanian-Turonian.

The upper age limit for the Boltysh ejecta is determined by the age of overlying Tertiary sediments. Although the upper contact of the breccia layer with the overlying deposits is not exposed in the outcrops, it has been traced in numerous drillholes by Bryansky et al. (1978) and also by I.M. Etingof, V.K. Ryabchun, N.F. Poddubnii, and others (unpublished data). The extensive erosion of the impact breccia and deposition of the Middle Eocene basal conglomerate on its surface has been established in numerous drill holes. The overlying near-shore sediments are sands and marls with Middle Eocene fauna (Stratigraphy of the Ukrainian SSR 9 1963). Thus, the age of the Boltysh fall-out ejecta is between the Cenomanian-Turonian and Middle Eocene periods.

A better constraint on the upper stratigraphic age of the Boltysh crater formation comes from studies of the well-preserved post-impact sediments in the structure (Fig. 5). The oldest series, which overlie the impact-melt sheet and fall-back suevites, are siltstones and sandstones with thin intercalations of sedimentary breccia with a total thickness of ~120 m. No paleoflora seems to have been preserved in this series.

Those sandstones and siltstones are overlain by an around 300-m-thick series of shales, argillites, and oil shales with abundant remnants of ostracods, gastropods, fishes, and a variety of flora. Investigations of the macrofossils were undertaken by Stanislavsky (1968) using samples from core 1715, who determined a probable Thanetian (i.e., Late Paleocene) age for the lowermost series. The flora in the upper part of the series from core 1715 is coeval with the flora of the Ypresian and Thanetian stages of the Parisian Basin. Thus, the age range of the series is from the Paleocene to the Early Eocene. It is important to note again the different ages of the oil shales in the Boltysh and the Rotmistrovka depressions: they are Aptian-

Albian in the Rotmistrovka crater and Paleocene-Eocene in the Boltysh structure (Stanislavsky 1968). A 40-m-thick series of brown siltstones of Middle Eocene age (Stratigraphy of the Ukrainian SSR 9 1963) overlies the shales and oil shales in the Boltysh crater. Thus, the stratigraphic age of the Boltysh impact crater and its ejecta can be constrained to lie between the Cenomanian-Turonian and the Late Paleocene.

Until recently, radiometric dating of Boltysh impact-melt rocks was limited to determinations by the fission track and K-Ar methods. The earliest investigations of two samples of glassy impact-melt rocks by the fission track method gave ages of 96 ± 10 and 105 ± 13 Ma (Komarov and Raikchlin 1976). However, more recent dating of one core sample of impact-melt rock by the same method yielded an age of 65.0 ± 1.1 Ma (Kashkarov et al. 1998).

Various K-Ar age determinations of the impact melt of the Boltysh crater have yielded a wide range of apparent ages from ca. 56 Ma (Shcherbak et al. 1978) to 88 Ma and up to 173 Ma (Bass et al. 1967). The most recent K-Ar date for the Boltysh crater was 88 ± 3 Ma (Boiko et al. 1985), the age most often quoted for the Boltysh impact event (e.g., Grieve 1995). In fact, the Boltysh and Steen River craters have been cited as the only two known impact structures formed at or near the Cenomanian/Turonian boundary (Rampino and Haggerty 1996).

Recent ^{40}Ar/^{39}Ar dating analyses (Kelley and Gurov 2002) of Boltysh impact-melt rocks, using a laser step-heating technique, help to constrain the crater age. Although a detailed description of this work is beyond the scope of the present paper, we will summarize the results. Seven core samples of impact-melt rocks from the drill-hole 50 were selected for investigations (Fig. 5c). The impact-melt sheet occurs in the interval from 593.5 m depth to the bottom of the hole at 736 m. Two sub-types of the impact-melt rocks, namely those with microcrystalline matrix and those with glassy matrix, are found in the depth-intervals from 595 to 657 m and 657 to 736 m, respectively. The light brown glass of the fresh glassy matrix melt rocks is completely isotropic, while dark brown glass of partly devitrified melt rocks displays microspherulitic textures (Grieve et al. 1987). Three samples of the microcrystalline melt rocks, and four samples of the glassy matrix melt rocks were analyzed.

Three core samples of the microcrystalline impact-melt rocks yield plateau ages, although one contained some excess argon. Of the four sampled melt rocks with a glassy matrix, only one sample yielded a plateau age. Several samples were repeatedly analyzed to check the results. A weighted mean age of three samples yielding plateaus (excluding the analysis that may contain excess argon) is 65.17 ± 0.64 Ma (the error is the 95% confidence level, using the square root of the MSWD). Note that this age was calculated assuming an age of 98.8 Ma for the international biotite standard GA1550 (Renne et al. 1998). This age is in good accordance with the fission track age of 65.04 ± 1.10 Ma of one glassy impact-melt rock sample measured by Kashkarov et al. (1998). The stratigraphic position of the Boltysh ejecta, as discussed also by Valter and Plotnikova (2003), is also in good concordance with the radiometric dating of the impact-melt rocks. This age for the Boltysh impact melt also coincides, within experimental errors, with the age of the Chicxulub structure and the age of the Cretaceous-Tertiary (K/T) boundary.

5
Discussion and Conclusions

The ejecta of the Boltysh impact crater form a thick layer of allochthonous breccias and fallback suevites within the structure. Within the crater the suevites are well preserved under post-impact sediments, but the ejecta outside the crater rim have been extensively eroded; however, remnants of the ejecta still occur within an area of about 6500 km^2 within the central part of the Ukrainian Shield. The extent of the erosion is evident from drill-core studies. The Boltysh crater is of about the same diameter as the Ries crater in southern Germany, for which only few patches of ejecta are preserved at distances of more than one crater diameter from the crater rim (e.g., Pohl et al. 1977, Hörz and Banholzer 1980).

Estimates of the initial volume and thickness of the Boltysh ejecta were made using the relationships between thickness of the fall-out rocks and their distance from the crater center (McGetchin et al. 1973; Stöffler et al. 1975). The area in which breccia is presently preserved probably corresponds to the area originally covered by a fallout blanket with an average thickness of up to 10 m. The area covered by an ejecta blanket 1 m thick may have been more than 25,000 km^2 (for details see Gurov and Khmelnitsky 1996).

The age of the Boltysh impact can be constrained from the stratigraphy of its extensive ejecta layer. Boltysh impact ejecta overlie Cenomanian-Turonian post-impact sediments within the nearby Rotmistrovka impact crater. This may be the only known terrestrial occurrence where the (proximal) ejecta from one crater occur within another crater. The upper age limit of the Boltysh ejecta is determined by the presence of Middle Eocene deposits covering the (eroded) breccia layer. A more precise upper age for Boltysh is obtained from post-impact sediments within the crater. Whereas the lowermost series of those deposits has no determinable fossils, a Paleocene age was determined for the basal layers of the crater-fill shales and oil shales (Stanislavsky 1968). Thus, stratigraphy constrains the age of Boltysh and its ejecta to the period between the Cenomanian-Turonian and the Paleocene.

The most commonly quoted age for the Boltysh age is 88 ± 3 Ma, determined by K-Ar dating of clast-free impact-melt rock (Boiko et al. 1985). However, this age is problematic when interpreting the crater history. It is difficult to see how such a large crater (24 km diameter) could have remained free of any fossil evidence for about 28 Ma, from the Turonian to the Paleocene. Indeed, recent ^{40}Ar/^{39}Ar and fission track ages of glassy material from Boltysh impact melt rocks are in better agreement with the stratigraphic position of the ejecta, implying only ~5 Ma of post-impact sedimentation in the crater prior to the preservation of the first fossils. Further detailed paleontological, especially paleofloral, investigations of the Boltysh post-impact sediments may yield more information (see also Valter and Plotnikova 2003).

The age of the Boltysh crater from $^{40}Ar/^{39}Ar$ plateau ages of three specimens of impact-melt rocks is 65.17 ± 0.64 Ma (95% confidence level). This age for the Boltysh impact coincides with the ^{40}Ar-^{39}Ar ages of glasses from the Chicxulub crater with plateau ages of 65.36 ± 0.11 and 65.42 ± 0.08 Ma (recalculated to an age of 98.9 Ma for the international biotite standard GA1550) (Montanari and Koeberl 2000; Swisher et al. 1992), and with ages of the K/T boundary (e.g., Pillmore and Miggins 2000). These data indicate a simultaneous or almost simultaneous formation age for the Boltysh and Chicxulub impact structures. Although it might be suggested that the Boltysh impact event provided some contribution to the catastrophic events at K/T boundary, the energy released during the Boltysh crater formation was about 10^6 Mt TNT equivalent (Gurov et al. 1999), which is less than 1% of the energy released during the Chicxulub impact event. However, further investigations of the Boltysh impact-melt rocks, ejecta and post-impact sediments may yield important information clarifying their relation to the K/T boundary. Further study of the Boltysh impact crater and its ejecta may also help with a better understanding of the Chicxulub impact event and the catastrophic events at the end of the Mesozoic.

Acknowledgments

Part of this work was supported by the Austrian Science Foundation, project Y58-GEO (to CK). We are grateful to the reviewers, Drs. H. Dypvik and M. Rampino, for their constructive comments that helped to improve this manuscript.

References

Bass YB, Ghalaka AI, Grabovsky VI (1967) The Boltysh oil shales. Razvedka i ochrana nedr 9: 11-15 (in Russian)

Boiko AK, Valter AA, Vishnyak AF (1985) About the age of the Boltysh depression. Geologichesky Zhurnal 45 (5): 86-90 (in Russian)

Bondarchuk VG (ed) (1960) Atlas of the paleogeographic maps of the Ukrainian and Moldavian SSR (1960) Academia Nauk Ukrainian SSR Press, Kiev, Ukraine, 78 maps (in Ukrainian)

Bryansky VP, Zlobenko VG, Ryabchun VK (1978) Breccia rocks of the Paleogene from the Boltysh basin area. Geologichesky Zhurnal 38 (2): 135-138 (in Russian)

Golubev VA, Karpov GM, Popovichenko VA (1974) About the meteorite-impact origin of the Boltysh depression in Kirovograd region. Doklady Academii Nauk UkrSSR series B, No 1, pp 10-13 (in Russian)

Grieve RAF, Reni G, Gurov EP, Ryabenko VA (1987) The melt rocks of the Boltysh impact crater, Ukraine, USSR. Contributions to Mineralogy and Petrology 96: 56-62

Grieve RAF, Rupert J, Smith J, Therriault A (1995) The record of terrestrial impact cratering. GSA Today 5: 193-196

Gurov EP (1993) The Acraman impact crater structure: Estimation of the crater diameter by the ejecta thickness [abs]. Lunar and Planetary Science 24: 589-590

Gurov EP (1996) The Boltysh impact crater: lake basin with heated bottom [abs]. In: Drobne K, Gorican C, Kotnik B (eds) The role of impact process in the geological and biological evolution of Planet Earth. Ljubljana, Slovenia, p 30

Gurov EP, Babina NV (2000) Formation of the Boltysh impact structure: Catastrophe of regional scale [abs]. In: Catastrophic Events and Mass Extinctions: Impacts and Beyond, Lunar and Planetary Institute Contribution No. 1053, Lunar and Planetary Institute, Houston, p 62

Gurov EP, Gurova EP (1985) Boltysh astrobleme: impact crater pattern with a central uplift [abs]. Lunar and Planetary Science 16: 310-311

Gurov EP, Gurova EP (1991) Geological Structure and Composition of Rocks in Impact Structures. Naukova Dumka Press, Kiev, Ukraine, 160 pp (in Russian)

Gurov EP, Khmelnitsky AF (1996) Distribution and preservation of the ejecta from the impact structures by example of the Boltysh and Acraman craters. Astronomichesky Vestnik 30 (1): 19-24 (in Russian)

Gurov EP, Valter AA (1977) Ejecta of the Boltysh impact crater in the Ukrainian Shield. Geologichesky Zhurnal 37 (6): 76-81 (in Russian)

Gurov EP, Val'ter AA, Rakitskaya RB (1980) Coesite in rocks of the meteorite explosion craters of Ukrainian Shield. International Geological Review 22: 329-332

Gurov EP, Gursky DS, Gurova EP, Rakitskaya RB, Yamnichenko AY (1999) The Boltysh impact crater: the main parameters and estimations of consequences of impact. Mineralny resursi Ukrainy 4: 20-25 (in Russian)

Hörz F, Banholzer GS (1980) Deep seated materials in the continuous deposits of the Ries crater, Germany. In: Papike JJ, Merrill RB (eds) Proceedings of Conference on Lunar Highlands Crust, Pergamon Press, New York, pp 211-231

Kashkarov LL, Nazarov MA, Kalinina GV, Lorenz KA, Kononkova NN (1998) Fission track dating of the Boltysh impact crater, Ukraine [abs]. Lunar and Planetary Science 29: abstract # 1257 (CD-ROM)

Kelley SP, Gurov EP (2002) The Boltysh, another end-Cretaceous impact. Meteoritics and Planetary Science 37: 1031-1044

Komarov AN, Raykhlin AI (1976) Comparative study of impactite age by fission track and K-Ar methods. Doklady Academii Nauk USSR 228: 673-676 (in Russian)

Masaitis VL (1973) The geological consequences of the crater-forming meteorite falls. Nedra, Leningrad, 17 pp (in Russian)

Masaitis VL (1974) Some ancient meteorite craters in the territory of USSR. Meteoritica 33: 64-68 (in Russian)

Masaitis VL, Danilin AN, Mashchak MS, Raykhlin AI, Selivanovskaya TV, Shadenkov YM (1980) The geology of astroblemes. Nedra, Leningrad, 231 pp (in Russian)

McGetchin TR, Settle M, Head JW (1973) Radial thickness variation in impact crater ejecta: implications for lunar basin deposits. Earth and Planetary Science Letters 20: 226-236

Montanari A, Koeberl C (2000) Impact Stratigraphy: the Italian record. Lecture Notes in Earth Sciences 93, Springer, Heidelberg-Berlin, 364 pp

Moroz SA, Sovyak-Krukovsky YV (1993) The Luzanov stratoregion of the Paleogene of Europe. Paleontologitchesky Zbirnik 29: 65-72 (in Ukrainian)

Pillmore CL, Miggins DP (2000) A new $^{40}Ar/^{39}Ar$ age determination of the K/T boundary interval: possible constraints on the timing of the K/T event and sedimentation rates of the K/T sequence [abs.]. In: Catastrophic Events and Mass Extinctions: Impacts and Beyond. Lunar and Planetary Institute Contribution No. 1053, Lunar and Planetary Institute, Houston, p 166

Pohl J, Stöffler D, Gall H, Ernstson K (1977) The Ries impact crater. In: Roddy DJ, Pepin RO, Merrill RB (eds) Impact and Explosion Cratering. Pergamon Press, New York, pp 343-404

Rampino MR, Haggerty BM (1996) Impact crises and mass extinctions: A working hypothesis. In: Ryder G, Fastovsky D, Gartner S (eds) The Cretaceous-Tertiary Event and Other Catastrophes in Earth History. Geological Society of America, Special Paper 307, pp 11-30

Renne PR, Swisher CC, Deino AL, Kamer DB, Owens TL, DePaulo DJ (1998) Intercalibration of standards, absolute ages and uncertainties in $^{40}Ar/^{39}Ar$ dating. Chemical Geology 145: 117-152

Shcherbak NP (ed) (1983) Geological map of the crystalline basement of the Ukrainian Shield. Scale 1: 500,000. Kiev: Ministry of Geology of UkrSSR Press

Shcherbak NP, Zlobenko VG, Zhukov GV, Kotljvskaya FI, Polevaya NI, Komlev LV, Kovalenko NK, Nosok GM, Pochtarenko VI (1978) Catalogue of isotopic data of the Ukrainian Shield. Naukova Dumka, Kiev, Ukraine, 224 pp

Shoemaker EM, Shoemaker CS (1996) The Proterozoic impact record of Australia. AGSO Journal of Australian Geology and Geophysics 16: 379-398

Stanislavsky FA (1968) Age and stratigraphy of the sapropelites of the Boltysh depression. Geologichny Zhurnal 28 (2): 105-110 (in Russian)

Stöffler D, Gault DE, Wedekind J, Polkovski G (1975) Experimental hypervelocity impact into quartz sand: Distribution and shock metamorphism of ejecta. Journal of Geophysical Research 80: 4062-4077

Stratigraphy of the Ukrainian SSR 9. Paleogene (1963) Syabryay VT (ed), Academia Nauk Ukrainian SSR Press, Kiev, Ukraine, 319 pp (in Ukrainian)

Stratigraphy of the Ukrainian SSR 8. (1971) Cretaceous Period. Kaptarenko-Tchernousova OK (ed), Academia Nauk Ukrainian SSR Press, Kiev, Ukraine, 320 pp (in Ukrainian)

Swisher CC, Grajales-Nishimura JM, Montanari A, Margolis SV, Claeys P, Alvarez W, Renne P, Cedillo-Pardo E, Maurrasse FJMR, Curtis GH, Smit J, McWilliams MO (1992) Coeval $^{40}Ar/^{39}Ar$ ages of 65.0 million years ago from Chixulub crater melt rock and Cretaceous-Tertiary boundary tektites. Science 257: 954-958

Valter AA, Plotnikova L (2003) Biostratigraphic indications of the age of the Boltysh impact crater, Ukraine. This volume.

Valter AA, Ryabenko VA (1977) The impact craters of the Ukrainian Shield. Naukova Dumka, Kiev, Ukraine, 156 pp (in Russian)

Valter AA, Dobryansky YP, Lazarenko EE, Tarasyuk VK (1982) Shock metamorphism of quartz in basement rocks and estimation of the form modification of the Boltysh and Ilyinets astroblemes of the Ukrainian Shield. Doklady Academii Nauk USSR 264: 132-136 (in Russian)

Williams GE (1986) The Acraman impact structure: Source of ejecta in late Precambrian shales. Science 233: 200-203

Williams GE, Schmidt PV, Boyd DM (1996) Magnetic signature and morphology of the Acraman impact structure, South Australia. AGSO Journal of Australian Geology and Geophysics 16: 431-442

Yurk YY, Yeremenko GK, Polkanov YA (1975) The Boltysh depression – a buried meteorite crater. International Geology Review 18: 196-202

Zaritsky AI (ed) (1992) Structural-facies map of crystalline basement of the western part of the East-European Platform. Scale 1:1,000,000. Kiev: Goskomgeologiya Ukrainy Press

Stratigraphy and Sedimentology of Coarse Impactoclastic Breccia Units within the Cretaceous-Tertiary Boundary Section, Albion Island, Belize

David T. King, Jr.[1] and Lucille W. Petruny[2]

[1]Department of Geology, Auburn University, Auburn, Alabama 36849-5305, USA. (kingdat@auburn.edu)
[2]Astra-Terra Research, Auburn, Alabama 36831-3323, USA and Department of Curriculum and Teaching, Auburn University, Auburn, Alabama 36849, USA. (lpetruny@att.net)

Abstract. At Albion Island in northern Belize, Cretaceous-Tertiary boundary deposits, also known as the Albion formation, rest upon karsted and fractured Maastrichtian dolostones. These deposits consist of a basal impactoclastic clay layer (~ 1 to 2-m thick) and an upper carbonate-rich, coarse impactoclastic breccia layer (up to 15-m thick). The focus of this paper is the upper layer, the Albion impactoclastic breccia.

The Albion impactoclastic breccia shows several important sedimentary structures, including development of discrete sedimentation units (2 to 7-m thick), which are strata that have been enhanced by horizontal shearing, and other sedimentary structures such as normal and reverse size grading, clast imbrication, flow lamination, and isolated and linked aggregates of clasts (i.e., clast clustering). Most carbonate clasts within the coarse impactoclastic unit show a broad range of angularities and shapes, with the most common being subangular and compact-bladed to compact-elongated, respectively. Surface texture analysis of carbonate clasts shows several types of surface markings, which display a gross sequential order (i.e., facets, polish, striations, cryptographic markings, bruises and pits, and chips).

In-situ, apparent-diameter measurements of the carbonate clasts, which ranged in size from 10 to 300-mm (or -3.3 to -8.2 ϕ), resulted in cumulative grain-size (ϕ) frequency curves with similar shapes through the interval -3.3 ϕ and -6.25 ϕ (i.e., 10 to 76-mm). Matrix, the total area comprised of less-than-10 mm (< -3.3 ϕ) particles, ranged from approximately 71 to 82 percent. Modified moment measures of these curves show these breccias are "extremely poorly sorted." The matrix content increases upward through the entire coarse impactoclastic layer, but is slightly lower near its top.

The Albion impactoclastic breccia has sedimentary structures and sedimentologic characteristics suggesting its mode of emplacement during the impact aftermath was similar to that of a very large volcanic debris avalanche. Sedimentation units show evidence of early turbulent flow and a more conspicuous later stage of laminar flow with shearing accompanying emplacement of most breccia sedimentation units. Clasts within these debris flows are not locally derived for the most part. We speculate that each sedimentation unit at Albion may represent a separate emplacement event during the process of ejecta-curtain collapse, perhaps owing to variations in atmospheric interaction with the debris.

1
Introduction

In northeastern Belize near the Belizean border with México and in closely adjacent areas of the Mexican state of Quintana Roo, the discontinuous Albion formation (Ocampo et al. 1996), lies between Maastrichtian dolostones and Paleocene limestones (King 1996a). The Albion formation represents direct ejecta deposits from Chicxulub impact crater (Ocampo et al. 1996), which is centered nearly 360 km away in the Mexican state of Yucatán (Sharpton et al. 1996; Fig. 1). The Albion formation consists of a basal impactoclastic[1] clay layer and an overlying coarse impactoclastic deposit of carbonate breccia (also known as the informal Albion spheroid bed and Albion diamictite, respectively; Ocampo et al. 1996; Pope et al. 1999). In the study area in northeastern Belize, the Albion formation ranges from 8 to 15-meters thick (Fig. 2). In most places, the Albion formation is exposed to weathering at ground level and a tropical soil zone is developed at its top.

Less than 50 km away to the southeast, in central Belize, an un-named lateral equivalent of the Albion formation is several meters thick. This un-named unit consists of a basal impactoclastic clay layer (resting upon Maastrichtian bedrock) and an overlying coarse impactoclastic unit (a carbonate-pebble conglomerate informally called the Pook's pebble bed by Ocampo et al. 2000). At this locale, the basal impactoclastic clay layer and overlying conglomerate are in contact without intervening breccias as at Albion Island (Ocampo et al. 2002).

In order to account for their stratigraphic and geographic occurrence, both Pope et al. (1999) and Ocampo et al. (2000) have suggested different mechanisms of emplacement for the Cretaceous-Tertiary ejecta facies of Belize and adjacent México. Specifically, they interpret the basal impactoclastic clay layer as having

---

[1] The term *impactoclastic layer, deposit, or unit* is used here in the sense of Stöffler and Grieve (1994), who define this term as "consolidated or unconsolidated sediment resulting from ballistic excavation, transport, and deposition of rocks at impact craters; may contain particles of impact melt rock."

formed directly upon the karsted and fractured Maastrichtian bedrock surface by action of a very hot, rapidly expanding vapor cloud. Overlying coarse impactoclastic deposits are interpreted as secondary, ballistic sediments principally derived from a more slowly moving, volatile-rich vapor cloud associated with the collapsing ejecta curtain. Lastly, they interpret the rounded carbonate pebbles within the Pook's pebble bed as high-altitude (sub-orbital?) ejecta originating mainly from shallow target-rock material, which fell in Belize beyond the limit of the Albion impactoclastic breccia deposits (Pope and Ocampo 2000).

The focus of this paper is upon the Albion formation's coarse impactoclastic deposit. Specifically, this paper deals with the stratigraphy and sedimentology of this breccia unit and what our additional data can tell us about formative processes for this coarse ejecta. Various mechanisms have been proposed for development of the coarse impactoclastic unit (reviewed by Pope et al. 1999). Our goal is to add to the basic knowledge of this unit and its attributes. Our work was conducted at Albion Quarry (site AQ in Fig. 1), which is located on Albion Island (near San Antonio), in the Orange Walk District, Belize.

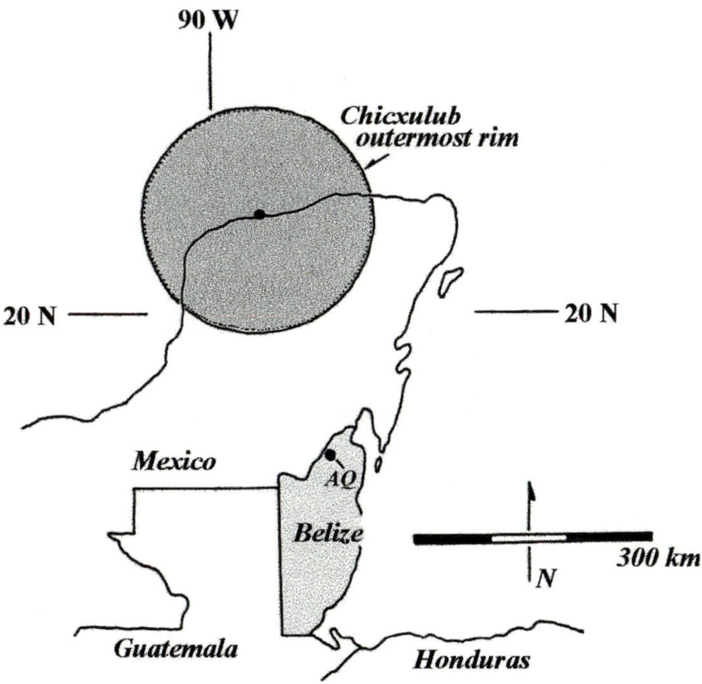

Fig. 1. Location map showing the Yucatán peninsula of México. Chicxulub outermost crater rim and the study site at Albion Quarry (*AQ*) are indicated.

2
Stratigraphy

Cretaceous-Tertiary boundary stratigraphy on Albion Island and vicinity consists of the Maastrichtian Barton Creek Formation and overlying Albion formation, which is capped by a thin, modern soil. A direct upper contact between Albion formation and overlying Paleocene rocks was not observed on Albion Island, but such a contact occurs 30 km farther south in the Cayo District of central Belize (King 1996a) and in adjacent Quintana Roo, México (Claeys 2001). At these adjacent sites, thickness of the Albion formation is not much different from what is noted at Albion Island. Therefore, we assume that we have an essentially complete section of the Albion formation within Albion Quarry.

2.1
Barton Creek Formation

The Maastrichtian Barton Creek Formation, a formal stratigraphic unit of northern Belize (Cornec 1985; 1986), is a thick sequence of tan to medium gray, highly dolomitized shallow-water carbonates. At Albion Island, the Barton Creek consists of cyclically arranged groupings of strata, approximately 5-m thick, each comprised of two main lithic types (or facies). In each cycle, the upper facies consists of 8 to 15-cm thick beds of planar cross-stratified, relatively coarsely crystalline dolomitic packstone and grainstone. Within coarsely crystalline dolostones near the top of some cycles, voids (vugs) several centimeters in diameter suggest the former presence of nodular evaporite minerals. The lower facies in each cycle consists of several 3 to 5-cm thick beds of cross-stratified, dolomitic wackestone and packstone. At Albion Island, a 1 to 2-m thick unit of vuggy, coarsely crystalline dolostone with thin lenses and layers of evaporite minerals comprises a distinctive altered zone in most places directly beneath the impactoclastic clay layer (Fig. 3).

The Barton Creek is largely devoid of substantially intact macrofossils, and the few that may be discerned are typically moldic voids within the dolostone. However, in some places these voids are sufficiently similar to macrofossil shape that it is possible to identify shapes of rare gastropods, pelecypods, and brachyuran crabs (Fouke et al. 2002). Abundant microfossils, especially foraminifers, were present in original sediment texture, but are now reduced to fine moldic voids or textural ghosts in the rock as well.

The age of the Barton Creek is established by two very rare, but key macrofossils: a late Maastrichtian carcineretid crab (*Carcineretes planetarius* Vega) and a nerineid gastropod with infolding wall structures (Vega et al. 1997; Pope et al. 1999). A late Maastrichtian age of upper Barton Creek is supported by $^{87}Sr/^{86}Sr$ ratios ranging from 0.70786 to 0.70796 (Ocampo et al. 1996).

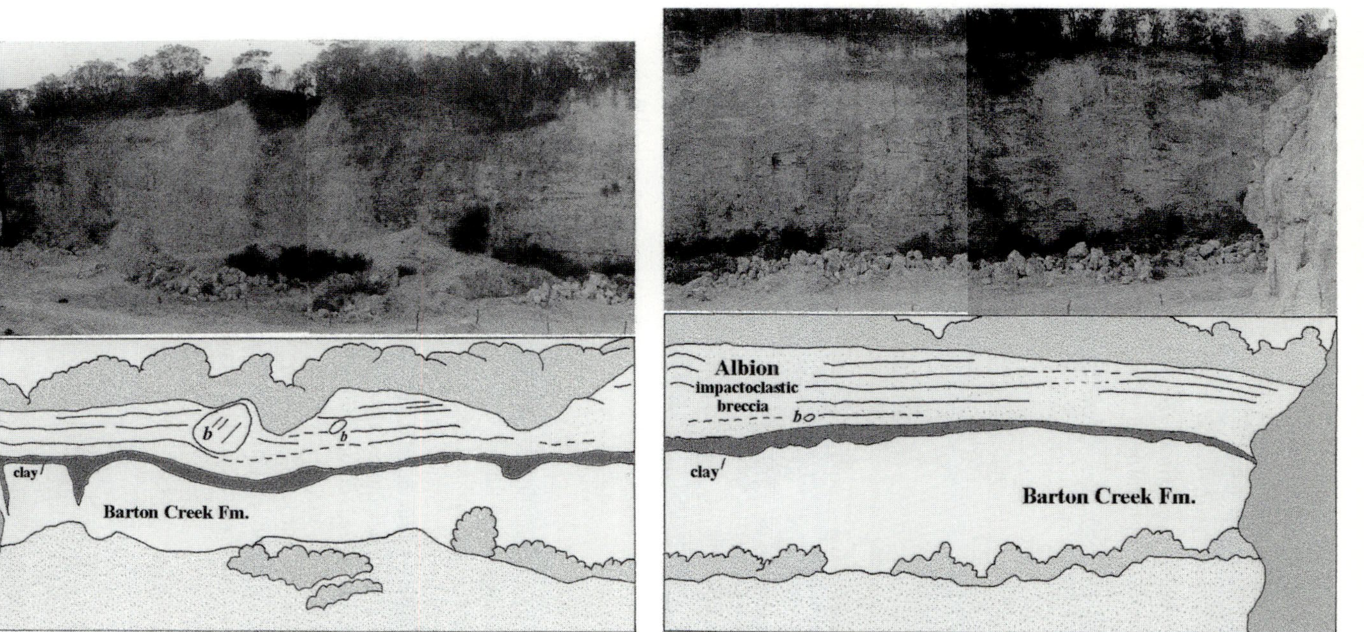

Fig. 2. Panoramic view of the western wall of Albion Quarry, Albion Island, Belize, in two parts. Breccia bedding contacts, which are accentuated by weathering, are marked in each sketch. Wall is approximately 25 m high. LEFT: Impactoclastic clay layer is 1 to 2.5-m thick; impactoclastic breccia, 8 to 13-m thick. On left side, impactoclastic clay fills two 6-m deep fissures. b = carbonate boulder and block ('boulders with bedding'). Block at left is 9-m diameter. RIGHT: All six layers may be seen near center of this half. b = carbonate boulder ('boulder with bedding').

The upper surfaces of outcrops of Barton Creek on Albion Island are highly irregular, including solution pits and pipes and deep fractures (Fig. 2). Pope et al. (1999) report patches of caliche composed of angular Barton Creek fragments with iron-oxide crusts cemented by calcite and dolomite lying upon the upper surface of the Barton Creek. Fouke et al. (2002) cites negative co-variation trends in $\delta^{13}C$ and $\delta^{18}O$ within uppermost Barton Creek as evidence of subaerial exposure during time of emplacement of overlying Albion impactoclastic clay layer.

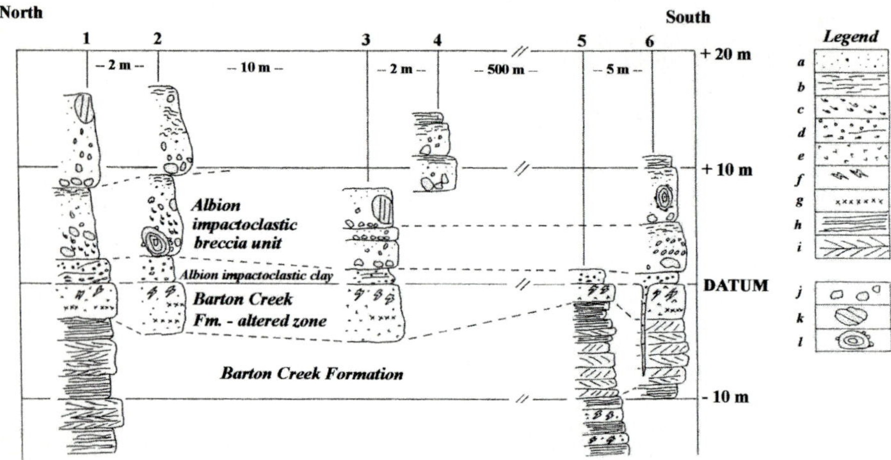

Fig. 3. Correlation among six measured sections within Albion Quarry. Top of all sections is the ground surface. Datum is the Barton Creek-Albion impactoclastic clay layer contact. Spacing between sections is indicated at top. Clasts are shown to relative scale; smallest clast shown (dot) is pebble size. Spheroid symbol shapes indicate degree of flattening. Legend: (*a*) impactoclastic breccia matrix; (*b*) flow laminations; (*c*) red and green clay clasts; (*d*) spheroid-bearing impactoclastic clay; (*e*) coarse dolostone; (*f*) vugs; (*g*) evaporite minerals; (*h*) planar cross-stratified, dolomitic packstone and grainstone; (*i*) cross-stratified, dolomitic wackestone and packstone; (*j*) carbonate clasts; (*k*) boulders with bedding; and (*l*) coated boulders. Sections: 1 = pit road, north end, west side; 2 = loading road, north end, upper part; 3 = pit road, north end, east side; 4 = loading road, north end, lower part; 5 = pit road, south end, east side; and 6 = pit road, south end; west side.

2.2
Albion Formation

The Albion formation[2] is an informal stratigraphic unit that was first described by Ocampo et al. (1996). They place the lower boundary at the basal disconformity,

---

[2] The use of lower case f in formation in this paper reflects the unit's continued informal status (see Salvador 1994 regarding formal versus informal units). Ocampo et al. (1996)

which marks the Cretaceous-Tertiary boundary horizon in this area. The Albion formation has been subdivided into two informal members at Albion Island: a basal impactoclastic clay layer and overlying coarse impactoclastic breccia unit.

2.2.1
Basal Impactoclastic Clay Layer

The basal impactoclastic clay layer at Albion rests directly upon Barton Creek dolostones (Fig. 3) and consists of dense, brown to rust-colored clay, which is locally rich in oblate, pebble-size dolomitic spheroids[3]. Near this layer's base, it is also rich in angular, centimeter-sized, red or green clay clasts. This clay layer ranges from 15-cm to 1.5-m thick, averaging about 90-cm. This layer tends to be thicker in topographically low areas of the irregular Barton Creek surface (Fig. 2). The layer's basal contact is sharp and is usually marked by a hematitic rind (< 1-mm thick) on the underlying Maastrichtian dolostone. The clay layer is highly deformed, as attested to by numerous internal glide planes marked by slickensides. Pope et al. (1999) recognized four individual beds within this clay layer, but we have not been able to consistently delineate any such primary layers.

Spheroids comprise about 5 to 25 percent by volume of this layer and are its most unusual constituents. Pope et al. (1999) supposed that the internal structure of such a spheroid may have consisted of core and crudely layered coating and thus was once much like an accretionary lapillus. However, Albion spheroids typically lack a clearly defined internal structure (a possible effect of diagenesis according to Fouke et al. 2002) and their origin remains enigmatic. Ocampo et al. (1996) conducted petrographic and isotopic analyses of several spheroids and concluded that they are likely high-temperature vapor-phase condensates. After further study, Fouke et al. (2002) concluded that the spheroids may be either (1) devitrified and dolomitized glass spherules, (2) accretionary lapilli originally composed of quicklime that rapidly hydrated after deposition (thus losing internal structure), or (3) dolomitized 'chrondrule-like objects.'

Close study of the angular, centimeter-sized red and green clay clasts suggest that they are vesicular glass fragments that have been altered since impact formation (Ocampo et al. 1996). Green clay clasts are smectites with Si:Al ratios higher than normal, but similar to other Cretaceous-Tertiary boundary glasses

---

describe the clay and breccia layers at Albion as "informal units" and mention the name Albion Formation (*sic*) for the first time in their paper. However, they do not say they intend the Albion formation as a formal unit and name no stratotype. Further, this unit's mappability has not been demonstrated. For these reasons, Albion formation should remain informal until established by additional publication. It is worthy of note that the term Albion formation has not been used by the Belize Geology and Petroleum Office of Petroleum in any publications.

[3] The term spheroid is used here in the same sense as Ocampo et al. (1996), Pope et al. (1999), and Fouke et al. (2002), namely that these are impact-generated, sub-spherical objects that lack good internal structure and clear evidence of exact mode of origin.

(Pope et al. 1999). Green clay clasts have compositional similarities with palagonites formed during alteration of glass spherules from the Cretaceous-Tertiary boundary in Haiti (Pope et al. 1999). Red clasts are thought to be similar to green, but perhaps more highly oxidized due to hydrothermal alteration or surface-weathering effects.

2.2.2
Coarse Impactoclastic Breccia Unit

The coarse impactoclastic breccia unit at Albion rests directly upon the impactoclastic clay layer (Figs. 2 and 3). This breccia consists of light tan, moderately indurated and matrix-rich, poorly sorted carbonate rock fragments. The matrix-rich nature of this breccia (i.e., framework grain-to-matrix ratio < 4:1) classifies it as a parabreccia (*sensu* Prothro and Schwab 1996). In a parabreccia, hypothetical "removal of the matrix collapses the framework" (Prothro and Schwab 1996). At Albion Quarry, the whole breccia unit is approximately 13-m thick and is capped at ground surface by an erosional surface.

At Albion Quarry, the coarse impactoclastic unit contains at least 6 individual sedimentation units[4] (Figs. 2 and 3). The thickness of these sedimentation units ranges from 2 to 7-meters vertically, but thickness ranges considerably over several hundred meters in a lateral direction. Individual sedimentation units appear to pinch and swell and cannot be correlated accurately unless each unit is directly traceable between measured sections (e.g., in Fig. 3, compare sections 1 and 2, 2-m apart, versus sections 2 and 3, 10-m apart; see also King 1996b). In at least one instance, a breccia unit pinches out on the outcrop (Fig. 4). Bases and tops of breccia units are relatively sharp, but do not show good evidence of significant erosion between units.

Sedimentation units within the coarse impactoclastic breccia were delineated initially by small slope breaks on cliff faces within Albion quarry (Figs. 2 and 4). These slope breaks, which are the result of weathering that has occurred since quarry excavation, appears to be quite sensitive to changes in bulk grain size (especially matrix content), clast sorting, and clast density.

Within individual sedimentation units, three styles of size grading were noted. Each is about equally common. Units that generally fine upward display crude normal grading (e.g., most units in Fig. 3). Some units, which have crude normal grading, also have large floating boulders near or at their tops (e.g., the upper units in sections 1 and 3, Fig. 3). Units that generally coarsen upward display crude reverse grading (Fig. 5). In some instances, beds that have size grading in one place do not show good evidence of size grading in another place at Albion Quarry. In other words, size grading may not be a laterally persistent characteristic of some units. In addition to size grading in the main size ranges, some units have

---

[4] Sedimentation unit is used here in the sense of Otto (1938), who defined this as "that thickness of sediment which was deposited under essentially constant physical conditions." The term is used rather than stratum or bed because it is more generic.

rare, carbonate boulders and blocks[5] (i.e., 'boulders with bedding,' 1 to 9-m in diameter) located near their bases or their tops (Figs. 2 and 4; see section 1 and 3, Fig. 3). There are at least six large boulders with bedding at Albion Quarry.

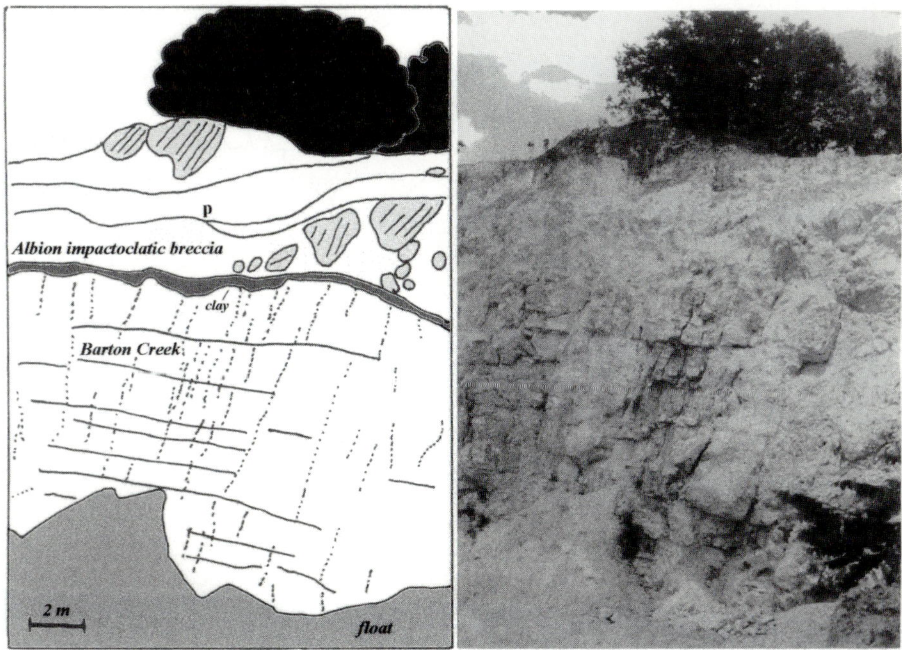

Fig. 4. Fractured Barton Creek dolostones overlain by the Albion impactoclastic clay and impactoclastic breccia units, north-end high wall, Albion Quarry. Sketch shows contact between four sedimentation units within the breccia. Large carbonate boulders ('boulders with bedding') are shown shaded. **p** = pinch out of breccia sedimentation unit. Scale (2 m) is indicated.

---

[5] Size terms are from Blair and McPherson (1999), who present a useful classification for coarse sedimentary particles larger than -8 ϕ (i.e., boulders, blocks, slabs, monoliths, and megaliths). *Boulders* are defined as a sedimentary particle between 0.25 and 4.1 m in diameter and *blocks*, between 4.1 to 65.5 m in diameter. In their scheme, large particles at Albion are classified as "coarse boulders to fine blocks."

Fig. 5. Sketch from photograph of a rectangular area approximately 2.1 m x 1.4 m, showing outline of all clasts larger than 5-cm diameter. Long axis of figure is parallel to bedding and area is located entirely within one sedimentation unit. Gross clast size distribution shows crude reverse grading (general coarsening upward) and various other features pertaining to clast aggregates. Aggregate types: *1* = clasts with jigsaw cracks; *2* = linked aggregates of clasts; *3* = isolated aggregates of clasts. See text for discussion and reference to these terms. Hammer is 31.5 cm long. Pit road, north end, near section 3.

Other sedimentary structures of note in the breccias include clast imbrication, flow lamination, and isolated and linked aggregates of clasts (i.e., clast clustering). Platy or bladed boulders, which are located at the base of some units, display imbrication. This imbrication occurs with clast support and, less commonly, as isolated imbrication (i.e., the clasts are not supporting one another but are aligned nearly parallel to one another). Flow laminations, which are located within the upper 1 to 2-meters of many sedimentation units, are also rather common features. Flow lamination is displayed by parting lineation within matrix and/or by preferential alignment of platy or bladed clasts (Fig. 6). Isolated and linked aggregates of clasts (term of Laznicka 1988), also known as clast clustering (term of Glicken 1996), is also rather common within Albion's sedimentation units. In these aggregates or clusters, groups of adjacent clasts are apparently slightly disintegrated from a larger, original rock mass (Fig. 5). Examples of non-disintegrated (fractured) rock masses (i.e., 'clasts with jigsaw cracks' *sensu* Ui 1983; Fig. 5) occur along with aggregates and clusters, thus allowing us to make this connection.

Some units contain rare matrix-coated boulders (up to 3-m in diameter), i.e., boulders surrounded by concentric layers of impactoclastic matrix, in some places armored by small clasts, which evidently formed by rolling (Fig. 7). Locations of some of these boulders within different units are shown diagrammatically in Figure 3.

Impactoclastic breccia matrix is mainly highly comminuted dolomite (in the silt-clay size range) that is mixed with minor amounts of clay. Matrix commonly makes up approximately 60 to 85 percent of the rock, and matrix is more abundant near the tops of most sedimentation units. Matrix contains some red and green clay clasts, similar to those found within the basal clay layer. These clay clasts are thought to be altered impact-glass fragments like the clay clasts found within the basal impactoclastic clay layer (Ocampo et al. 1996). Red and green clay clasts are much more common within the matrix of the lowermost 2 to 3 meters of the impactoclastic breccia layer.

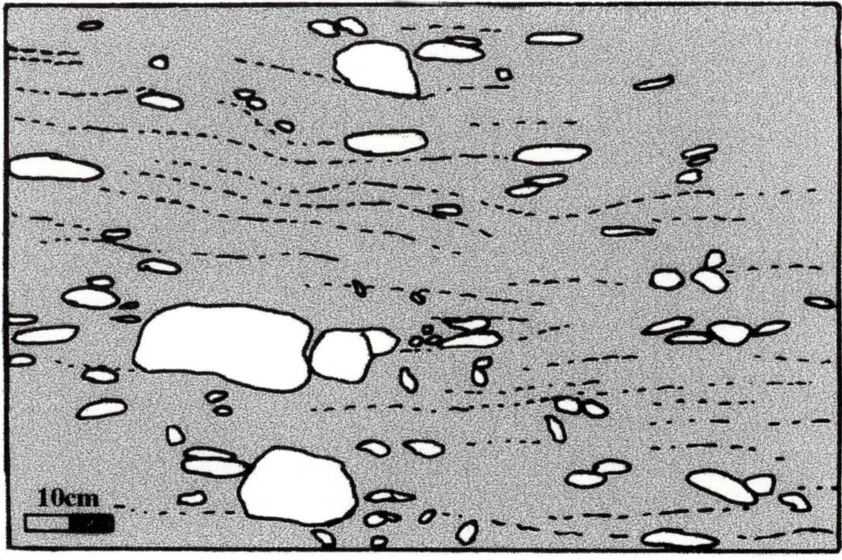

Fig. 6. Sketch from photograph of a rectangular area, approximately 90 cm x 60 cm, showing outline of all clasts larger than 2.5-cm diameter. Long axis of figure is parallel to stratification and sketch area is located entirely within one sedimentation unit. Top of rectangle is approximately at the upper boundary of this sedimentation unit. Clast distribution shows preferential alignment of platy and bladed clasts. Dashed lines indicate weathering along parting lineation in the matrix. Scale (10 cm) is indicated. West side, north end, Albion Quarry.

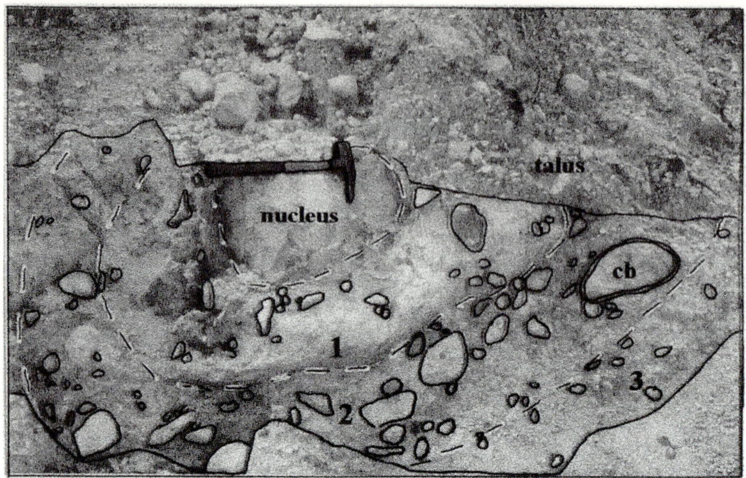

Fig. 7. Partially exhumed coated boulder, section 6, Albion Quarry. Hammer (31.5 cm long) rests on nucleus (tan dolostone boulder). There are three thick matrix coating layers (numbered 1 – 3), which are delineated by dashed lines. On the right side, layer 1 shows some evidence of armoring (i.e., small clasts adhering to its surface (base of layer 2)). Layer 2 contains a smaller coated boulder (**cb**) within it. Talus covers upper half of coated boulder, above solid line.

3
Sedimentology

3.1
Shape Analysis

The coarse impactoclastic breccia contains abundant carbonate clasts, which show a broad range of angularities and shapes. The most common degree of angularity is subangular (Fig. 8) and the most common shape is compact-bladed to compact-elongated (angularity and shape terms from Folk 1968; Fig. 9). Extremes within clastic angularity are highly abraded and rounded (i.e., subspherical) clasts and, on the other end of the scale, highly angular bladed and platy slabs of rock. Angular and subangular clasts tend to retain basic bladed and platy shapes (Fig. 9), which likely were derived from comminution of layering within fragments of target rock (Petruny and King 2000). Subrounded and rounded clasts, which are more highly chipped, bruised, and pitted (as defined in the next section) comprise a subpopulation of compact shapes (Fig. 9). This indicates a high level of mechanical erosion upon more equidimensional target rock fragments (Petruny and King 2000).

3.2
Surface Texture Analysis

Many carbonate clasts within the coarse impactoclastic breccia unit have several types of distinctive surface markings that are related to their genesis and transport history. A clast with such markings is shown in Figure 10. These surface markings include: facets; polish; striations; cryptographic markings; bruises and pits; and chips. Facets are primary, polygonal bounding surfaces of the clastic blocks, which are up to several hundred square centimeters in area. Polish is composed of compound, parallel ultrafine-size striations (several microns across) that together comprise a smooth and shiny surface. Striations are parallel grooves of a fine to coarse nature (0.01 to 0.2-mm across) that give the clast's surface a distinctive "ruled" appearance. Cryptographic markings (sensu King et al. 1997) are branching, dendritic channel-like networks of scratches. These scratches are composed of many segments, joined at angles, which are in turn composed of closely spaced parallel scratches (0.1-mm across) or individual relatively wide scratches (0.2 to 0.5-mm across). Bruises and pits are indentations in a clast, probably made by grain-to-grain impacts. Pits are damaged areas that are generally circular (or concentric) brittle impact marks with sharp, fractured rims. Whereas bruises are light-colored, shallow, rimless, irregularly shaped areas of impact damage on the surface of a clast. Pits are less than 0.5-cm in diameter, whereas bruises are larger than 0.5-cm in diameter. Chips are the result of brittle deformation of at the edges or margins of clasts. Some chips display minor conchoidal fracture. Most chips are less than 3-cm^2 in area. Rampino et al. (1996) noted some of these same surface features and concluded a general similarity with surface markings on other proximal ejecta deposits of carbonate-target craters.

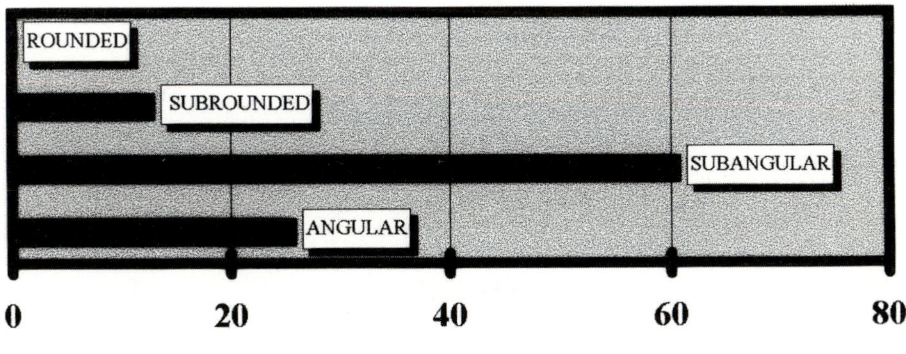

Fig. 8. Graph showing distribution of standard classes of grain angularity among Albion clasts in the size range from 10 to 300-mm (-3.32 to -8.23 φ). Data from all four locations studied for grain-size analysis at Albion Quarry was pooled for this graph. N = 1,574 clasts.

The surface markings listed above are arranged in an approximate order of occurrence (Fig. 11) with facets having formed first, followed typically by polish and striations. Cryptographic markings, if present, are superimposed on polish and facets. Bruises and pits follow cryptographic markings (or they follow polish and striation, if cryptographic markings are absent). Lastly, the chips occurred. Not all clasts show this complete order of markings, but study of numerous clastic surfaces has established this sequence.

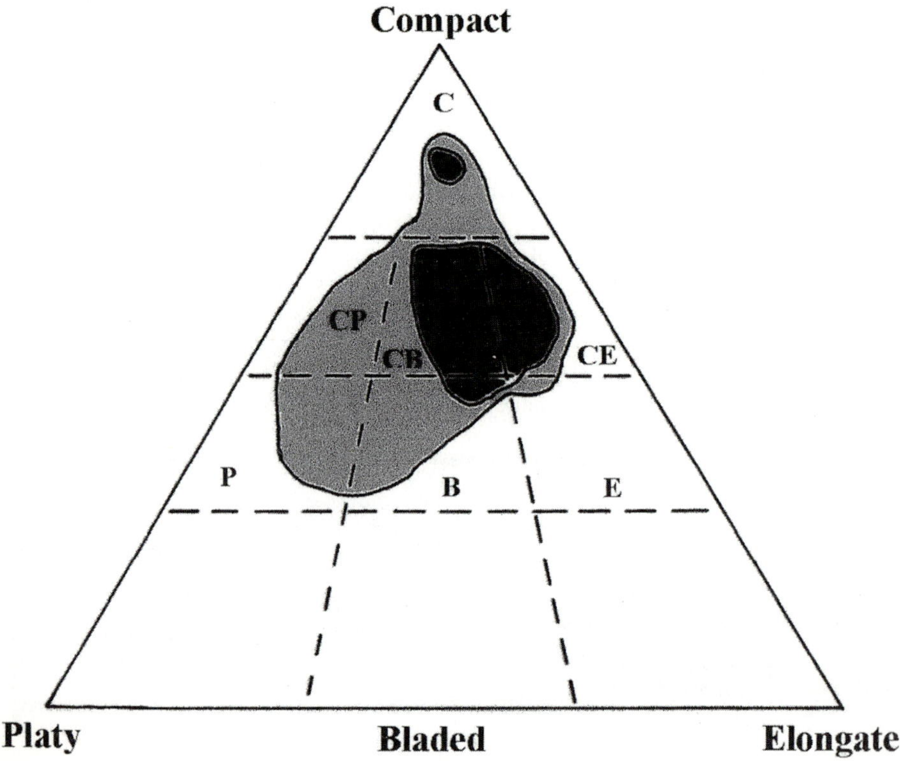

Fig. 9. Ternary diagram showing results of *in-situ* (field-based) shape analysis of clasts from impactoclastic breccia layer. Shading: white = no data points; medium gray = 30 % of data points; black = 70 % of data points. Ternary diagram poles (i.e., end-member shapes) were defined by Folk (1968). Internal subdivisions of the diagram: C = compact; CP = compact platy; CB = compact bladed; CE = compact elongated; P = platy; B = bladed; and E = elongated. Data is from various sites within Albion Quarry. N = 93.

Fig. 10. Representative clast surface (facet) from impactoclastic breccia at Albion Quarry, which shows several of the typical surficial markings. Markings include: *P* = polish (first); *S* = striations; *cm* = cryptographic markings; *b/p* = bruises and pits; and *C* = chips (last). In this instance, cryptographic markings and chips are most evident on the left side, and striations and bruises and pits are more evident in the right. In this instance, polish is more evident in the lower half of the clast surface. Scale (upper left) is 3-cm long. Dark material at lower right is iron-oxide coating.

Surficial features of clasts	**Suggested origin of features**
Chips *Bruises and pits* *Cryptographic markings*	*Debris-flow interactions* *and* *compaction effects*
Striations *Polish* *Facets*	*Excavation effects* *and* *vapor-cloud interactions*

Fig. 11. Matrix showing order of surface markings (facets = oldest; chips = youngest) and related stages in their origin (as explained in the text).

3.3
Grain-size Frequency Analysis

Using meter-square grids, we made *in-situ*, apparent-diameter measurements (in 5 mm-increments) of carbonate clasts in the coarse impactoclastic unit. We measured all carbonate clasts, which ranged in size from 10 to 300-mm (or -3.3 to -8.2 ϕ). These apparent-diameter measurements were then converted to equivalent-area statistics in order to calculate volume (area) percents within each size class. The following is a synopsis of each step involved in this process.

1. Using a string grid, we delineated a one-meter square area on the scraped surfaces of selected exposures of the coarse impactoclastic breccia. We divided this one-meter square area into 10 vertical strips, each 10-cm wide.

2. Within each successive vertical strip, beginning at the top and moving downward, we individually measured *in-situ* diameter of all particles (i.e., their long-axis aspect presented in outcrop) to the nearest 0.5-cm. All measurements were thus assigned to a bin comprising a 0.5-cm interval, e.g., 0.5 to 1.0-cm, 1.0 to 1.5-cm, 1.5 to 2.0-cm, etc. Particle size was consistently *rounded up* to the nearest 0.5-cm increment, e.g., a particle 1.7-cm in diameter was placed in the 2.0-cm bin. All particles under 0.5-cm were considered as part of the finer matrix, and thus not measured separately.

3. We used 'percent of area' to determine grain-size frequency distribution by first converting all particle measurements to cross-sectional areas of spheres with diameters equal to the long-axis aspect measured in step 2. Then, individual counts within each 0.5-cm bin were multiplied by cross-sectional area to determine the total amount of area assigned to each grain-size bin.

4. We computed 'cumulative percent finer than' by adding the total cross-sectional area in each bin, beginning with the largest particle in each grid's grain-size distribution.

5. We converted grain sizes originally measured in centimeters to ϕ values for making standard cumulative grain-size (ϕ) frequency curves. The ϕ value is the negative $\log_2$ of the grain size in millimeters (Folk 1968). The ϕ value of the larger end of each grain-size bin is used for this purpose (e.g., for the 1.5 to 2-cm size bin, 2-cm (= -4.32 ϕ) was used).

Cumulative grain-size (ϕ) frequency curves display distinctive, similar shapes in the range from -3.32 to -6.25 ϕ (Fig. 12). However, in the coarse tails of these size distributions (i.e., among grain sizes coarser than -6.25 ϕ), there is considerable variation owing to the relatively small number of grains from each grid area found within the larger size bins. We should also note that very rare boulders with bedding and matrix-coated boulders, within the size range of -8.32 to -12.30 ϕ, also occur at Albion Quarry (Figs. 4 and 7). However, these larger boulders are not part of our present analyses because they could not be effectively treated using this method and did not occur in places where accurate direct measurement was feasible.

The curves show a range in matrix area (i.e., matrix content) from approximately 71 to 82 percent among the four grid locations analyzed (Fig. 13). The four grid locations (A through D, which are meter-square grids with bases at 1.5, 3, 8, and 11.5 meters above the base of the impactoclastic breccia), are arranged so that a vertical trends for the whole impactoclastic unit could be studied. Distribution of matrix percent suggest a fining upward within the whole impactoclastic breccia unit, except near the top where perhaps some winnowing may have occurred during much later soil development (Fig. 13). An overall fining upward trend is suggested by decrease upward in other grain-size parameters, such as number of small clasts (0.5 to 1.0 cm in diameter) and average size of five largest clasts (Fig. 13).

Because such a high percent of matrix is involved in our measurements, and because we could not disaggregate the matrix for purposes of compatible detailed size-frequency analysis, we are limited as to the type of statistical analysis that can be performed upon these grain-size data. Specifically, we are unable to compute a mean grain-size value (*sensu* Folk 1968), as defined in standard sedimentology texts, owing to the absence of graphic mean and other key graphic points on the curve within the fine tail. However, using a modified 'moment method' computation (modified from Folk 1968), which allows comparisons of standard deviations within grain-size populations without computing a mean (i.e., where mean is unknown), we were able to compute an approximation of standard deviation (σ_ϕ) for comparative purposes (see Folk 1968, p. 50).

Grain-size frequency curves from Locations A through D yield modified σ_ϕ values approximately equal to $\sim 4\ \phi \pm 0.1\ \phi$ indicating "extremely poorly sorted sediment" (term of Folk 1968). This sorting measure (combined with data showing widely variable angularities and grain shapes and observations of very rare extremely large boulders within breccia deposits at Albion) supports the idea of mixed populations of breccia clasts. A simple way to have such a distribution is for there to have been multiple sources for carbonate breccia clasts from within the region (Petruny and King 2000).

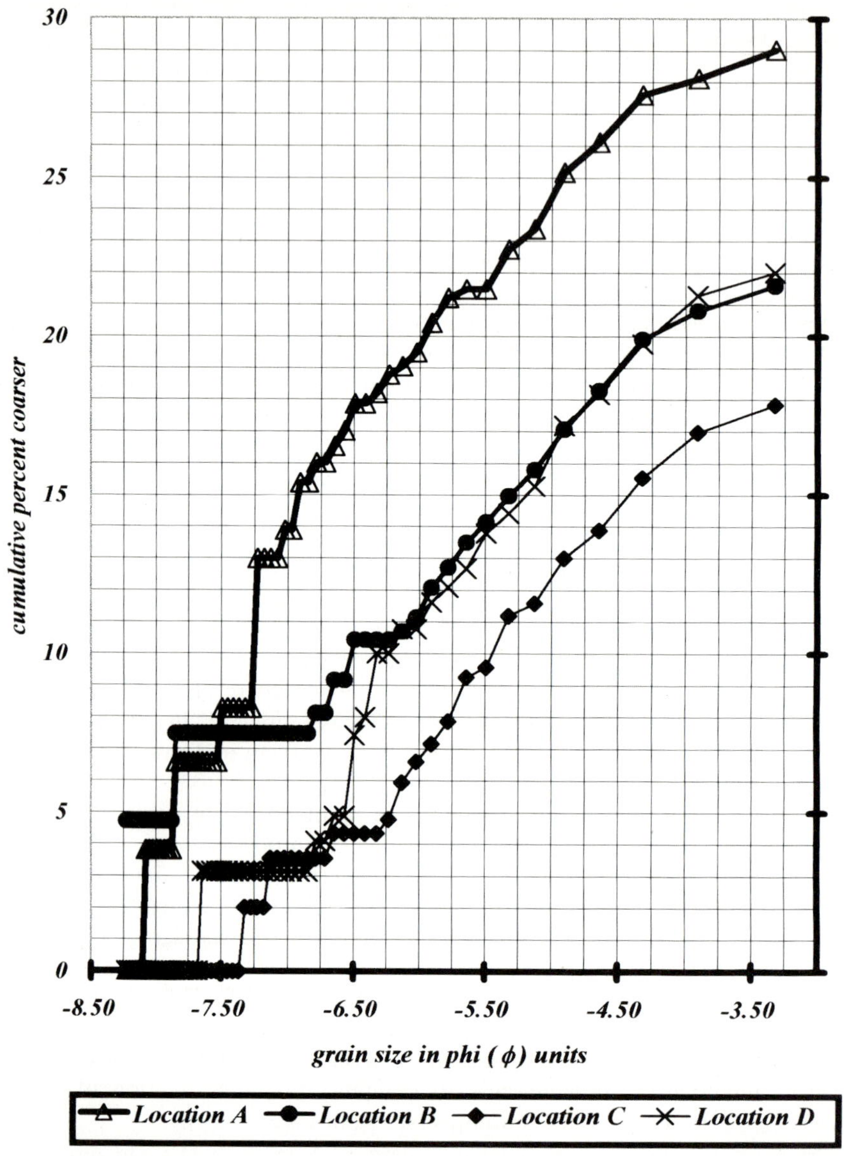

Fig. 12. Cumulative grain-size (φ) frequency curves for grid locations A, B, C, and D, at Albion Quarry. The approximate location above base of the impactoclastic breccia for each grid location is as follows: A (1.5-m); B (3.0-m); C (8.0-m); and D (11.5.0-m). The number of clasts measured in our analyses at each grid location is as follows: A (413); B (421); C (323); and D (319). Clasts smaller than -2.32 φ (5-mm) were not measured in the field and are considered as part of the matrix. Rare boulders larger than -8.23 φ (300-mm) were not part of our analyses.

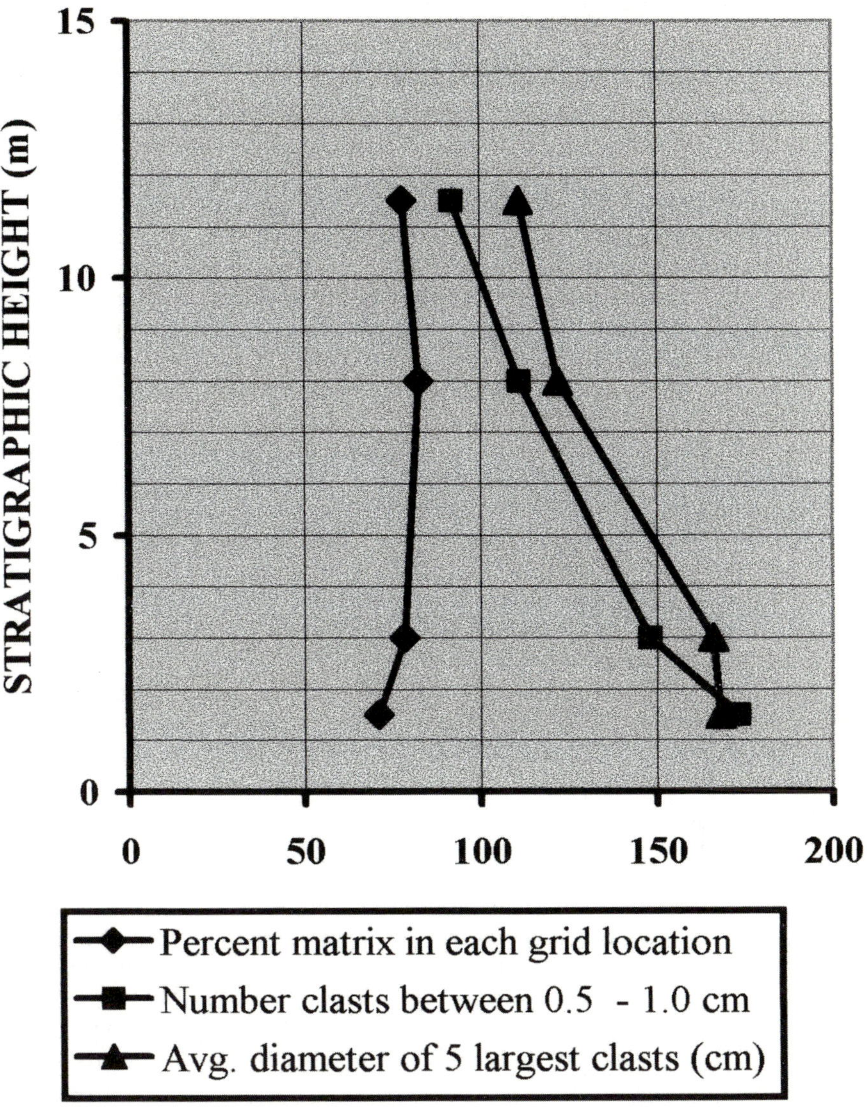

Fig. 13. Vertical trends in three grain-size parameters at grid locations A, B, C, and D (respectively, 1.5, 3, 8, and 11.5-m above base of impactoclastic breccia unit at Albion Quarry). Percent matrix is total surface area of grains under 0.5-cm diameter. Other grain-size parameters plotted are (1) number of clasts between 0.5 and 1.0 (i.e., number of grains in smallest size bin) and (2) average diameter of five largest grains. All parameters indicate fining upward in the breccia unit.

4
Discussion and Conclusions

We regard individual sedimentation units within the Albion coarse impactoclastic unit as debris-flow deposits, which show evidence of effects of initial turbulence and later laminar flow conditions (cf. Glicken 1996). Initial turbulence is demonstrated by the mixing of various clast types within sedimentation units and other features such as size-grading effects and presence of matrix-coated boulders. Later laminar flow conditions are indicated by flow laminations, flow-oriented clasts (within flow laminations), and the presence of isolated and linked aggregates of clasts (i.e., clast clustering) and intact clasts with jigsaw cracks. These latter features are taken as strong evidence of laminar flow in a highly viscous medium where turbulence was at a minimum (otherwise aggregates would be dispersed and clasts with jigsaw cracks would be disintegrated; Glicken 1996). Similar features suggesting initial turbulence and later laminar flow has been carefully documented for volcanic debris-avalanche deposits by Glicken (1996) who used similar field methods to uncover physical evidence like that found at Albion. Our conclusion contrasts sharply with the interpreted sedimentation conditions for the Bunte Breccia of Ries crater, Germany, according to Hörz et al. (1983), wherein the sedimentologic aspects of that unit support only "a highly turbulent depositional environment" for deposition.

Individual sedimentation units at Albion Quarry have peculiar characteristics that make their complete origin difficult to fully explain. Whereas each unit has its own set of depositional characteristics such as grain size and sorting and its own style of sedimentary structure such as clast aggregation and clustering, imbrication, and size grading, it is not absolutely clear the extent to which these features result from primary sedimentation processes alone. For example, upper and lower contacts between sedimentation units may also have been affected by shear between units during and after emplacement. Such shearing may account for grain-size differences between sedimentation units at their contacts (i.e., "shear sorting" described by Stone 1967) and for some aspects of grain orientation at contacts as well (i.e., imbrication at contacts due to shear, as noted by Glicken 1996).

We view each sedimentation unit as an indicative of a separate emplacement event, whether due entirely to sedimentation processes within debris flows or sedimentation plus subsequent shearing. Our view is supported by evidence from volcanic debris-avalanche (flow) deposits wherein post-depositional shearing is an effect of subsequent emplacement of overlying flow lobes (Glicken 1996). Sedimentation units at Albion Island may represent sequential stages in the collapse of Chicxulub's ejecta curtain and/or phases of interaction between that ejecta curtain and Earth's atmosphere (cf. Schultz 1992). Alternatively, individual sedimentation units may represent separate debris-flow lobes that followed slightly different paths or traveled at slightly different speeds before arriving at Albion Island (cf. Glicken 1996).

Some aspects of the Albion impactoclastic breccias may shed some light on the nature of debris-flow fluidization. Fluidization (*sensu* McSaveney 1978) may be gaseous or mechanical. Gaseous fluidization requires gas to effect necessary dilation and thus the loss of strength needed to permit flow (Wilson 1980). Mechanical fluidization is a state where the myriad of interparticle collisions causes dilation and loss of strength (McSaveney 1978). This is really a type of grain flow (*sensu* Bagnold 1954) in which interparticle collisions create dispersive stresses that act normal to the flow's movement. Gaseous fluidization can be ruled out as an important aspect of sedimentation in the present instance because of the extremely poor sorting of the material (Wilson 1980; Glicken 1996). Thus, by default, mechanical fluidization is supported as the main mechanism for maintaining flow (cf. Glicken 1996).

A peculiar kind of mechanical fluidization was at work in the present instance, however, as many intact clasts with jigsaw cracks are present and many closely spaced clast aggregates and clusters also occur. Thus, the mechanical fluidization operative in the Albion flows was one in which the grains were more often in contact than not (a phenomenon also characteristic of some mechanical acoustic fluidization situations; Melosh 1983). Savage (1984) refers to such flows where particles rarely lose contact as "quasi-static plastic flows."

Comparison of the underlying Barton Creek facies at Albion Island with lithologies of coarse impactoclastic breccia clasts shows significant differences. For example, none of the cross-stratified, coarsely crystalline dolomitic packstone and grainstone, the most common Barton Creek lithic type at Albion, is present in the overlying coarse impactoclastic unit. Thus, we think that the underlying Barton Creek at Albion Island was *not* the main source for clasts composing the impactoclastic breccia in this study (i.e., the source was probably excavated from within Chicxulub transient crater). Our interpretation is supported by two recent studies at Albion Quarry. Fouke et al. (1996) showed that cathodoluminescent signatures of carbonate cements in impactoclastic breccia clasts were substantially different from those of bedrock samples taken from the underlying Barton Creek. And, Fouke et al. (2002) showed that there were significant upward decreases in bulk-rock $^{87}Sr/^{86}Sr$ within the coarse impactoclastic unit (suggesting a deep excavation of clasts found in the impactoclastic breccia). This interpretation of ejecta provenance is very different from the German Ries crater, widely cited as a somewhat analogous large carbonate-target crater, wherein far more locally derived clastic material seems to be incorporated in the Bunte Breccia with increasing distance from crater center (Hörz 1983).

A possible exception to the distant-source notion for carbonate clasts may be found in the six Albion 'boulders with bedding' (i.e., coarse boulders to fine blocks, 1 to 9-m in diameter). These very coarse sedimentary clasts are enigmatic because their size nearly matches or exceeds the thickness of the sedimentation units with which they are associated (i.e., their transport by such flows is more difficult to explain than if they were smaller clasts). Further, we wonder why these 'boulders with bedding' are still intact whereas there are many disaggregated clasts within the impactoclastic breccia. Considering that volcanic debris flows (a good analogue to this situation) show a profound decrease in the volume of largest particle with increasing distance from source (Glicken 1996), we perhaps should look to a more local source for these huge clasts. The Albion 'boulders with bedding' are rather similar lithologically to local Barton Creek bedrock, and thus the interpretation that they may represent locally excavated bedrock (from secondary cratering?) seems a possible explanation for their seemingly anomalous presence.

Grain-shape analysis suggests that many impactoclastic breccia clasts display shape characteristics derived directly from sedimentary bedding in target rocks. However, some clasts are highly abraded and perhaps were transported within the debris flow for significant distances in order to become rounded (cf. conclusions of Fouke et al. 2002). The variety of grain shapes, angularities, and surface textures generally supports the idea of mixing of carbonate clast populations during the early turbulent phase of movement. Further, cumulative grain-size (ϕ) frequency curves showing a high percentage of matrix, a substantial coarse component, and extremely poor sorting could also mean that grain-size distribution within the impactoclastic breccia represents a mixing of several clast populations from different excavated target layers (as in volcanic debris flow deposits; Glicken 1996).

Clasts within the impactoclastic breccia layer display surficial evidence of a sequence of events that may have been initially very energetic and on the whole indicates intensive interaction between clasts (as expected in mechanical fluidization, as noted above). King et al. (1997) interpreted facets, polish, and striations as representing effects of hypervelocity interactions among clasts during ballistic excavation and ejection (cf. Ries polished ejecta described by Chao 1976). Cryptographic markings, bruises and pits, and chips are interpreted as later particle interactions within turbulent and laminar debris flows. Also, some of these later features may have formed during compaction of the breccias shortly after deposition.

Acknowledgments

We thank the owner of the Albion Quarry at Albion Island, D. Grijalva, for kindly allowing us to work there during March 2000 and Director E. L. Wade of the Belize Geology and Petroleum Office for her assistance with access and permits. We appreciate helpful discussions with Kevin Pope, Adriana Ocampo, and Al Fischer, who introduced us to these unusual breccias. We are grateful for the comments of our reviewers, which helped improve this manuscript.

References

Bagnold RA (1954) Experiments on a gravity-free dispersion of large solid spheres in a Newtonian fluid under shear. Proceedings of the Royal Society of London (Series A) 225: 49-63

Blair TC, McPherson JG (1999) Grain-size and textural classification of coarse sedimentary particles. Journal of Sedimentary Research 69: 6-19

Chao ECT (1976) Mineral-produced high-pressure striae and clay polish: Key evidence for non-ballistic transport of ejecta from Ries crater. Science 194: 615-618

Claeys P (2001) The Chicxulub proximal ejecta blanket [abstract]. In: Smelror M, Dypvik H, Tsikalas F (eds) 7th Workshop of the ESF IMPACT Programme, Submarine Craters and Ejecta-Crater Correlation and Icy Impacts and Icy Targets (Longyearbyen, Svalbard), 29 August – 3 September, 2001, Abstracts and Proceedings of the Norwegian Geological Society No. 1, pp 19-20

Cornec JH (1985) Notes on the provisional geological map of Belize at the scale 1:250,000. Petroleum Office, Belmopan, Belize, 22 pp.

Cornec JH (1986) Provisional geological map of Belize. Petroleum Office, Belmopan, Belize

Folk RL (1968) Petrology of sedimentary rocks. Hemphill's Company, Austin, Texas, 170 pp

Fouke BW, Alvarez W, Claeys P, Ocampo AC, Pope KO, Smit J, Vega FJ (1996) Cathodoluminescence study of carbonate growth phases in bedrock and KT ejecta from the Chicxulub impact, Albion Island quarry, Belize [abstract]. Geological Society of America Abstracts with Programs 28(7): A183

Fouke BW, Zerkle AL, Alvarez W, Pope KO, Ocampo AC, Wachtman RJ, Grajales Nishimura JM, Claeys P, Fischer AG (2002) Cathodoluminescence petrography and isotope geochemistry of KT impact ejecta deposited 360 km from the Chicxulub crater, at Albion Island, Belize. Sedimentology 49: 117-138

Glicken H (1996) Rockslide-debris avalanche of May 18, 1980, Mt. St. Helens Volcano, Washington. United States Geological Survey Open File Report 96-977, 98 pp

Hörz F, Ostertag R, Rainey DA (1983) Bunte breccia of the Ries: continuous deposits of large impact craters. Review of Geophysics and Space Physics 21: 1667-1725

King Jr DT (1996a) Stratigraphy: layers of evidence (at Albion Island). Planetary Report 16(3): 10-11

King Jr DT (1996b) Cretaceous-Tertiary boundary stratigraphy near San Antonio, Orange Walk District, Belize, Central America. Gulf Coast Association of Geological Societies Transactions 46: 213-217

King Jr DT, Demick JD, Pope KO, Ocampo AC (1997) Surface markings on Chicxulub ejecta clasts from Belize, Central America [abstract]. Geological Society of America Abstracts with Programs 29(3): 12

Laznicka P (1988) Breccia and Coarse Fragmentites. Petrology, Environments, Associations, Ores. Elsevier, Amsterdam, 832 pp

Melosh HJ (1983) Acoustic fluidization. American Scientist 71: 158-164

McSaveney MJ (1978) Sherman glacier rock avalanche, Alaska. In: Voight B (ed) Rockslides and Avalanches 1, Natural Phenomena: Elsevier, Amsterdam, pp 197-258

Ocampo AC, Pope KO, Fischer AG (1996) Ejecta blanket deposits of the Chicxulub crater from Albion Island, Belize. In: Ryder G, Fastovsky D, Gartner S (eds) New

Developments Regarding the KT Event and Other Catastrophes in Earth History. Geological Society of America, Special Paper 307, pp 75-88

Ocampo AC, Pope KO, Vega F, Fouke BW (2000) Depositional facies of the Chicxulub ejecta blanket in the southern Yucatán Peninsula [abstract]. Eos, Transactions of the American Geophysical Union 81(48): F798

Ocampo AC, Pope KO, King Jr DT, Fischer AG (2002) A new impact ejecta deposit from Chicxulub crater in central Belize [abstract]. In: Jakes P (ed) 8th Workshop of the ESF IMPACT Programme, Impacts: a Geological and Astronomical Perspective (Prague, Czech Republic), 12-16 October, 2002, Abstracts and Proceedings volume, p 47

Otto GH (1938) The sedimentation unit and its use in field sampling. Journal of Geology 46: 569-582

Petruny LW, King Jr DT (2000) Grain-size frequency distributions of Cretaceous-Tertiary boundary breccia deposits in Belize, Central America [abstract]. Geological Society of America Abstracts with Programs 32(7): A451-452

Pope KO, Ocampo AC (2000) Chicxulub high-altitude ballistic ejecta from central Belize [abstract]. Lunar and Planetary Science 31: abstract number 1419 CD-ROM

Pope KO, Ocampo AC, Fischer AG, Alvarez W, Fouke BW, Webster CL, Vega FJ, Smit J, Fritsche AE, Claeys P (1999) Chicxulub impact ejecta from Albion Island, Belize. Earth and Planetary Science Letters 170: 351-364

Prothro DR, Schwab F (1996) Sedimentary Geology. Freeman and Company, New York, 575 pp

Rampino MR, Ernston K, Fischer AG, King Jr DT, Ocampo AC, Pope KO (1996) Characteristics of clasts in K/T debris-flow diamictites in Belize compared with other known proximal ejecta deposits [abstract]. Geological Society of America Abstracts with Programs 28(7): A-182

Salvador A (ed) (1994) International Stratigraphic Guide, 2nd ed. International Union of Geological Sciences and Geological Society of America, Boulder, Colorado, 214 pp

Savage SB (1984) The mechanics of rapid granular flows. Advances in Applied Mechanics 24: 289-366

Schultz PH (1992) Atmospheric effects on ejecta emplacement. Journal of Geophysical Research 97: 11623-11662

Sharpton VL, Marín LE, Carney JL, Lee S, Ryder G, Schuraytz BC, Sikora P, Spudis PD (1996) A model of the Chicxulub impact basin based on evaluation of geophysical data, well logs, and drill core samples. In: Ryder G, Fastovsky D, Gartner S (eds) The Cretaceous-Tertiary event and other catastrophes in Earth history. Geological Society of America, Special Paper 307: 55-74

Stöffler D, Grieve RAF (1994) Classification and nomenclature of impact metamorphic rocks: A proposal to the IUGS subcommission of the systematics of metamorphic rocks. In: Montanari A, Smit J (eds) Post-Östersund Newsletter, European Science Foundation (ESF) Network on Impact Cratering and Evolution of Planet Earth, Strasbourg, pp 9-15

Stone RO (1967) A desert glossary. Earth-Science Reviews 3: 211-268

Ui T (1983) Volcanic debris avalanche deposits – identification and comparison with non-volcanic debris stream deposits. Journal of Volcanology and Geothermal Research 18: 135-150

Vega FJ, Feldman RM, Ocampo AC, Pope KO (1997) A new species of Late Cretaceous carcineretid crab (Brachyura: Carcineretidae) from Albion Island, Belize. Journal of Paleontology 71: 615-620

Wilson CJN (1980) The role of fluidization in the emplacement of pyroclastic flows: an experimental approach. Journal of Volcanology and Geothermal Research 8: 231-249

New Geochemical Insights from Electron-Spin-Resonance Studies of Mn^{2+} and SO_3^- in Calcites: Quantitative Analyses of Chicxulub Crater Ejecta from Belize and Southern México with Comparison to Limestones from Distal Cretaceous-Tertiary-Boundary Sites

David L. Griscom,[1,2,3*] Virgilio Beltrán-López[3,7], Kevin O. Pope[4] and Adriana C. Ocampo[5,6]

[1]Laboratoire de Minéralogie et Cristallographie de Paris, Université de Paris 6, 4 place Jussieu, 75252 Paris, France. (dlgriscom@netscape.net)

[2]Materials and Structures Laboratory, Tokyo Institute of Technology, 4259 Nagatsuta, Yokohama 226, Japan.

[3]ICN, Universidad Nacional Autónoma de México, 04510 México D.F., México.

[4]Geo Eco Arc Research, Inc., 16305 St. Mary's Church Rd., Aquasco, MD 20608, USA. (kpope@starband.net)

[5]European Space Agency, ESTEC, Planetary Division, code SCI-SB, Keplerlaan 1, 2200 AG Netherlands. (Adriana.Ocampo@rssd.esa.int)

[6]Jet Propulsion Laboratory, Pasadena, CA 91109, USA.

[*]Fulbright-García Robles Fellow at Universidad Nacional Autónoma de México (UNAM), México D.F.

[7]deceased

Abstract. The solid-state-physics technique of electron spin resonance (ESR) has been employed in an exploratory study of marine limestones and impact-related deposits from Cretaceous-Tertiary (KT) boundary sites including Spain (Sopelana and Caravaca), New Jersey (Bass River), the U.S. Atlantic continental margin (Blake Nose, ODP Leg 171B/1049/A), and several locations in Belize and southern Mexico within ~600 km of the Chicxulub crater. The ESR spectra of SO_3^- (a radiation-induced point defect involving a sulfite ion substitutional for CO_3^{2-} which has trapped a positive charge) and Mn^{2+} in calcite were singled out for analysis because they are unambiguously interpretable and relatively easy to record. ESR signal strengths of calcite-related SO_3^- and Mn^{2+} have been studied as functions of stratigraphic position in whole-rock samples across the KT boundary at Sopelana, Caravaca, and Blake Nose. At all three of these sites, anomalies in SO_3^- and/or Mn^{2+} intensities are noted at the KT boundary relative to the corresponding background levels in the rocks above and below. At Caravaca, the SO_3^- background itself is found to be lower by a factor of 2.7 in the first 30,000 years of the Tertiary relative to its steady-state value in the last 15,000 years of the Cretaceous, indicating either an abrupt and quasi-permanent change in ocean chemistry (or temperature) or extinction of the marine biota primarily responsible for fixing sulfite in the late Cretaceous limestones. An exponential decrease in the Mn^{2+} concentration per unit mass calcite, $[Mn^{2+}]$, as the KT boundary at Caravaca is approached from below (1/e characteristic length =1.4 cm) is interpreted as a result of post-impact leaching of the seafloor.

Absolute ESR quantitative analyses of proximal impact deposits from Belize and southern Mexico group naturally into three distinct fields in a two-dimensional $[SO_3^-]$-versus-$[Mn^{2+}]$ scatter plot. These fields contain (I) limestone ejecta clasts, (II) accretionary lapilli, and (III) a variety of SO_3^--depleted/Mn^{2+}-enriched impact deposits. Data for the investigated non-impact-related Cretaceous and Tertiary marine limestones (Spain and Blake Nose) fall outside of these three fields. With reference to these non-impact deposits, fields I, II, and III can be respectively characterized as Mn^{2+}-depleted, SO_3^--enhanced, and SO_3^--depleted. It is proposed that (1) field I represents calcites from the Yucatán Platform, and that the Mn^{2+}-depleted signature can be used as an indicator of primary Chicxulub ejecta in deep marine environments and (2) field II represents calcites that include a component formed in the vapor plume, either from condensation in the presence of CO_2/SO_3-rich vapors, or reactions between CaO and CO_2/SO_3 rich vapors, and that this SO_3^--enhanced signature can be used as an indicator of impact vapor plume deposits.   Given these two propositions, the ESR data for the Blake Nose deposits are ascribed to the presence of basal coarse calcitic Chicxulub ejecta clasts, while the finer components that are increasingly represented toward the top are interpreted to contain high-SO_3^- calcite from the vapor plume.   The apparently-undisturbed Bass River deposit may contain even higher concentrations of vapor-plume calcite. None of the three components included in field III appear to be represented at distal, deep marine KT-boundary sites; this field may include several types of impact-related deposits of diverse origins and diagenetic histories.

1

Introduction

Electron spin resonance (ESR) is useful for non-destructive identifications and quantitative analyses of both paramagnetic ions and radiation-induced point defects in minerals and can even be used as a means of dating the latter (Ikeya 1993).   In the most common cases, "point defects" comprise vacancies or interstitials in a crystal lattice. But to be detected by ESR requires that the point defects in question each contain an unpaired electron; (thus, "self trapping" of radiation-generated electrons or holes can sometimes give rise to point defects not involving atomic displacements). The ESR technique is also capable of detecting fine-grained metallic iron or magnetite precipitates in natural glasses (Griscom

1984). Accordingly, ESR was recently employed to quantify the amounts of single-domain (SD) ferrite particles (i.e, ferrimagnetic spinels in sizes <60 nm: Dunlop 1972) in Cretaceous-Tertiary (KT) boundary impact layers (Griscom et al. 1999). The object of this SD-ferrite research was to explore the efficacy of using ESR-detected <60-nm spinel particles as an impact marker analogous to, but simpler than, the process of Robin et al. (1991), which involved magnetic separation, automated scanning electron microscopy, and microprobe analyses to identify cosmic spinels in sizes >1 μm. As is commonly true when ESR is applied to complex materials systems, numerous additional resonance signals were recorded that were not the objects of the original study—and some of these suggested themselves as potential objectives for future study.

In fact, unpublished research performed concomitantly with the SD-ferrite study of Griscom et al. (1999) revealed that the ESR intensities of Mn^{2+} and the sulfite radical, SO_3^-, both present in the calcite fractions of limestone samples from two Spanish KT-boundary sections, each displayed its own peculiar anomaly coinciding with the boundary. Given these discoveries, we embarked on a broader ESR study of these two paramagnetic species in KT-boundary calcites to determine the potential of this technique for characterizing deposits related to the Chicxulub impact.

Two other prior applications of ESR to KT-boundary materials should be mentioned. Both Miura et al. (1985) and Premovič et al. (2000, 2002) recorded ESR spectra of paramagnetic centers in KT-boundary fish clays from Stevns Klint, Denmark. While Premovič et al. (2002) confined the ESR component of their extensive geochemical investigation to Cu^{2+} and V^{4+} present as organic complexes in the "Fiskelers clay", Miura et al. (1985) focused on Mn^{2+} and a radiation-induced point defect (probably CO_2^-) present in the calcite fraction of these materials. The latter authors also incidentally reported, but did not quantify, a spectrum that was later to be identified as that of SO_3^- in calcite.

The SO_3^- radical ion results from trapping of an electron "hole" (induced by the ionizing effects of natural irradiations, mostly ^{238}U, ^{232}Th, and ^{40}K decays) at the site of a sulfite ion (SO_3^{2-}) substitutional for a CO_3^{2-} ion in the calcite and aragonite crystalline forms of $CaCO_3$ (Kai and Miki 1991 and 1992, Ikeya 1993). [Note that the SO_3^- ion is a "radical" because it possesses an unpaired electron, in this case a "hole" in the normally-filled P shells of the three oxygens (Ikeya 1993). Note too that charge neutrality requires that each missing electron must be stably trapped elsewhere in the crystal lattice in order for measurable numbers of SO_3^- radicals to remain stable.]

All measurements to be reported here are unambiguously specific to calcite. The ESR spectrum the SO_3^- species in calcite is distinctively different from that of the same species in aragonite (Kai and Miki 1992). In addition, the ESR spectrum of Mn^{2+} in dolomite (e.g., Shepherd and Graham 1984) is easily distinguishable from that of Mn^{2+} calcite. (The Mn^{2+} ion is substitutional for Ca^{2+} in calcite but it substitutes for Mg^{2+} in dolomite, leading to differently shaped ESR spectra due to the differing electrostatic interactions of the unpaired D-shell electrons of Mn^{2+} with local crystalline electric fields specific to the crystal structures of calcite and dolomite.) While some Mn^{2+} is known to be present in clays, Griscom et al.

(1999) found its ESR spectrum to be about two orders of magnitude weaker in the separated clay fractions of KT-boundary deposits from Caravaca (Spain) and Bass River (New Jersey) than is the spectrum of Mn^{2+} in calcite which typifies whole-rock samples of these same two KT-boundary materials. Above and below the KT boundary, the calcite fractions in the marine limestones are much larger, so interferences from Mn^{2+} in the clay fraction are expected to be even less in non-KT-boundary rocks. The SO_3^- species has not been reported in dolomite (see Ikeya 1993 for a review of applications of ESR to limestones) nor was it observed in six samples from the Albion Island (Belize) spheroid bed investigated in the course of the present study—all of which displayed dolomite-type Mn^{2+} ESR spectra. Thus, even though whole-rock samples were studied here and some of these were predominantly dolomite, the data selected for presentation below pertain exclusively to samples displaying ESR spectra dominated by the signals of Mn^{2+} and SO_3^- in *calcite*.

2
Samples

This paper reports the aforementioned unpublished Mn^{2+} and SO_3^- ESR results for two Spanish KT-boundary sections (Caravaca and Sopelana) together with new data for samples from the Ocean Drilling Program (ODP) Leg 171B/1049/A (Blake Nose) drill core (Norris et al. 1998). All three of these sections are more distant than five crater radii from the Chicxulub structure (greater than ~500 km) and hence are termed "distal" (Melosh 1989). In addition, absolute concentrations of Mn^{2+} and SO_3^- in calcites have been determined for individual spheroids and bulk samples from proximal Chicxulub impact deposits in Belize and southern México (Ocampo et al. 1996, Pope et al. 1999, Pope et al. 2002). Fig. 1 provides a map showing the proximal sites where the present samples were collected. Fig.s 2 and 3 respectively show a photograph of several oblate spheroids from El Guayal, México, and a thin-section photomicrograph of spheroids from Ramonal, México.

In all, we analyzed 60 samples from 11 sites in North America, Central America, and Europe (Table 1). With only two exceptions (mentioned below), all samples for ESR measurement were ground to fine powders in agate mortars. Most samples were dried for at least an hour at 100 or 150 °C before measurement; however, two significant exceptions are mentioned when they arise in context.

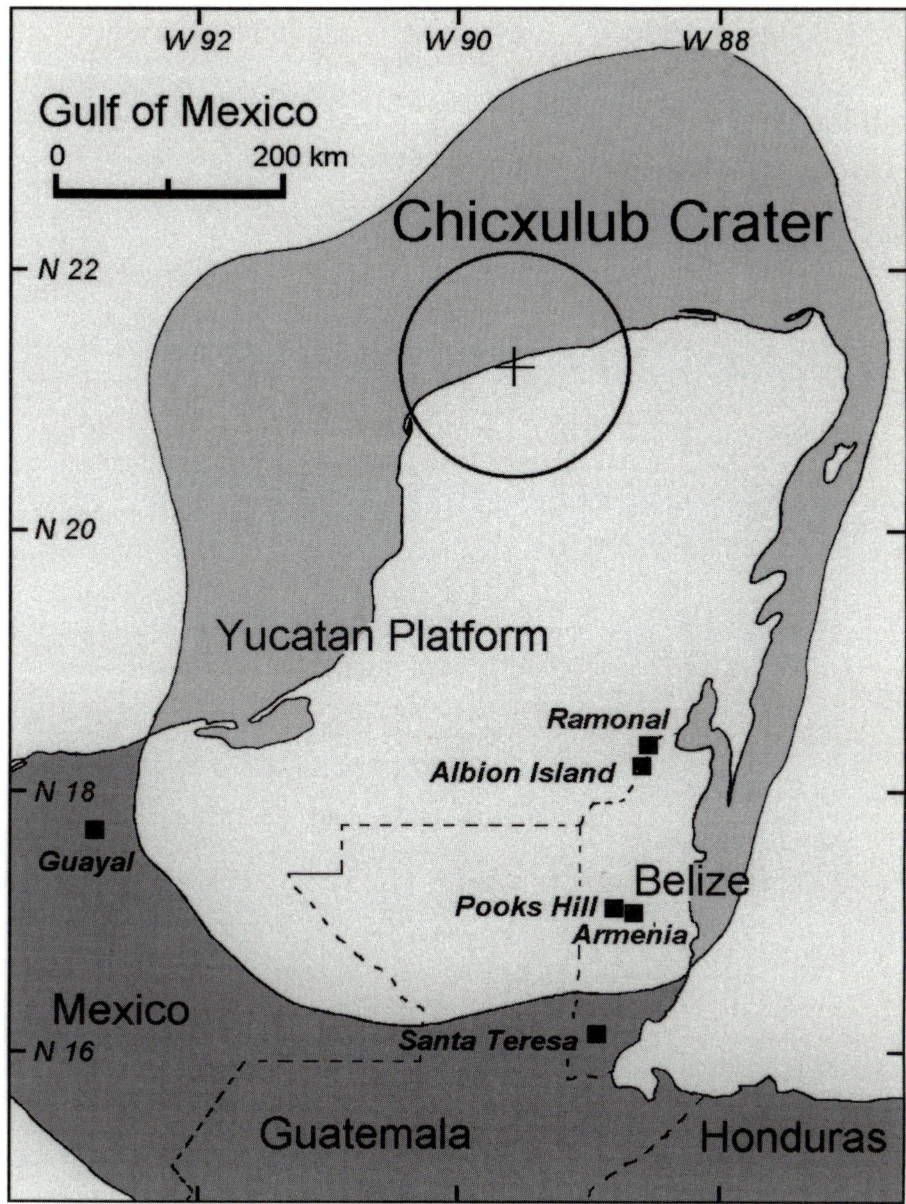

Fig. 1. Map of the Chicxulub impact crater and the proximal sites with impact ejecta studied by ESR. At the time of impact, Ramonal (North and South), Albion Island, Pook's Hill, and Armenia were all located on the Yucatán Platform (lightly shaded area), and Santa Teresa and El Guayal were just off the edge of the platform in deep water.

Table 1. Characteristics of KT boundary samples analyzed by ESR.

Site	Number of Samples	Stratigraphic Unit	Description
Caravaca, Spain	11	Cretaceous/Tertiary	Marly limestone/marly clay
Sopelana, Spain	15	Cretaceous/Tertiary	Marly limestone/marly clay
Bass River, New Jersey	1	KT-boundary spherule bed	Gray spherules plus matrix, mostly smectite
ODP Leg 171B/ 1049A core, Atlantic Ocean (Blake Nose)	8	Upper Maastrichtian/ ejecta layer/fireball layer/ lower Danian	Calcareous ooze/greenish spherules/red capping layer/clay-rich ooze
Guayal, Tabasco, México	1	Spheroid bed ~2 m below fireball layer	Chert-permeated 6×12 mm^2 oblate spheroid
Santa Teresa, southern Belize	1	KT-ejecta-bed slump, re-deposited in Paleocene	Silicate glass microtektites altered to smectite, calcite
Ramonal North, Quintana Roo, México	4	Albion Formation spheroid bed	Accretionary lapilli 1-2 cm in diameter; also bulk samples with mm-sized lapilli and calcite matrix
Ramonal South, Quintana Roo, México	2	Albion Formation spheroid bed	Bulk samples of calcite and clay matrix from base, no lapilli
Albion Island, northern Belize	5	Albion Formation spheroid bed	Weakly-consolidated chalky calcite layers 1-4 cm thick, no lapilli

Table 1. (Cont.)

Site	Number of Samples	Stratigraphic Unit	Description
Albion Island, northern Belize	2	Albion Formation diamictite bed	Calcite spherules 1 cm in diameter with radial crystal habit, some with lithic cores
Pook's Hill, central Belize	1	Albion Formation Pook's pebble bed	Pook's pebble: rounded micritc limestone clast
Armenia, central Belize	8	Albion Formation Pook's pebble bed	Pook's pebble: rounded micritc limestone clasts ~1-5 cm; also bulk samples of coarse calcite matrix
Armenia, central Belize	2	Albion Formation spheroid bed	Accretionary lapillus ~1 cm; bulk sample of fine calcite matrix

Fig. 2. Chert-permeated spheroids from El Guayal, Tabasco, México, showing impressions left in the matrix material in which they were found. (Samples made available by J. Urrutia-Fucugauchi.)

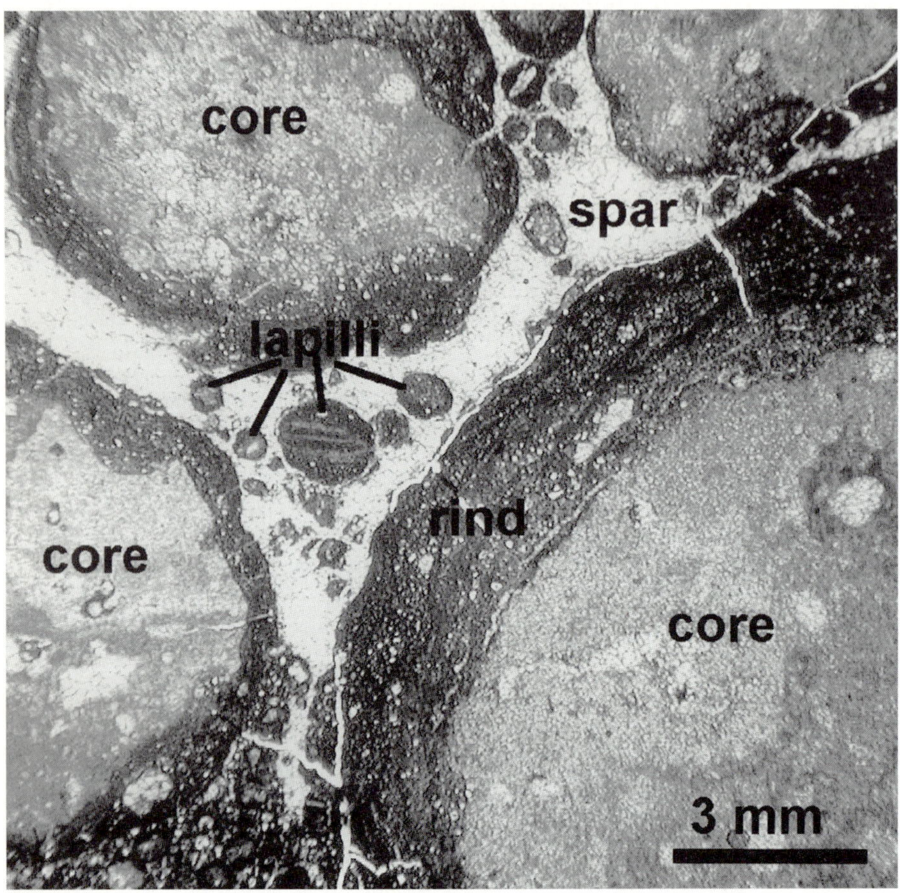

Fig. 3. Photomicrograph of accretionary lapilli in the Albion Formation spheroid bed, Ramonal North, Quintana Roo, México. Transmitted, plain polarized light. The heterogeneous lapilli cores are composed of fine-grain calcite and small limestone lithic fragments. The rinds enveloping the cores have multiple layers and in these examples are mostly stained red (dark) by iron oxides. Note the smaller lapilli in the sparry calcite matrix, some of which are also accretionary and have lithic cores.

3
Experimental Methods, Sample Encapsulation, and Data Analysis

3.1
Electron Spin Resonance

All commercial electron spin resonance spectrometers measure the absorption of fixed-frequency microwaves by unpaired electrons as a function of the magnitude, H, of an externally applied laboratory magnetic field, $\boldsymbol{H}$. For *Zeeman interactions* alone, the resonance condition is given by

$$h\nu = g\beta H \tag{1}$$

where h is Planck's constant, ν is the frequency of the microwaves, β is the Bohr magneton, and g is a dimensionless number related to the projection of the electron magnetic moment along the direction of $\boldsymbol{H}$ (see, e.g., Weil et al. 1996, Griscom 2001). The g value of the free electron (having quantum-mechanical spin $S=1/2$) is isotropic and given by $g_e=2.00232$. However, the g matrices of unpaired electrons embedded in solids are strongly dependent on local crystalline electric fields, which are in turn determined by the crystal structure. Thus, for crystals the g value depends on the direction cosines of the angles between $\boldsymbol{H}$ and the crystallographic axes. When, as in the present case, the samples are polycrystalline, all angles are represented with equal probability, giving rise to mathematically well-defined line shapes called "powder patterns", from which the principal-axis g values are extracted by computer line-shape simulations (Weil et al. 1996, Griscom 1990 and 2001, Poole, 1983).

Unpaired electrons are also sensitive to other interactions besides the Zeeman interaction with the laboratory field. For example, under ideal conditions the SO_3^- species should display a *hyperfine interaction* with the magnetic moment of the 0.75 %-abundant ^{33}S nucleus (nuclear spin $I=3/2$); this weak quartet of satellite lines was not detected in the present work but has been reported by Kai and Miki (1992). By contrast, Mn^{2+} is always subject to a hyperfine interaction with the magnetic moment of the ^{55}Mn nucleus (100% abundant, $I=5/2$), which splits each component of the spectrum into $2I+1=6$ nearly-equally spaced hyperfine lines. Because the paramagnetic nature of Mn^{2+} arises from five unpaired electrons in $3d$ orbitals vectorially adding to a spin of $S=5/2$, its ESR spectrum always consists of $2S=5$ *fine-structure* components, each of which is split into six lines by the ^{55}Mn hyperfine interaction (e.g., Weil et al. 1996, Griscom 1990 and 2001, Poole 1983). In powder samples, the six hyperfine lines of the central ($M_S=1/2 \leftrightarrow M_S=-1/2$) fine-structure component are the strongest spectral features; (these are the ones apparent in Fig. 4).

All ESR spectra reported here were recorded at X-band frequencies ($\nu \approx 9.1$-9.5 GHz) on continuous-wave spectrometers, including Varian E-Line (Universidad Nacional Autónoma de México, México, DF) and Bruker ER200 (Naval Research Laboratory, Washington, DC) analog instruments and computer-controlled Bruker

ED300 spectrometers (Universités de Paris 6 & 7). In each case, the magnetic field was modulated at 100 kHz and lock-in signal detection was employed. Thus, the recorded spectra were in the form of the first derivatives of the absorption curves as functions of laboratory magnetic-field strength.

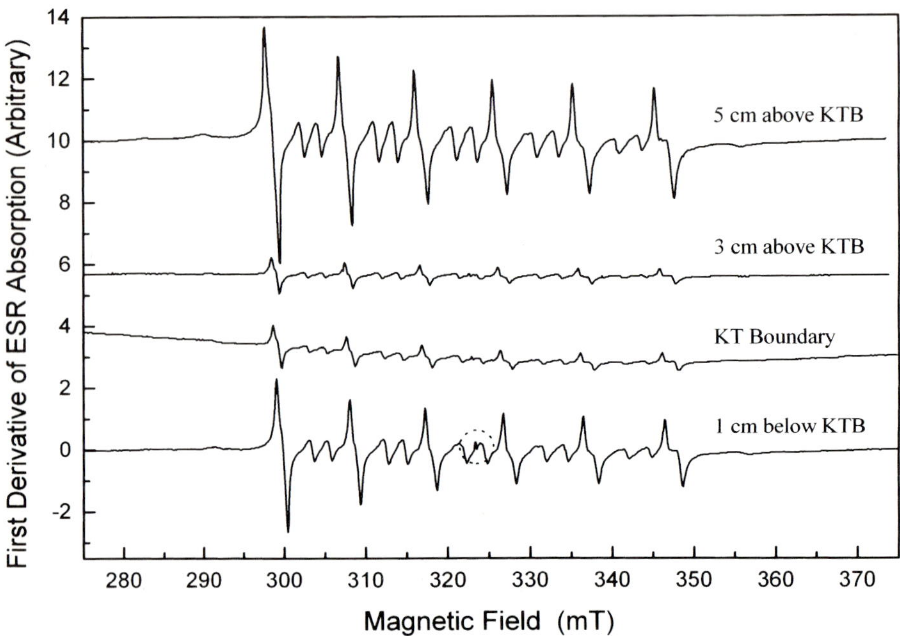

Fig. 4. X-band (9.1 GHz) ESR spectra recorded at ambient temperature and fixed gain for a succession of volumetrically-equal whole-rock limestone samples traversing the KT boundary at Caravaca, Spain. The principal six-line spectra are due to the "allowed" ESR transitions of Mn^{2+} in calcite; the weaker pairs of lines between the members of this ^{55}Mn hyperfine sextet are "forbidden" transitions of the same spectrum. The sloping "baseline" underlying the KT-boundary spectrum is actually the central portion of a much broader ESR signal due to single-domain ferrite particles (Griscom et al. 1999). A circle encloses the position of the ESR signal of the radiation-induced defect SO_3^{-} in calcite. The intensity differences seen here include the effect of calcite depletion in the boundary clays, which are ~3 cm thick in the sampled section. (Samples provided by E. Robin.)

3.2
Sample Encapsulation

Except where explicitly noted, all samples were in the form of homogeneous fine powders, which were loaded into ~25-cm-long, 3-mm-bore silica-glass tubes, generally to heights of 3 cm or more. The tubes (if not irradiated) gave no ESR signal. These were inserted vertically into the 3-cm-high TE_{102} microwave cavities common to all three spectrometers so that the filled part of the sample tube extended all the way through cavity—or, in cases where the fill height was <3 cm, the filled part was vertically centered in the cavity. The measurement is effectively sensitive to the 1-cm-high part of the sample column exactly centered in cavity (Hyde 1960). In all cases, the cavity also contained coaxial double-walled fused-silica glassware for temperature control by flowing N_2 gas. Although those measurements reported here were performed in the range ~290-300 K, the presence of the glassware significantly enhances ESR sensitivity by concentrating the oscillating microwave magnetic field at the sample position. With the exception of the Sopelana series, all samples were weighed and the height of the sample in the tube was measured (just before insertion) in order to determine the linear mass density ρ in units of g/cm (at the time of measurement).

3.3
Data Reduction and Determination of Absolute Spin Concentrations

Electron spin resonance spectral amplitudes are recorded in arbitrary "spectrometer units" versus magnetic field strength, H. For any series of samples (run on the same spectrometer) for which the ESR line shapes are essentially identical and only *relative* intensity differences are important, the amplitudes of the first derivative spectra can be used as the measure of intensity after normalization to the product of spectrometer gain, the magnetic-field modulation amplitude, and the square root of the microwave power, P (provided that P was set low enough that a linear dependency on $\sqrt{P}$ prevailed—or the identical value of P was used in all cases). In the present study, the Sopelana and Caravaca stratigraphic series were analyzed in this way. In the case of the Caravaca series, however, small sample-to-sample differences in the shapes of the SO_3^- line were approximately accounted for by multiplying the derivative amplitude by the width-at-half-amplitude of the strongest derivative feature.

When line shapes are varying significantly on a sample-to-sample basis or when *absolute* spin concentrations are desired, raw ESR intensities in the form of (spectrometer units)×(field strength)^2/(sample height in cm) were determined as the areas under the absorption curves by double numerical integration of the first-derivative spectra, normalized as before to the product of spectrometer gain, the magnetic field modulation amplitude, and $\sqrt{P}$. (As explained in the following section, corrections for "microwave power saturation" were made where needed.) Baseline corrections were made as described by Poole (1983).

To obtain *absolute* spin concentrations, these raw intensities were first divided by the identically obtained raw intensity of a Varian strong-pitch linear standard sample containing $(3\pm0.75)\times10^{15}$ spin-1/2 electrons per linear centimeter of sample height (Hyde 1960), and the resulting dimensionless number was then multiplied by 3×10^{15} cm^{-1} and divided by the linear mass density ρ of the sample to obtain the formal equivalent of the number of spin-1/2 states per gram of whole rock. Division of the latter number by the carbonate fraction, if known, yields the number of spin-1/2 states per gram calcite. Further multiplication by the formula weight of $CaCO_3$ and division by Avogadro's Number gives the absolute number of spin-1/2 states per calcite formula unit. These were exactly the desired results in the case of the SO_3^- radical, which, like the carbon pitch in the standard sample, happens to be an $S=1/2$ species.

In the case of Mn^{2+}, a theoretical correction [multiplication by factor of 0.1196 for T=290-300 K (Griscom et al. 1999)] was applied to convert the corresponding spin-1/2 equivalent to the number of spin-5/2 states. Quantitative analysis of the Mn^{2+} spectra in limestones is further complicated by the fact that they overlie broader spectral components associated with Fe^{3+} and ferrite phases in clay minerals (Ikeya 1993, Griscom et al 1999), thus precluding reliable integration of the full spectrum. For this reason, only the low-field ^{55}Mn hyperfine line of the central ($M_S=1/2\leftrightarrow M_S=-1/2$) fine-structure component was integrated here. The intensity of the full absorption curve was then assumed to be this integral multiplied by a factor of $2S(2I+1)=30$ (the total number of fine structure/hyperfine structure lines).

4
Accuracy, Precision and Possible Systematic Errors

4.1
Accuracy

The accuracies of all absolute spin concentrations are limited foremost by the accuracy of the standard sample calibration, given by the manufacturer as ±25 % (Hyde 1960). This factor affects the concentrations of SO_3^- and Mn^{2+} equally. There is an additional uncertainty of approximately ±10 % in the Mn^{2+} spin concentrations due to the necessary, but untested method of extrapolating the overall intensity from that of a single hyperfine line (see above). A third uncertainty of ±15 % in the absolute accuracy of our results arises from issues related to inter-laboratory calibrations (see below). While in the case of Mn^{2+}, these three *a priori* uncertainties combine (as the square root of the sum of the squares) to ±30 %, *a posteriori* comparisons with the results of a more conventional analytic technique suggest the actual accuracy of our results to be closer to ±10 % (see below).

The SO_3^- spectra gathered on the analogue spectrometers in Mexico City and Washington DC were recorded at a microwave power of 50 μW, while those acquired on computer-driven spectrometers in Paris were recorded at 10 μW after discovery that spectra obtained on the Parisian spectrometers at 50 μW were distorted in a way that prevented accurate numerical integrations. (For reasons not yet understood, spectra recorded in México and Washington were not distorted in this way.) Thus, we were faced with the unwanted problem of reconciling intensity data recorded at two different microwave powers for a paramagnetic species, SO_3^-, subject to microwave power saturation (the intensity dependence on $\sqrt{P}$ is non-linear). This problem was addressed in the following way: Data for the SO_3^- species acquired in México and Washington were corrected by multiplication by a nominal "saturation factor" of 3.0, as scaled from a plot of SO_3^--in-calcite ESR intensity versus microwave power (microwave-saturation curve) provided by Ikeya (1993). Ordinarily, the scale-factor error implicit to such a round-number correction would systematically bias the México/Washington SO_3^- data points relative to the ones recorded in Paris. However, this outcome was averted by calibrating the Paris SO_3^- data, not to another standard sample, but to several of the KT samples previously calibrated in Washington (using a standard sample, but applying the nominal factor-of-3 correction). In consequence, an additional uncertainty (± 10 %) in absolute *accuracy* accrues to *all* Mn^{2+} and SO_3^- concentrations, but systematic bias between clusters of data recorded in different locations is avoided.

4.2
Precision of Absolute Measurements

The present ESR spectra were obtained on four different spectrometers in three different parts of the world, requiring repeated cross calibrations involving the running several samples multiple times on two or more of these instruments. In general, a single spectrometer in good repair can yield reproducible results (within ~ 2-3 %) for a period of years. Even though one of the present spectrometers was repaired in the middle of a sample run, re-running a few samples was sufficient to correct for the consequent gain change. The net result of the intensive cross calibrations has led to the discovery and correction of a few discrepancies in absolute intensities outside of the expected 2-3 % precision that were ultimately traced to errors in parameter entry. However, several variances in the relative SO_3^- ESR intensities of *pairs* of samples run both in Washington and Paris persisted at levels of up to $\pm 15\%$. These variances are most likely due to sample-dependent variations in microwave power saturation, which were not accounted for by the flat factor-of-3 correction applied to all SO_3^- spectra recorded in México and Washington (see above). Thus, a precision of $\pm\ 15$ % is assigned to all of the present SO_3^- spin concentrations *when they are displayed on an SO_3^--versus-Mn^{2+} scatter plot including data recorded at different locations*. This uncertainty (large for ESR) turns out to be small in comparison with the sample-dependent scatter in the experimental data to be presented below (spanning four orders of magnitude in SO_3^- and a factor of about 200 in Mn^{2+}).

4.3
Precision of Relative Measurements for a Single Stratigraphic Column

Data for the three studied stratigraphic columns are not affected by the larger of the above-mentioned uncertainties in precision (i.e., ±15 %) because each data set was recorded on a single spectrometer under a single set of conditions in a single day. Any errors in absolute accuracy are irrelevant, since the principal object was determination of the *relative* intensities as functions of column height. Only short-term spectrometer drift and weighing errors, both of the order of ±1 %, affect these measurements—except in the case of un-dried KT-boundary clays as will be discussed below. The Sopelana samples were not weighed but were loaded into fused quartz tubes to equal heights, making them volumetrically equal to within a variance of about ±4 % owing to variations in the inside diameters of the tubes.

Five separate sets of data were recorded for the Caravaca section to optimize instrumental settings. Each of these five data sets was very similar to the others but, rather than average them, only the last set is presented below because the methods were deemed significantly improved in the process. The Blake Nose spectra were recorded twice (both times in Paris) with superficially similar results; however, those spectra obtained at a microwave power setting of 50 μW were disregarded because of the line-shape distortions mentioned above. The second set of Blake Nose spectra, recorded at 10 μW, was instrumentally integrated, providing absolute spin concentrations suitable for display on Mn^{2+}-SO_3^- scatter plots. Data for the Sopelana and Caravaca sections were recorded as spectral amplitudes.

4.4
A Possible Source of Systematic Error

A potential source of non-random error was recognized in the course of performing an isochronal annealing experiment (10 minutes at temperature) on a sample of Bass River, New Jersey, KT-boundary spherules (Olsson et al. 1997). When the annealing temperature applied to the sample in the bottom part of the silica tube was raised to 200 °C, water was observed to condense inside the cool top part. Concomitantly, when the ESR spectra were re-recorded after quenching back to room temperature, the spectral intensity of SO_3^-, which should have decreased progressively as the annealing temperature was raised (Ikeya 1993), had actually increased by ~50% after annealing to 200 °C (Fig. 5a). Since this sample was determined to contain 35 wt % carbonates (Griscom et al. 1999), it is reasonable to suppose that the source of the evolved water was the clays that comprise most of the non-carbonate fraction (Olsson et al. 1997). At the conclusion of the isochronal annealing experiment the sample was "re-irradiated" with γ rays to a dose (30 kGy), equivalent to a 65-Ma paleo-dose calculated using the algorithm of Rink and Odom (1991) and, absent information on the radioisotopes present in the Bass River materials, the ^{238}U and ^{232}Th contents of Stevns Klint KT-boundary calcites reported by Alvarez et al. (1980). The resulting component of "re-induced" SO_3^- ESR intensity stable after 168 h at room temperature (see Fig. 5b) turned out to be 5.6 times greater than that recorded for the as-received sample.

It is well known that non-resonant absorption of microwaves by water molecules can degrade the sensitivity of an ESR experiment. Thus, subject to future verification, it is tentatively supposed that ESR intensities measured in the as-received Bass River samples could have been artificially lowered by this mechanism by as much as a factor of 5.6 (the ratio of the intensity after "re-irradiation" to that in the as-received material). Alternatively, if H_2O molecules were to have had negligible effect, it would have to be assumed that the paleo-dose delivered to the Bass River samples was ~5.6 times lower than that calculated using Stevns-Klint radionuclide data and/or significant unstable components with decay times >>168 h (but <<65 Ma) were present in the "re-irradiated" sample. Some puzzling results of an early attempt to date the Stevns Klint fish clays by the ESR additive-dose method (Miura et al. 1985) might possibly be explained by the effects of radiation-induced components with decay times too long for practical laboratory controls.

When a comparable isochronal annealing experiment was performed on a gray-white KT-boundary calcite spheroid from southern México (Ramonal North), only a 6% increase in SO_3^- intensity was noted at 175 °C and progressively lower intensities were recorded for higher annealing temperatures (Fig. 6a). After the final, 325-°C annealing and γ irradiation to 40 kGy, the "re-induced" SO_3^- intensity recorded in the Ramonal North sample 1/2 h after irradiation was a factor of 1.6 times greater than in the as-received sample (Fig. 6b). In the case of the Bass River sample (Fig. 5b), short-lived "re-induced" SO_3^- defects were found to decay exponentially for a week (time constant 74 h determined from a four-point fit). But for the Ramonal North spheroid, data acquisition in Washington was discontinued 24 h after "re-irradiation", so possible decays with time constants >24 h were not probed in this case.

As also shown in Fig. 6a, the SO_3^- ESR signal strengths in two of the three well-dried synthetic calcites doped with SO_3^{2-} decreased monotonically with increasing annealing temperature following γ irradiation, while that of the third sample increased only by 2 % before monotonically declining at higher temperatures. These results suggest that moisture-free samples do not show increases in SO_3^- intensity upon isochronal annealing after irradiation. To explore the notion that the spectrometer sensitivity was degraded by the initial presence of water molecules in the Bass River sample during the first isochronal anneal, these data were divided by the corresponding data points for the isochronal annealing experiment following "re-irradiation" by γ rays. The results of this division (assuming that the data points at 275 and 300 °C are pairwise identical within experimental accuracy) are shown as the dashed curve in Fig. 5a. This curve seems to imply that the entire factor-of-5.6 "discrepancy" between the observed and "predicted" SO_3^- intensity in the as- received Bass River sample is due to

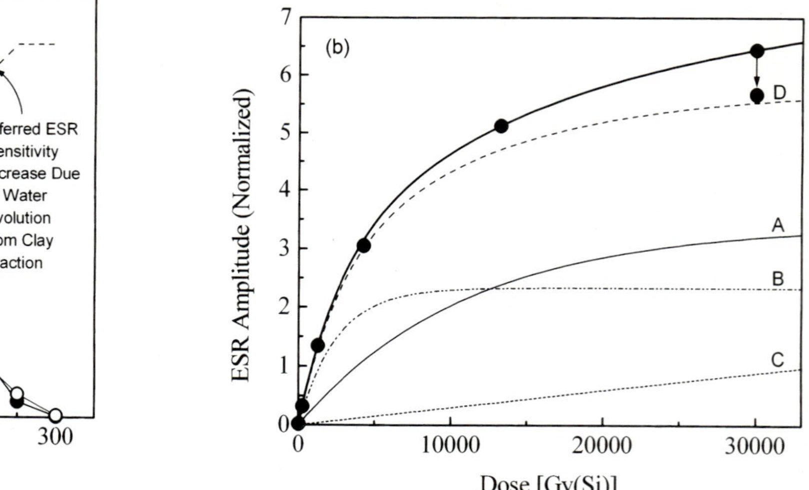

Fig. 5. (a) Isochronal annealing data (10 min at temperature) for ESR-detected SO3- ions in a Bass-River-spherule-bed bulk sample in as-received condition (open circles) and after annealing to 300 °C and γ-irradiating to ~30 kGy (solid circles). (b) Growth of SO3- ions in the annealed Bass River sample measured ~3-5 days after each 60Co γ irradiation (solid circles), except data point near 5 k Gy (30 days after). All data are normalized to the SO3- intensity in the as-received sample. After terminating irradiation at ~30 kGy, the room-temperature isothermal decay was monitored for 168 h. Four data points (only the first and last of which are plotted here) were well fitted by an exponential decay with time constant 74 h, leading to a no-adjustible parameters "postdiction" of the growth curve of metastable component C. The solid curve passing through the large solid circles is the sum of saturating exponential functions A and B, determined by cut-and-try methods, plus curve C. Dashed curve D is the sum of stable components A + B. The dashed curve in (a) is a hypothetical increase in ESR sensitivity due to evolution of water during the annealing experiment, calculated by dividing the lower curve (open circles) by the upper curve (solid circles) and normalizing to the "as-received" data point.

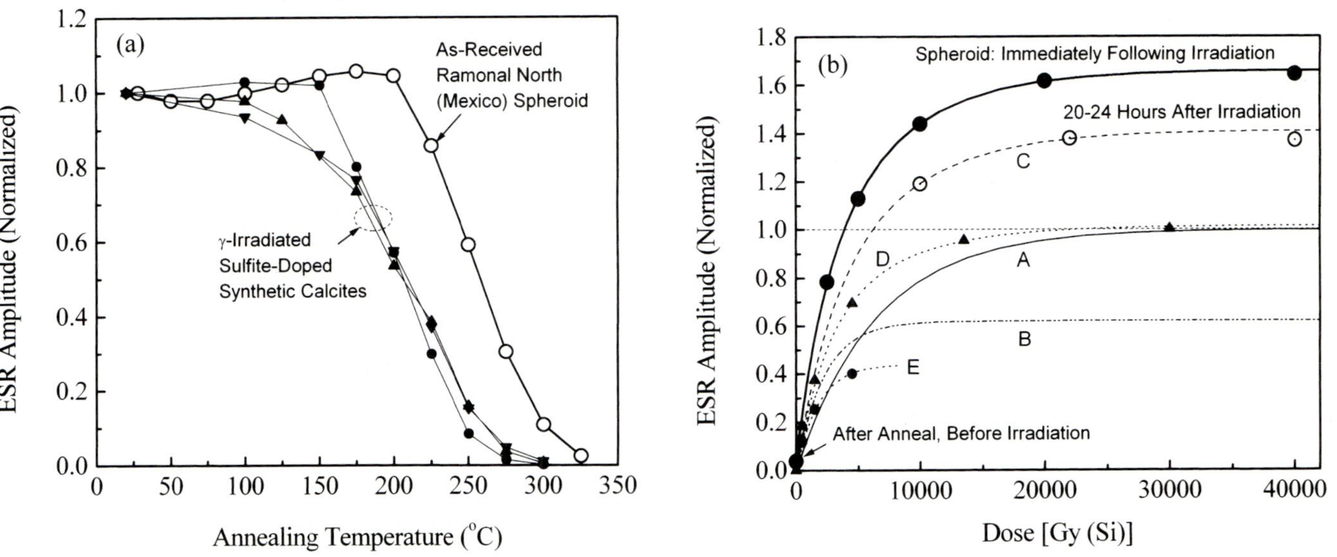

Fig. 6. (a) Isochronal annealing data (10 min at temperature) for ESR-detected SO_3^- ions in an as-received Ramonal North (RN) KT-boundary spheroid (large open circles) and in γ-irradiated, sulfite-doped synthetic calcites (small solid symbols). (b) Regeneration of SO_3^- ions in the annealed RN spheroid by ^{60}Co γ irradiation (large solid circles) and first creation of these species in the synthetic calcites (small symbols). The RN spheroid data are normalized to the SO_3^- intensity in the virgin RN sample; the other data are arbitrarily normalized. The bold curve passing through the large solid circles is the sum of saturating exponential functions A and B, which were determined by cut-and-try; the other fitted curves were accomplished as linear combinations of curves A and B. The small solid circles pertain to calcite singly doped with sulfite; the upward and downward solid triangles pertain to calcites co-doped with SO_3^{2-} plus CrO_4^{2-} and Pb^{2+} ions, respectively

spectrometer sensitivity degradation due to water molecules that are not fully removed until the annealing temperature reaches ~250 °C. If this interpretation is correct, this effect may have led to significant underestimations of both Mn^{2+} and SO_3^- contents in distal KT-boundary clays, especially those from Caravaca, which were not dried. Although all Blake Nose samples were dried at 100 °C before any ESR spectra were recorded, any illite present is likely to have retained its structural water molecules (Deer et al. 1975). This issue must be resolved in future work.

5
ESR Results and Technical Considerations

5.1
Identification, Simulation, and Interpretation of the Measured ESR spectra

Figure 4 illustrates a set of ESR spectra recorded under identical conditions for four members of the Caravaca series immediately spanning the KT boundary. The magnetic field scan range and spectrometer gain were optimized for observation of the ^{55}Mn hyperfine sextet and each spectrum has been vertically offset from the others for convenient viewing. The sloping "baseline" of the KT-boundary sample is actually due to single-domain ferrites in the clay fraction, as reported in detail by Griscom et al. (1999). At the time these particular spectra were recorded (in México), the SO_3^- spectrum (circled) had not yet been "discovered" in KT-related samples. Only by narrowing the magnetic-field scan range, lowering microwave power and magnetic-field modulation amplitude, and increasing the spectrometer gain (requiring higher time constants and longer scan times) were SO_3^- spectra of sufficient diagnostic quality acquired.

The solid curve in Fig. 7a shows, not the circled feature of Fig. 4, but the stronger SO_3^- spectrum of material from a 1.2×0.6-cm oblate spheroid similar to those of Fig. 2 recovered from the KT boundary at El Guayal, México (Grajales et al. 1996, Griscom et al. 1999). This particular SO_3^- signature, appearing to the left of the vertical dash-dot line in Fig. 7 (low-magnetic-field side), happens to be accompanied by the signature of E′ centers in α quartz (on the high-field side). The E′ signal, due to holes trapped at oxygen vacancies in the α-quartz lattice (Weeks 1956, Silsbee 1960, Feigl and Anderson 1970, Odom and Rink 1988), was also present but more than 10 times weaker in all samples from the Spanish KT sections.

Figure 7b represents a "material simulation" of the Guayal spheroid spectrum accomplished in the following way: Sulfite-doped calcium carbonates were prepared by adding small amounts of $NaSO_3$ to aqueous solutions of $NaCO_3$ and precipitating $CaCO_3$ by addition of equimolar $CaCl$ solutions. After washing and drying, these precipitates were sealed into fused-silica ESR sample tubes and

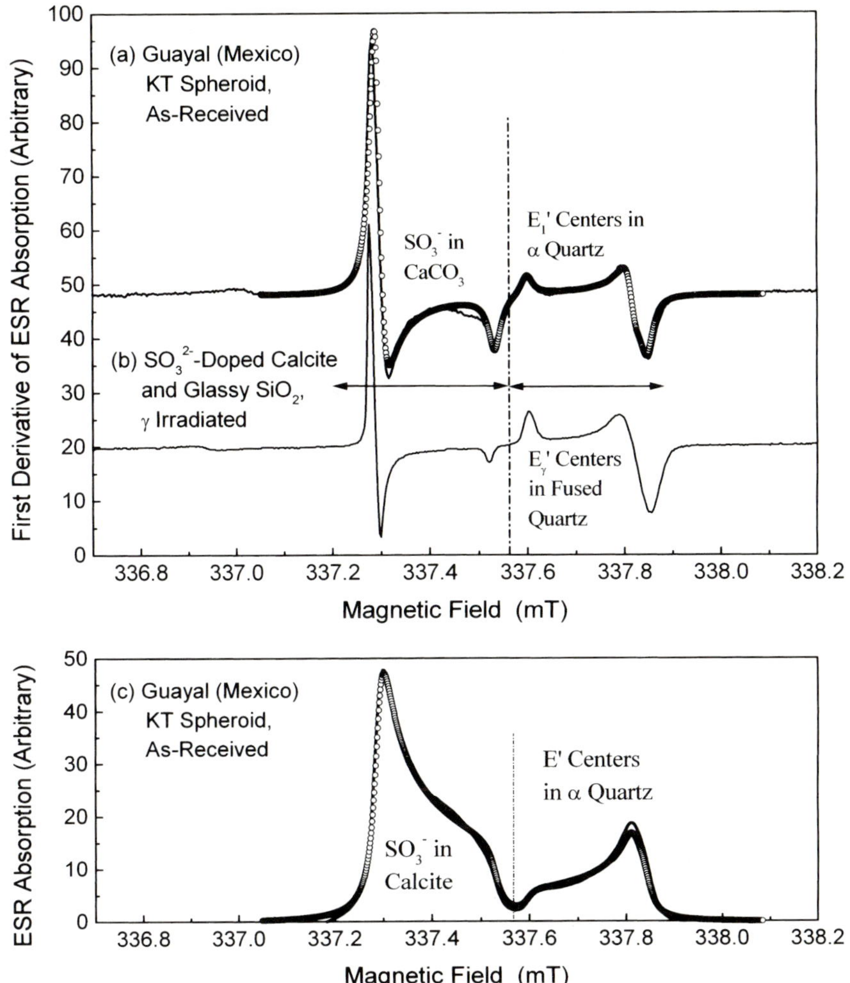

Fig. 7. (a) X-Band (9.5 GHz) ambient-temperate ESR spectrum (solid curve) and its computer simulation (circles) for an accretionary lapillus from El Guayal, México. (Microwave power 50 μW; magnetic-field modulation amplitude 0.01 milliTesla.) The simulated spectrum of (a) is a weighted sum of the computed spectra of SO^{3-} ions in calcite and E' centers in α quartz, using literature g values (Feigl and Anderson. 1970) for the latter. The quartz in this sample was in the form of authigenic chert (Griscom et al. 1999) which had either filled empty interstitial spaces or, more speculatively, replaced anhydrite. (b) "Material simulation" of the spectrum of (a) using a γ-irradiated hybrid calcite/SiO2 simulant consisting of SO_3^{2-}-doped calcite and the fused-quartz sample tube containing it. (c) The spectrum of (a) after numerical integration (solid curve) and its computer simulation (circles). The areas under the curves of (c) are used in quantitative measurements of the numbers of paramagnetic centers.

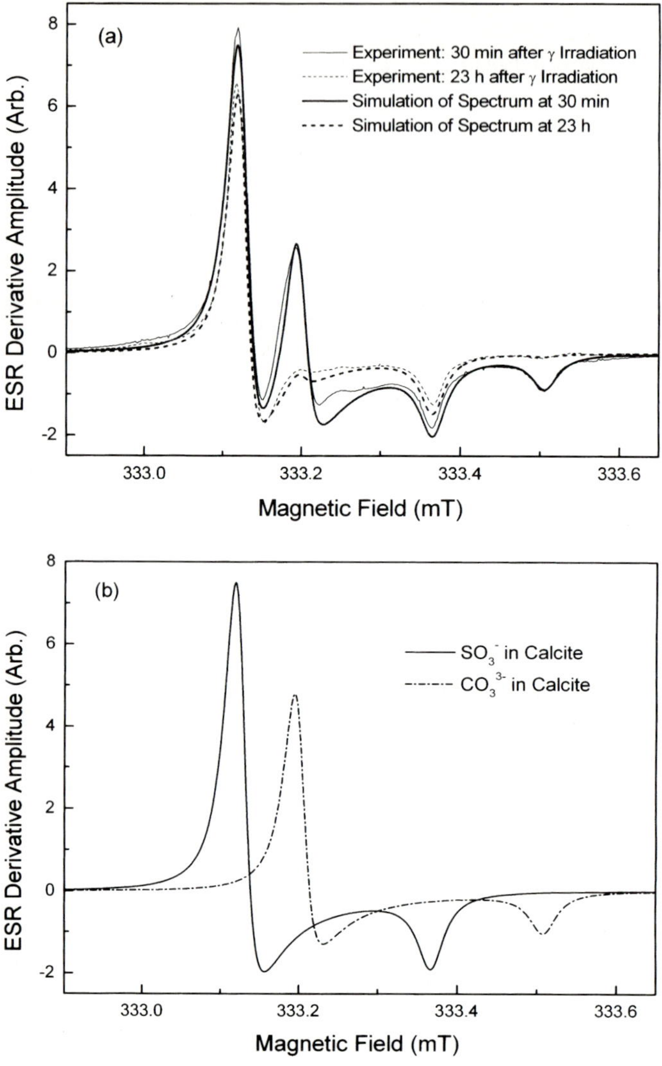

Fig. 8. (a) ESR spectra of SO_3^- and CO_3^{3-} radical ions in the annealed Ramonal North KT-boundary spheroid 30 minutes after ^{60}Co γ irradiation to a dose of 40 kGy (thin solid curve) and after an additional 23 hours at room temperature (thin dashed curve). The bold gray curves in (a) are computer simulations based on ESR powder-pattern theory (e.g., Griscom, 1990). Each simulation is a linear combination of the two simulated component spectra shown in (b), for which the g values exactly match those in the literature (Serway and Marshall 1967, Ikeya 1993). Both paramagnetic species are seen to thermally fade during 23 hours at room temperature, although fading of the CO_3^{3-} is nearly complete in this time span while a more stable SO_3^- component remains.

subjected to irradiation by ^{60}Co γ rays. The ESR spectrum of the resulting doped-calcite/silica-tube system was then recorded (Fig. 7b). Other synthetic samples were also prepared in the same way but doped (or co-doped) with other ions (Pb^{2+}, Ni^{2+}, Zn^{2+}, SO_4^{2-}, CrO_4^{2-}) selected as possessing (1) physical and chemical properties that might lead to their substitution for Ca^{2+} or CO_3^{2-} in the calcite lattice where they would exhibit an axially symmetric g matrices upon trapping an electron or hole and (2) isotopic abundances that would yield ESR spectra with weak-to-nonexistant hyperfine satellites..(Both of these properties are characteristic of the ESR spectrum attributed to SO_3^- in calcite.)

In fact, only doped, or co-doped, with sulfite were found to give this particular signal upon irradiation, in full agreement with the earlier evidence of Kai and Miki (1991 and 1992) for the attribution of this spectrum to the SO_3^- radical ion. Growth and isochronal-anneal data for the SO_3^- signal in some of these γ-irradiated synthetic samples are illustrated in Fig. 6

In addition to the "material simulation" of Fig. 7b, the ESR line shape of the Guayal spheroid was simulated theoretically (see, e.g., Griscom 1990); the optimized result is displayed as the small open circles in Fig. 7a. This simulation involved six spin-Hamiltonian parameters (three principal-axis g values per paramagnetic species), plus two "natural" Linewidths and the ratio of the strengths of the two signals (2.7:1 in favor of SO_3^-). However, the g values for E' centers in α quartz were taken from the literature (Feigl and Anderson 1966) and used without further adjustment, thus (i) confirming the ESR signature of E' centers and (ii) providing an "in-cavity g-value marker" to aid in determining the g values of the SO_3^- signal. That is, with reference to the E'-center signal "marker", the g values of the SO_3^- species could be determined to virtually the same accuracy as those of the E' center without performing *absolute* measurements of either the frequency v or the magnetic field strength H, which appear in eqn. (1). Rather, only an accurate knowledge of the *relative* magnetic-field separation on the recorder chart is required to determine any g value with respect to a known "marker" g value. Lorentzian convolution linewidths of 0.03 mT were found suitable for both spectra. The g values which optimized the SO_3^- spectrum were $g_\parallel$ = 2.00214 ±0.00005 and $g_\perp$ = 2.00357 ±0.00005, not only agreeing with the literature [2.0021 and 2.0036, respectively (Ikeya 1993)], but adding an additional significant figure.

Figure 7c shows the absorption curve (solid curve) determined by numerical integration of the experimental first-derivative spectrum of Fig. 7a, together with the theoretically simulated absorption curve (circles) before its differentiation to give the circles in Fig 4a. (The areas under the absorption curves are employed in ESR quantitative analyses.)

Figure 8a shows ESR spectra (thin curves) recorded for the Ramonal North calcite spheroid after annealing to 325 °C and "re-irradiation" to a dose of 40 kGy (see kinetics data in Fig. 6). This time the E' centers induced in the silica sample tube were annealed out in an H_2-O_2 flame while carefully keeping the sample itself at ambient temperature at the opposite end of the sealed tube. Nevertheless, another spectral feature still occurs to the high-field side of the SO_3^- spectrum.

This feature is part of the well-known spectrum of radiation-induced CO_3^{3-} centers in the calcite lattice (Serway and Marshall 1967, Ikeya 1993) and was also observed in the irradiated synthetic calcites of the present study before they were allowed to anneal at room temperature. The thick curves in Fig. 8a are computer simulations of the experimental curves as linear combinations of the theoretically simulated SO_3^- and CO_3^{3-} component spectra of Fig. 8b. The CO_3^{3-} g values that optimized these simulations (this time using the SO_3^- spectrum as the "in-cavity marker") are $g_\parallel = 2.00132 \pm 0.00005$ and $g_\perp = 2.00315 \pm 0.00005$, agreeing with literature values (Ikeya 1993, Serway and Marshall 1967) and again adding another significant figure to the precision.

It is principally the presence of metastable CO_3^{3-} "self-trapped electrons" in the γ-irradiated samples that accounts for the observed room-temperature decay of a fraction of the simultaneously induced SO_3^- species in times ~1 day (Figs. 5b, 6b, and 8a). This is because each remobilized self-trapped electron eventually meets and annihilates a trapped-hole center, some of which are SO_3^-. However, most of the annihilated hole centers turn out to be "self-trapped holes", CO_3^-, the ESR spectrum of which occurs at lower magnetic fields than shown in Fig. 8 (Ikeya 1993, Serway and Marshall 1967).

5.2
Relative and Absolute ESR Intensity Measurements

Figure 9a and b portray the ESR-determined *relative* concentrations of SO_3^- and Mn^{2+} in calcite (per gram of whole rock) as functions of column height in the Sopelana and Caravaca KT limestones, respectively. Equivalent data for the ODP Leg-171B/1049A core are given in Fig. 10..The corresponding ratios $[SO_3^-]/[Mn^{2+}]$ are shown in all three cases. The data for Figs. 9 and 10 are given in Table 2.

Figure 11 is a scatter plot of the *absolute* determinations of the same two paramagnetic species in samples of the KT-impact deposits from proximal sites in Belize and southern Mexico. By "absolute determinations" is meant the number of paramagnetic species (either SO_3^- or Mn^{2+}) per "molecule" of $CaCO_3$. Stated another way, Fig. 11 shows the number of SO_3^- ions per million CO_3^{2-} ions in calcite and plotted versus the number of Mn^{2+} ions per million Ca^{2+} ions in the same calcite. The positions of the large data points are based on measured carbonate mass fractions $f_{calcite}$. Since the carbonate mass fraction was not determined for the other investigated samples (small data points), $f_{calcite}$ was assumed for convenience to be equal to 1.0 in these cases. Thus, if $f_{calcite}$ subsequently proves to be less than 1.0, each small data point will move toward

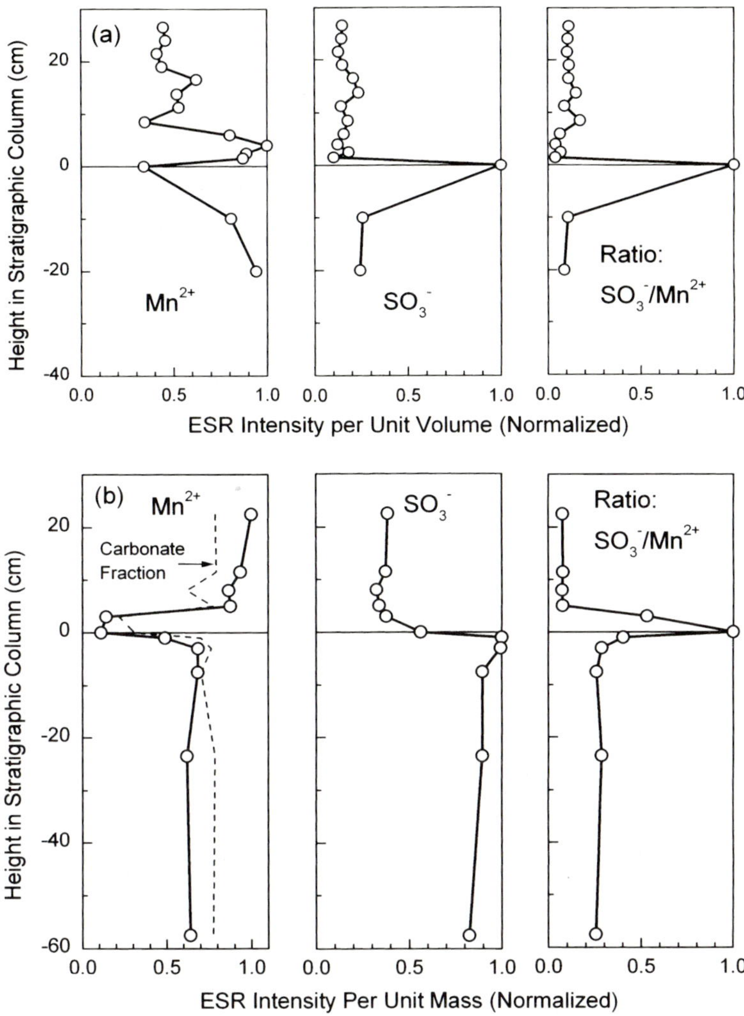

Fig. 9. Relative concentrations of Mn^{2+} and SO$_3^-$ in calcite per unit whole-rock mass or volume (open circles) determined by ESR for KT stratigraphic sections from (a) Sopelana and (b) Caravaca, Spain. The right-hand panel in each case is the SO3--to- Mn2+ ratio, which is independent of calcite fraction in these whole-rock samples and which is proposed as a potentially useful marker of KT-impact layers in marine limestones. The gradual "decay" of this ratio into the Maastrichtian has been argued to be a result of post-impact diagenesis (Griscom and Beltrán-López 2002). The KT boundary (base of ejecta layer) is at zero height in both cases. (Samples provided by E. Robin.)

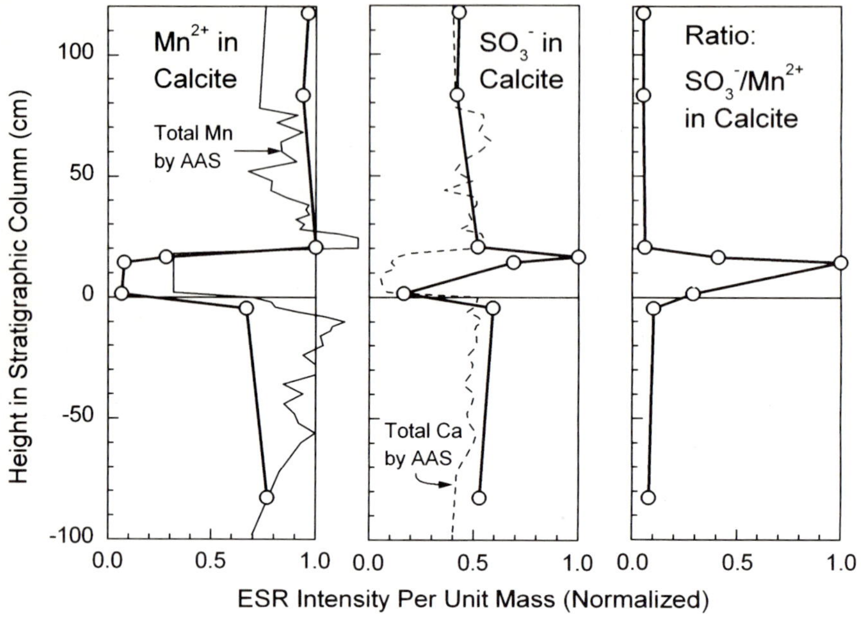

Fig. 10. Relative concentrations of Mn^{2+} and SO$_3^-$ in calcite per unit whole-rock mass (open circles) determined by ESR for a KT stratigraphic section from the Ocean Drilling Program Leg 171B/1049A core recovered from the U.S. Atlantic continental slope east of Florida (Blake Nose). The right-hand panel shows the ratio of [SO$_3^-$] to [Mn^{2+}]. Atomic absorption spectroscopy (AAS) data for total Mn and total Ca in more-closely-spaced samples from the same core (Martínez-Ruiz et al. 2001) are plotted in the left-hand and center panels, respectively; (they are arbitrarily normalized for comparison with the ESR data). The sharp maximum in SO$_3^-$ seen in the central panel corresponds to the "fireball" layer (Norris et al. 1998). The peak in the [SO$_3^-$]-to-[Mn^{2+}] ratio occurs 1 cm below this red capping layer. The base of the ejecta layer is at zero height.

the upper right along lines with unit slope by factors of 1/f$_{calcite}$. Fig. 12 illustrates absolute-concentration data recorded for both KT-boundary and non-KT-boundary limestones from several KT sections distal to the crater. The data of Fig. 12 are superposed on shaded fields taken directly from Fig. 11 for purposes of comparing the data for distal sites with the ranges of results for accretionary lapilli (gray-gradient) and limestone ejecta clasts (dark gray) recovered from the proximal sites. The data included in Figs. 11 and 12 can be found in Table 3.

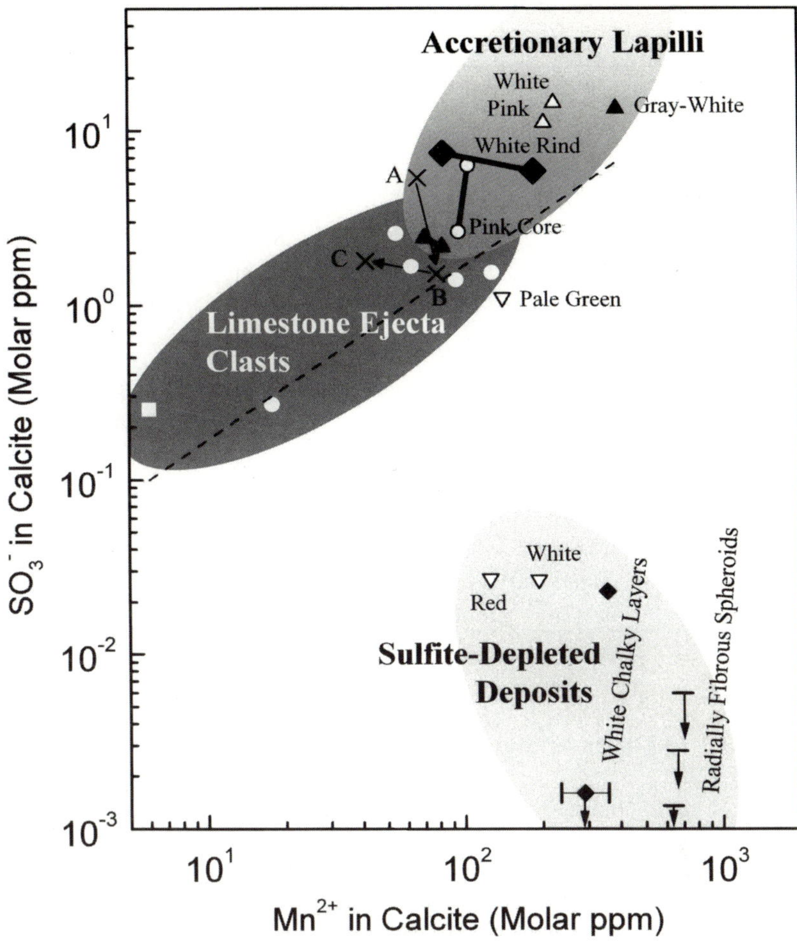

Fig. 11. Scatter plot of ESR-determined Mn^{2+} and SO_3^- concentrations in calcite in KT-impact deposits from México and Belize within ~600 km of the center of the Chicxulub structure. The white data points in the dark-gray field pertain to Chicxulub limestone ejecta clasts ("Pook's Pebbles") from central Belize. Solid data points and the open circles in the gray-gradient field refer to accretionary lapilli from México and central Belize. Solid diamonds in the pale-gray field represent white "chalky layers" in the Albion Island, Belize, quarry. Open triangles are spheroid-bed bulk samples or matrix materials mentioned in the text. The dashed straight line has unit slope and is provided as a guide to the eye. Crosses connected by arrows are matrix materials from a KT-ejecta outcrop in Armenia, Belize, representing a sequence from 1.3 m deep in the spheroid bed (A) upward to the base of the overlying Pook's pebble bed (B) and then upward 1 m into the Pook's pebble bed (C). The dark-gray, gray-gradient, and pale-gray elliptical fields define calcite types I, II, and III, respectively, as described in the text.

Table 2.   Relative ESR intensities of Mn^{2+} and SO_3^- in stratigraphic sequences from Sopelana (left) and Caravaca (right), Spain. The KT boundary (base of ejecta layer) is at zero height in both cases.

Sopelana			Caravaca		
Column Height (cm)	Mn^{2+}	SO_3^-	Column Height (cm)	Mn^{2+}	SO_3^-
27	0.447	0.150	22.5	1.000	0.385
24	0.459	0.146	11.5	0.936	0.374
21	0.414	0.129	8	0.868	0.325
19	0.439	0.150	5	0.877	0.339
16	0.623	0.207	3	0.140	0.377
13	0.519	0.236	0	0.111	0.562
11	0.529	0.143	-1	0.491	1.000
8	0.346	0.179	-3	0.686	0.994
6	0.803	0.157	-7.5	0.685	0.897
4	1.000	0.124	-23.5	0.621	0.895
2	0.891	0.183	-57.5	0.641	0.825
1	0.872	0.103			
0	0.341	1.000			
-10	0.809	0.257			
-20	0.942	0.243			

5.3
Comparison to Other Analytic Methods of Chemical Analysis

No chemical data obtained by other analytical methods are available for the particular Sopelana and Caravaca stratigraphic sequences examined here. However, in Fig. 10, the (normalized) ESR-determined Mn^{2+}-in-calcite data for the ODP Leg 171B/1049A core (circles) are compared with whole-rock Mn contents determined by atomic absorption spectroscopy (AAS) for samples from the same core (Martínez-Ruiz et al. 2001) (thin curve). In this Fig., the AAS data for Mn have been scaled by a factor arbitrarily selected to achieve approximate superposition with the ESR data. The two types of data appear to be mutually consistent but little else can be said, given the rapid oscillations in the AAS data with column height, the sparseness of the ESR data, and the unreliability of the AAS data obtained for the calcite-depleted ejecta and fireball layers between zero and +17 cm. (All AAS data for the ejecta and fireball layers were reported as 100 ppm but this number was in fact the detection limit in this experiment: F Martínez-Ruiz, private communication). By contrast, the ESR method is capable of measuring Mn^{2+} in calcite to levels down to1 ppm with precisions of ±5 % or better. The result of averaging the *absolute* whole-rock data for samples from +20 to +120 cm above the base of the ejecta layer (neglecting a significant spike in the AAS data between +22 and +26 cm that was not sampled by ESR) gives 280 ppm Mn by AAS versus 258 ppm Mn by ESR. Much of this 8 % difference could be due to Mn in the clay fraction that escaped ESR detection. Based on this high degree of agreement between ESR and AAS, the ESR measurements may in fact be far more accurate than implied by the *a priori* calibration uncertainties discussed above.

It can be seen in Fig. 11 that the ESR-determined SO_3^- concentrations in calcite range from ~20 ppm down to ~2 ppb. We know of no other analytical data with which these data can be compared, particularly given that we measure a specific tetra-atomic ionic impurity (sulfite) present in a particular mineral phase (calcite). Finally, we emphasize that the *precision* of the ESR intensity data pertaining to the stratigraphic sequences of Figs. 9 and 10 are of the order of ±1 % (and this precision can be maintained even at sub-ppm levels due to the experimenter's option to suppress noise by various means including signal averaging).

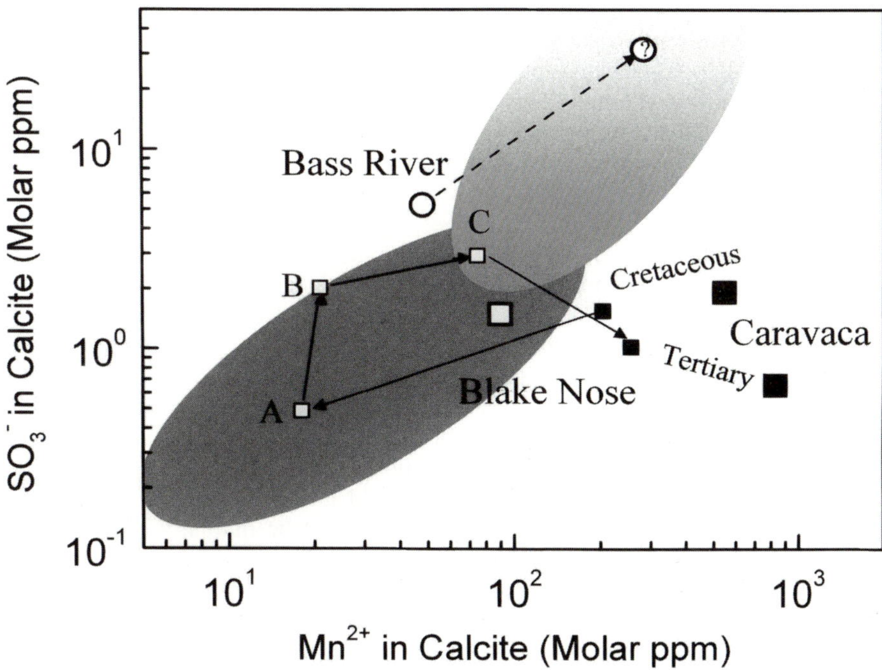

Fig. 12. ESR determinations of Mn^{2+} and SO_3^- concentrations in calcite in trans-KT-boundary materials from Blake Nose (small squares) and Caravaca (large squares), with comparison to Chicxulub impact deposits from México and Belize (shaded fields) and a bulk sample of Bass River spherules (circles). The dark-gray field recapitulates the compositional range of limestone ejecta clasts from central Belize as inferred in Fig. 11. The gray-gradient field, also taken from Fig. 11, represents data for accretionary lapilli from various sites in México and Belize. Arrows trace the stratigraphic sequence for the Blake Nose core: (A) base of ejecta layer, (B) top of ejecta layer, and (C) fireball layer. The large open square in the dark-gray field pertains to the KT boundary at Caravaca. The dashed arrow indicates the direction and maximum magnitude of a possible correction to the Bass River datum as discussed in the text.

6
Geological Issues

6.1
Phenomenological Interpretation of the Data

In Figs. 9 and 10, ESR measurements of both Mn^{2+} and the sulfite radical SO_3^- in the calcite fractions of *whole-rock samples* from three hemi-pelagic marine KT-boundary sections appear to provide stratigraphic markers of the KT impact event. Because these two paramagnetic species are specific to calcite, their intensities must naturally be proportional to the calcite mass fractions, $f_{calcite}$. However, the Mn^{2+} data of Fig. 10, when compared with the actual values of $f_{calcite}$ shown in superposition, clearly reveal that the Mn^{2+} concentration *per calcite formula unit*, $[Mn^{2+}]$, must itself vary with height in the stratigraphic column. Since the calcite fractions were presently determined only for the Caravaca samples, the ratio of $[SO_3^-]$ to $[Mn^{2+}]$ (right-hand panels in Figs. 9 and 10) is tested here as a stratigraphic marker which is insensitive not only to $f_{calcite}$ but also to all errors in determining the linear sample mass density ρ, *and* (very important; see above) to any possible spectrometer sensitivity variations arising from loading of the microwave cavity by large numbers of water molecules bound within the KT-boundary clays.

It is important, however, to attach some geochemical meaning to the $[SO_3^-]$: $[Mn^{2+}]$ ratio. There are primarily three possible explanations for shifts in this ratio across the KT boundary. One possibility is that there was a long-term change in the biologically-mediated incorporation of Mn^{2+} and/or SO_3^{2-} in calcite initiated by the KT event. Another is that post-impact diagenesis has altered the $[SO_3^-]:[Mn^{2+}]$ ratio by dissolving the original biogenic calcites and replacing them with authegenic calcites relatively depleted or enriched in sulfite or manganese. A third possibility is that the $[SO_3^-]:[Mn^{2+}]$ ratio has been affected by the influx of impact ejecta from Chicxulub; (this effect would mainly explain changes within the ejecta layer). These three explanations need not be mutually exclusive and indeed each may contribute in some segment(s) of a given stratigraphic column. Presumably, the concentrations of SO_3^- and Mn^{2+} in marine limestones deposited above the carbonate compensation depth were originally determined biogenically (e.g., Kastner 1999). Thus, the step-like decrease in the SO_3^- data of Fig. 9b from an average value 0.90 during the last ~15,000 years of the Cretaceous to an average value 0.36 during the first ~30,000 years of the Tertiary (times calculated from deposition rates estimated by Smit 1999) is likely to be a geochemical trace of a change in the biologically-mediated incorporation of SO_3^- in the deposited calcites. This change could relate to the turnover in marine species at the beginning of the Tertiary or to a change in ocean temperature or chemistry lasting at least 30,000 years.

With regard to diagenesis, it is known from ESR studies of cave rocks (speleothems) that Mn^{2+} is rejected in the process of dissolution and recrystallization of limestones, whereas there is some evidence that sulfite scavenged from environmental sources (e.g., "cave guano") may be incorporated into stalactites (Ikeya 1993). Consistent with this general result, we found by ESR that a large optical-quality crystal of natural calcite from a mine in northern México contained 50 times less SO_3^- *and at least 1000 times less* Mn^{2+} than typical non-impact limestones from Caravaca, Spain. On this basis, it can be suggested that dissolution and recrystallization of Caravaca limestones might lead to decreases in both SO_3^- and Mn^{2+} but *very likely to an increase in the [SO_3^-]:[Mn^{2+}] ratio*. In Fig. 9b, the gradual increase in $[SO_3^-]:[Mn^{2+}]$ approaching the KT boundary from below has been accurately fitted (± 1 %) with an exponential function that decays downward from its peak value at the KT boundary with a $1/e$ decay length of 1.4 cm. It is important to note that 1.4 cm is ten times larger than the *downward* range of bioturbation at Caravaca as determined by Robin et al. (1991) from high-stratigraphic-resolution counting of insoluble, micron-size cosmic spinels as a function of column height. Griscom and Beltrán-López (2002) discuss in further detail how the ESR results support the premise that the top few centimeters of the Maastrichtian at Caravaca have been diagenically altered—as is inferred from other evidence summarized by Smit (1999), who ascribed the effect to leaching by sulfuric acid produced by dissociation of pyrite framboids in the boundary clay.

As a first step toward determining the effects of the influx of Chicxulub ejecta in the KT-boundary sediments, we examine the $[Mn^{2+}]$ and $[SO_3^-]$ characters of known ejecta using the scatter plots of Figs. 11 and 12. These data are expressed as the number density of paramagnetic species per million of the ions for which they are substituents in calcite (Mn^{2+} replaces Ca^{2+}, while SO_3^- is formed at sites where SO_3^{2-} has substituted for CO_3^{2-}). The large data points in Figs. 11 and 12 are normalized to the measured calcite fractions, $f_{calcite}$, of the corresponding samples. The small data points in both figures pertain to whole rock samples for which $f_{calcite}$ has not yet been determined and has for convenience been taken to be 1.0. Thus, if $f_{calcite}$ is later determined to be <1, the small data points will move to the upper right along lines of unit slope (i.e., parallel to the dashed line in Fig. 11) by factors of $1/f_{calcite}$. However, from petrological inspection of the samples we know that they are generally >80% calcite, thus these correction factors are expected to be <1.25 in most cases.

Table 3. Absolute determinations of [SO$_3^-$] and [Mn^{2+}] in calcite, each expressed as number of ions per million CaCO$_3$ "molecules". All data were acquired using whole-rock samples. Those data captioned "Normalized to fcalcite" are corrected for known calcite mass fractions. The rest of these data assume that fcalcite = 1 and are therefore subject to future correction upward by a factor of 1/fcalcite.

Sample	Position	Mn^{2+} ppm	SO$_3^-$ ppm
Normalized to f$_{calcite}$:			
Caravaca, Spain	+11 cm	824	0.66
Caravaca KT boundary clay 0 cm		90	1.49
Caravavaca	-23 cm	546	1.93
Guayal, Mexico spheroid (rind)		188	5.86
(core)		84	7.41
Bass River, NJ, spherule bed, bulk		48	5.26
Type-I Calcites:			
Pook's pebbles (Armenia, Belize)			
#1		129	1.54
#2		94	1.39
#3		63	1.66
#4		55	2.58
#5		18	0.27
Pook's pebble, Teakettle, Belize (sawed from interior of cobble)		6	0.25
Type-II Calcites:			
Ramonal N. gray-white spheroid		395	13.31
Ramonal N. spheroid bed (bulk)		226	14.35
Ramonal N. spheroid bed (bulk)		207	11.04
Ramonal N. spheroid (interior)		83	2.16
... (brown crust)		71	2.44
Armenia spheroid: (pink core)		96	2.64
(white rind)		105	6.30

Sample	Position	Mn^{2+} ppm	SO$_3$ ppm
ODP Leg 171B/1049A Core:			
	+117.1 cm	257	1.26
	+83.2 cm	250	1.23
	+20.5 cm	267	1.53
fireball layer	+16.5 cm	75	2.95
top of ejecta layer	+14.3 cm	21	2.03
bottom of ejecta layer	+1.5 cm	18	0.49
	-4.5 cm	178	1.75
	-82.8 cm	205	1.56
Type-III Calcites:			
Albion Island chalky layers:			
SW wall #1 (top of layer)		284	<0.0010
SW wall #1 (bottom of layer)		358	0.0230
SW wall #2		284	<0.0010
SW wall #3		290	0.0016
NE wall		234	<0.0010
Albion Island radially fibrous Spheroids #1		695	<0.0060
#2		659	<0.0028
Ramonal S. spheroid bed (bulk, wht)		194	0.0268
Ramonal S. spheroid bed (bulk, red)		126	0.0270
Armenia, Belize matrix materials:			
Pook's pebble bed	+1.0 m	42	2.05
Base of Pook's pebble bed	0 m	79	1.73
Spheroid bed	-1.3 m	67	6.13
Santa Teresa, Belize:			
Altered microtektite layer		142	1.10

6.2
Proximal Chicxulub-Crater Deposits

Figure 11 portrays absolute ESR determinations of [Mn^{2+}] and [SO$_3^-$] for samples from several KT boundary deposits in southern México and Belize, all of which were collected within 600 km of the center of the Chicxulub structure. Only two of the sites (Guayal, México, ~530 km, and Santa Teresa, Belize, ~580 km) are farther than 500 km from the structure (see Fig. 1). We shall refer to all of these sites as "proximal" (c.f.., Melosh 1989).

Figure 11 demonstrates the present suite of proximal ejecta to include as many as three different types of calcites, each characterized by ESR data falling into a particular area on the [SO$_3^-$]-versus-[Mn^{2+}] scatter plot. As a guide to the eye, these three areas are highlighted by three differently-shaded elliptical fields; (the field to the upper right will be referred to as the "gray-gradient" area). While the upper pair of ellipses mutually overlap, the data defining them do not. Moreover, the materials yielding the data points within the dark-gray area are distinguishable from those of the gray-gradient area on the basis of lithology. That is, the non-eclipsed part of the dark-gray ellipse encompasses only limestone ejecta clasts, while the gray-gradient ellipse to the upper right contains only accretionary lapilli and spheroid bed bulk deposits that contain accretionary lapilli. The pale-gray ellipse at the bottom of Fig. 11 does not overlap either of the other two and contains three other lithologically-distinct materials that will be discussed below. Fig. 12 (data points) displays absolute ESR determinations of $[Mn^{2+}]$ and $[SO_3^-]$ for KT-boundary and non-KT-boundary materials from three marine sites distal to the crater; these are shown superposed on the two upper shaded areas defined in Fig. 11 in order to facilitate comparison of the data points for the distal samples with the areas occupied by two of the three calcite types present in the proximal ejecta. The natures of the samples yielding data points in the three highlighted fields of Fig. 11 will now be discussed in sequence:

Accretionary lapilli. The large solid diamonds in Fig. 11 pertain to a Guayal (México) KT-boundary oblate spheroid that was shown by thin-section microscopy to be an accretionary lapillus (Fig. 13 and Griscom et al. 1999); the left-hand data point pertains to the core and the right-hand one to material from within ~2 mm of the surface near the equator. As for all of the *large* data points in Figs. 11 and 12, those for the Guayal spheroid were determined by dividing the ESR species concentrations (number per gram whole rock) by the known value of $f_{calcite}$ to obtain number per gram calcite; this number was then converted to number per mole calcite Energy-dispersive spectroscopy (EDS) performed in conjunction with scanning electron microscopy (SEM) showed the Guayal spheroid to be an aggregate of calcite, dolomite, and aluminosilicate particles (Fig. 13c and Griscom et al. 1999). These particles range in size from ~1 to ~75 μm, with the interstitial spaces pervasively filled with fine-grained quartz (Fig. 13c and Griscom et al. 1999); (this interstitial quartz accounts for the high E′-center concentration in the ESR spectrum of Fig. 7a). Quartz and calcite were the only crystalline phases identifiable by X-ray diffraction. The aluminosilicate particles (which were found by EDS to contain Mg and Ca and possibly Na and Fe) appear black in Fig. 13b and are regarded as possible tektite glasses. However, no explicit evidence for their degree of (non-)crystallinity has yet been obtained. A value of $f_{calcite}$ =0.321 was eventually determined by several successive grindings, HCl treatments, and re-weighings, followed by re-measurement by ESR and performing an extrapolation based on the still-surviving Mn^{2+}-in-calcite ESR signals (Griscom et al. 1999). Thus, Griscom et al. (1999) inferred that much of the calcite was present as micrometer-sized particles hermetically encased in a pervasive matrix of authigenic chert.

The other data falling in the gray-gradient field in Fig. 11 refer to (chert-free) calcite spheroids from Albion Formation spheroid beds (Ocampo et al. 1996, Pope et al. 1999). In particular, the open circles belong to a single cm-size accretionary lapillus from the Armenia spheroid bed in central Belize (~475 km from the center of Chicxulub, Fig. 1). The lower open circle of this two-data-point grouping pertains to the coherent pink core of the lapillus; the upper circle was recorded for its friable snow-white rind. The solid triangles in the gray-gradient field represent

[SO$_3^-$] and [Mn^{2+}] determinations for individual accretionary lapilli from spheroid bed outcrops at Ramonal North, located in Quintana Roo, Mexico (~350 km from the center of the Chicxulub structure, Fig. 1). The lower-left-hand couplet of solid triangles in this field represents the white core (left) and gray rim with red-brown rind (right) of one of these cm-size Ramonal North spheroids. The small solid triangle at the far upper right pertains to the outermost ~3 mm of a ~1.5-cm-diameter grey-white accretionary lapillus also from the Ramonal North site. The hollow upward triangles represent bulk samples from the Ramonal North site containing smaller (~mm sized) accretionary lapilli; one of these samples was pink, the other white.

Chicxulub limestone ejecta clasts. Sub-rounded to rounded clasts of micritic limestone referred to as "Pook's pebbles" (the first exemplars were discovered near Pook's Hill Lodge in central Belize) are found in Chicxulub ejecta layers in Belize (Ocampo et al. 1997, Pope et al. 2000). Pook's pebbles are pink to white shallow-water limestones retaining their original shell fragments and microfossils, although they have been partially or totally recrystallized. The surfaces of the Pook's pebble ejecta are polished and striated and often exhibit surface-penetrating lithic fragments (Ocampo et al., 1997; Pope et al., 2000). The present study included a 2×3×26 mm^3 monolithic sample (not powdered) sawed from the interior of a Pook's pebble recovered near Pook's Hill, Belize; this sample gave rise to the data point (solid square) at the extreme left of Fig. 11

At a highway cut at Armenia, central Belize, a ~5-m-thick Pook's pebble bed overlies a ~5-m-thick spheroid bed, which in turn rests on dolomites of the Late Cretaceous Barton Creek Formation and soils derived there from (Pope et al. 2000). Five ~1-to-5-cm-diameter Pook's pebbles were sampled from within 1 meter of the base of the bed, and ESR spectra of SO$_3^-$ and Mn^{2+} in calcite were recorded for powdered samples (solid white circles in the dark-gray field of Fig. 11). The data points for four of these plot close to those of the lowest-SO$_3^-$ /lowest-Mn^{2+} accretionary lapilli of the gray-gradient field. These four points, taken together with the Pook's Hill Pook's pebble data point (solid square) and an intermediate one in the Armenia suite, define the part of the dark-gray elliptical field not overlapped by the gray-gradient field.

Armenia, Belize, matrix materials. Also sampled at Armenia were matrix materials from the impact deposits. The crosses (×) in Fig. 11 represent absolute ESR determinations of [SO$_3^-$] and [Mn^{2+}] in matrix material from: (A) the spheroid bed, 1.3 m below the top, (B) the base of the overlying Pook's pebble bed, and (C) within the Pook's pebble bed, ~1 m above its base. It can be seen that the matrix material of the spheroid bed bears a strong affinity for the single Armenia accretionary lapillus thus far analyzed by ESR, whereas the matrix materials from the base and interior of the Pook's pebble bed associate with the embedded pebbles. Thus, the large clasts and the lapilli are inferred to be of approximately the same compositions as the fine-grained matrices that respectively contain them.

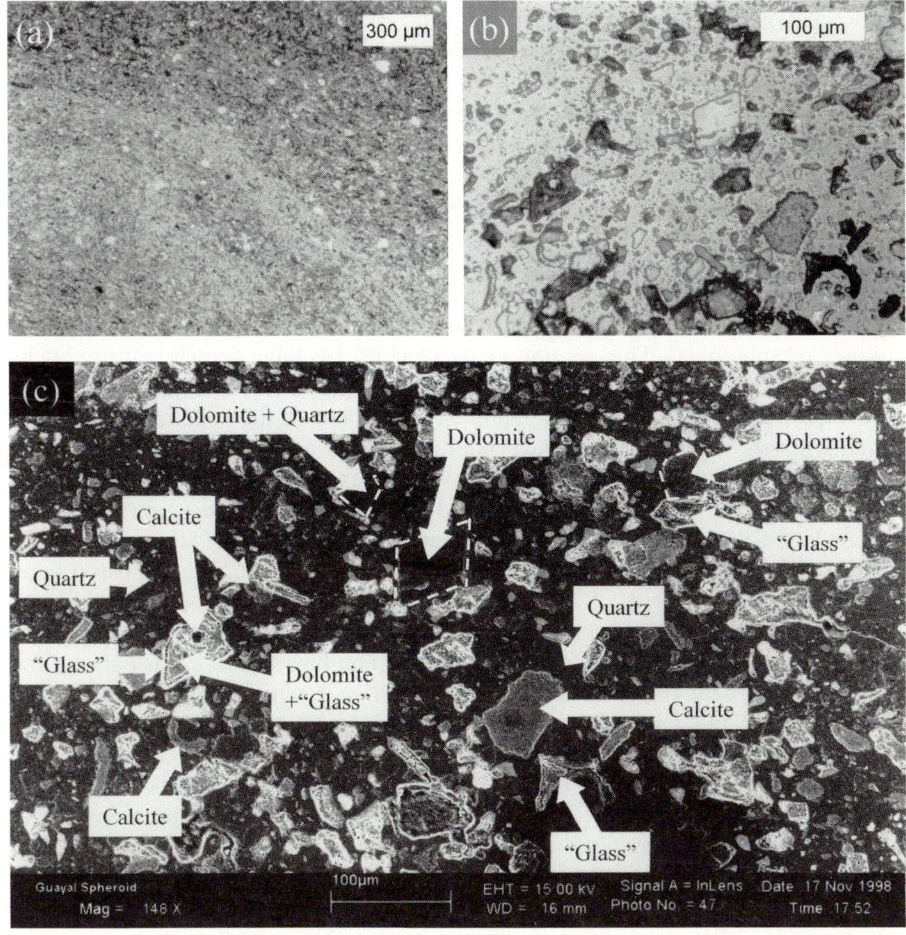

Fig. 13. Micrographs of a thin section of a 1.2×0.6 cm oblate spheroid from El Guayal, México: (a) viewed in transmitted plane-polarized light, (b) viewed under confocal illumination focused on the frontal surface, and (c) imaged by scanning electron microscopy encompassing the same field of view as (b). Various particles seen in both (b) and (c) are identified in (c) on the basis of energy-dispersive analyses performed under the electron microscope. Dolomite grains exhibiting low contrast in (c) have been highlighted by dashed lines. Objects appearing black in (b) and analyzed to contain Si, Mg, Al, and Ca are labeled "glass" but their degree of non-crystallinity has not yet been determined. Dark regions identified as quartz are areas of matrix material selected as being relatively free of small-particle inclusions. (See also Griscom et al. 1999.)

Sulfite-depleted proximal deposits. The pale-gray elliptical field in Fig. 11 includes data points for three distinctly different deposits from proximal Chicxulub ejecta beds. The open downward triangles within the pale-gray field represent matrix materials from the base of the Albion Formation spheroid bed in Ramonal South, Mexico (about 100 m south of Ramonal North). One of these samples was deep red in color, the other white. Petrographic analyses revealed that these samples are mostly limestone clasts and altered glass (smectite) in a coarse (~50 micron) calcite matrix. No accretionary lapilli were observed.

The solid diamonds in the pale-gray field of Fig. 11 belong to ~2-cm-thick "chalky layers" in the Albion Formation spheroid bed at Albion Island, Belize. These layers commonly lie along the upper and lower contacts of the spheroid bed and occasionally along diagonally-cross-cutting planes within the spheroid bed and they generally exhibit sharp slickensides contacts with both the spheroid bed and the overlying diamictite bed (Pope et al. 1999). The Albion Island "chalky layers" are uniform in their creamy white color, powdery texture, and apparent absence of spheroids. Thin section analyses (Fig. 14) have shown that the layers are composed mostly of coarse calcite and sparry calcite veins, which is consistent with their Mn^{2+}-in-calcite ESR signatures. By contrast, the spheroid bed proper contains mostly dolomite and clay spheroids and dolomite clasts in a dolomite matrix (Ocampo et al. 1996, Pope et al. 1999; Fouke et al., 2002). Indeed, all six samples from the Albion Island spheroid bed that have been studied by ESR revealed the ESR signature of Mn^{2+} in *dolomite*. Since dolomite does not manifest an SO_3^- signature, the Mn^{2+} data for KT dolomites are not further discussed here, except to mention that the [Mn^{2+}] value for one of the dolomite spheroids was determined by ESR to be 75 ppm (i.e., similar to the Mn^{2+} content of the limestone ejecta clasts from Armenia, Belize: Table 3). The horizontal "error bar" on the solid diamond near the bottom of Fig. 11 does not represent actual errors but rather delimits the entire measured range of Mn^{2+} contents in four different "chalky-layer" samples from diverse locations, including the northeast and southwest corners of the ~0.3-km long Albion Island quarry. On the other hand, the [SO_3^-] measurement pertaining to this lowest data point is precise only to within a factor of ~3 due to the difficulty of separating the extremely weak signal from a noisy and curving baseline. The SO_3^- signals, if any, of other three investigated samples were totally lost in the noise, thus restricting the possible concentrations of this paramagnetic species to one part per billion or less in these three samples (symbolized by downward arrow emanating from the lower solid diamond).

The ESR intensity data for the radially-fibrous calcite spheroids, found only in two small clay pockets in the diamictite bed at Albion Island (Ocampo et al. 1996, Pope et al. 1999), revealed the highest Mn^{2+} concentrations of any proximal KT-boundary calcites (see Table 3). On the other hand, their SO_3^- ESR signals were too weak to be measured under the prevailing experimental conditions. ESR spectra were recorded for two separate spheroids in this category, one of these on two different spectrometers. The horizonal bars in the pale-gray field of Fig. 11 mark the measured Mn^{2+} concentrations determined in these three measurements and also the noise-floor-imposed upper limits on the possible amounts of SO_3^-.

The downward arrows emanating from these bars indicate that the actual SO_3^- concentrations must lie below the bars.

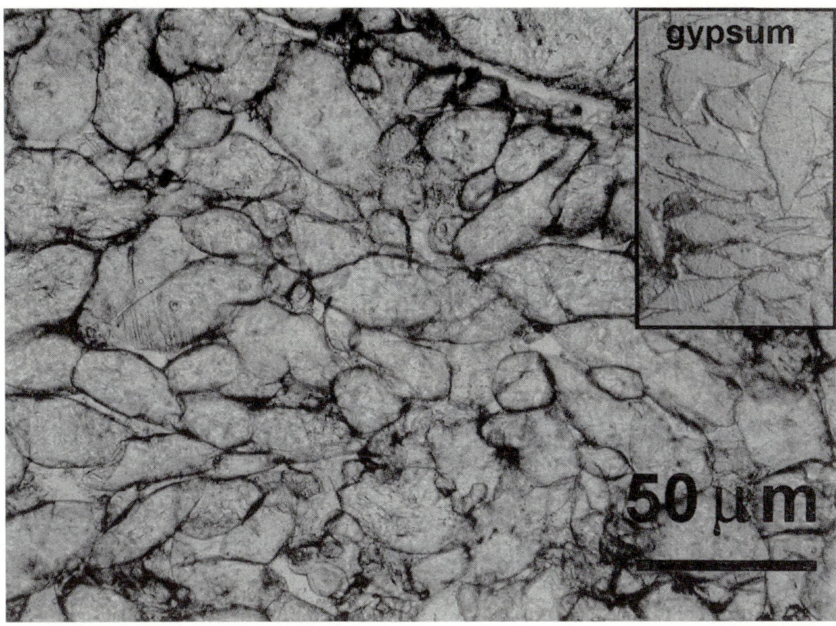

Fig. 14. Thin-section photomicrograph of a sample of material from the Albion Formation "chalky layers", Albion Island quarry, Belize. Transmitted plain-polarized light. Note the lozenge shaped calcite crystals and their similarity in form to the gypsum crystals in the inset (same scale and viewing parameters). The calcite in the Albion Formation "chalky layers" is interpreted as a secondary replacement of gypsum.

Another unusual sample. The open downward triangle falling slightly to the right of the dark-gray area in Fig. 11 represents a pale-green sample with vitreous luster from Santa Teresa in southern Belize, a site located about ~580 km from the crater center (Fig. 1). Thin section analyses reveal that this sample is composed of vesicular glass spherules (microtektites) and shards that have been altered to clay. The samples come from slump blocks of KT ejecta remobilized in the early Tertiary (Smit 1999). The ESR results suggest that none of the original glass is preserved [see Griscom (1990) for a review of possible signatures]; only the usual spectra of SO_3^- and Mn^{2+} in polycrystalline calcite were recorded for a sample selected for its glassy appearance.

7
Interpretations and Conclusions

7.1
Proximal Chicxulub Ejecta Layers

The ESR analyses of the proximal Chicxulub ejecta indicate that there are at least three types of calcite in these deposits. Calcite type I includes all data points used to define the dark-gray elliptical area in Figs. 11 and 12 and comprises limestone fragments ejected from the Yucatán platform. These limestones have been partially to wholly recrystallized. We suspect that much of this recrystallization occurred prior to impact; however, shock effects in limestones are notoriously difficult to detect. Therefore the type I calcites may typify Chicxulub clastic limestone ejecta.

All data points used to define the gray-gradient elliptical area in Figs. 11 and 12 thus far represent only accretionary lapilli and bulk samples of the beds containing such lapilli. This field touches or slightly overlaps that of calcite type I. We propose that this close proximity indicates that some of the accretionary lapilli contain some clastic limestone grains of type I. Thin-section analyses of some the accretionary lapilli confirm that they do contain common clastic limestone grains (e.g., Figs. 3 and 13, and Griscom et al. 1999). Nevertheless, the fact that many of the spheroid data are located far to the upper right of the gray-gradient field confirms that another type of calcite is present in all of the lapilli, and in a few samples abundant. We speculate that this "type-II" calcite is the product of calcite crystallization in the impact vapor plume, either directly by condensing vapors, or by the back reaction of CaO and CO_2 produced by the vaporization of carbonate target rocks (e.g., Pope et al. 1997). CaO condenses at 3500°C (Gupta et al. 2001) and thus large amounts of CaO must have condensed near the crater upon the initial expansion and cooling of the vapor plume. Presumably, the calcites forming under such conditions would trap part of the SO_3^{2-} (or SO_3) and perhaps also Mn^{2+} co-present in the vapor. Large amounts of sulfate were vaporized by the impact (e.g., Pope et al. 1997, Pierazzo et al. 1998). Moreover, impact vapor plume simulations with lasers and theoretical calculations both predict that the Chicxulub vapor plume contained abundant SO_3 (Gerasimov et al. 1994, Ohno et al. 2002), and presumably the ionized state SO_3^{2-} was present in the hottest vapors. The upward lightening of the gray gradient in the upper elliptical field of Figs. 11 and 12 reflects our speculation that the fraction of clastic limestone grains in the accretionary lapilli is decreasing upward, concomitant with an upwardly-increasing fraction of vapor-plume condensates.

The type-III calcites (data points in the pale-gray ellipse in Fig. 11) are characterized by very low contents of SO_3^- and in fact probably represent more than one type of calcite given their varied geological contexts. First, thin-section examinations of the Ramonal South spheroid-bed samples (open downward triangles) have shown them to comprise calcite and clays but no lapilli. Second, petrographic work with the white "chalky layers" from Albion Island (solid diamonds) indicates that these calcites are secondary replacements of gypsum (the calcite crystal forms seen in Fig. 14 are psuedomorphs after gypsum forms). And, third, petrographic analyses, including cathodluminescence studies, of the radially fibrous calcite spheroids (horizontal bars) indicate that they are primary crystal growths and not replacements of preexisting forms (Pope et al., 1996; Fouke et al., 2002).

Despite the similarity of their SO_3^{2-} and Mn^{2+} ESR signatures to those of the secondary Albion Island "chalky layers", we continue to favor the speculation of Pope et al. (1999) that the radially-fibrous spheroids may have crystallized from molten calcite inclusions in silicate glass melts, given that they have been found only within masses of smectite that originally may have been glass. We rule out recrystallization from aqueous solutions because Mn^{2+} is known to be expelled during recrystallization of speleothems (Ikeya 1993) and given our own observation that a primary solution-derived calcite crystal was depleted in Mn^{2+} by more than three orders of magnitude below the level (~700 ppm) that characterizes the radially fibrous spheroids (see above). Moreover, we view the presence of dolomite cores in many of the radially-fibrous spheroids (Pope et al. 1999) as inconsistent with a model of primary crystal growth inside of vugs. We therefore propose the radially-fibrous spheroids to be partially melted and partially digested dolomite xenoliths which recrystallized within cooling masses of silicate glass. In evoking this process, we assume without proof that MgO was incorporated into the glass more rapidly than was CaO (with evolution of CO_2) and that the melting point of calcite is lower than that of dolomite; (the melting points of both minerals are presently unknown since in unconstrained situations they decompose). In this view, molten calcite spheroids would have formed (by surface tension with the surrounding glass) with some of these still including undigested dolomite grains at their centers. It would be necessary to assume further that any initially-present sulfite ions evolved, perhaps as SO_3, while the molten calcite retained, or even concentrated, Mn^{2+}. This model should be amenable to verification by experimental glass melting.

7.2
Distal Chicxulub Ejecta Layers

The data point for the Caravaca KT-boundary layer (large open square at the center of Fig. 12) is coincident in SO_3^--Mn^{2+} space with the high end of the dark-gray "target-rock" regime. One possible reason for this result is that KT-boundary layer at Caravaca contains significant amounts of fine carbonate ejecta from the Yucatán platform, transported there along with the shocked quartz that is also present (e.g., Izett, 1990). The marked contrast between the proposed clastic ejecta and the pre- and post-impact limestones may relate to the fact that the ejecta come from a shallow-water platform environment (evaporites are common), whereas normal calcite deposition at Caravaca was dominated by deep water processes. Nevertheless, to the extent that the ESR spectrometer may have been desensitized by the presence of water molecules in the un-dried KT-boundary clay sample (see above), the "true" position of the Caravaca KT-boundary datum will lie somewhat to the upper right of its present position (along a line parallel to the dashed arrow connecting the as-measured Bass River datum to its possible location, as described above). Thus, the calcite component of the ejecta layer at the KT boundary at Caravaca might equally well be interpreted as containing vapor-plume condensates. More work will be needed to resolve this issue. In any event, the as-measured $[SO_3^-]$ and $[Mn^{2+}]$ values in calcite, with or without possible future correction, clearly link the Caravaca KT-boundary impact layer with possible Chicxulub ejecta.

Data points corresponding to the bottom and top of the ejecta layer intercepted by the ODP Leg 171B/1049A core (points A and B, respectively, Fig. 12) also fall within the calcite type-I field, suggesting that the ejecta layer at Blake Nose contains limestone clastic ejecta from Chicxulub.   Point C in Fig. 12, corresponding to the fireball layer, falls within the calcite type-II field and thus may be indicative of a contribution by vapor-plume condensates. Taking into account evidence for gravity-flow grading of the Blake Nose ejecta layer (Smit 1999), the present data are consistent with identifying the coarser fraction of the calcite ejecta as comminuted Chicxulub ejecta clasts.   The finer calcite components, especially those present in the fireball layer, are then interpreted as including a significant admixture of vapor-plume condensates.

In Fig. 10, the SO_3^--in-calcite ESR data (open circles) are compared with AAS data for Ca in the ODP Leg171B/1049A core (Martínez-Ruiz et al. 2001: dashed line), which are here normalized to bring the two forms of data into approximate coincidence.   The Ca AAS data are a proxy for the calcite mass fraction $f_{calcite}$. The ESR data represent the product of the SO_3^- concentration in calcite times the calcite mass fraction, i.e., $[SO_3^-] \times f_{calcite}$.   Thus, it can be seen that $[SO_3^-]$ is approximately constant in the basal Tertiary and also constant (but with a 10 % higher value) in the terminal Cretaceous, thus mimicking the larger step-like change in $[SO_3^-]$ across the KT boundary at Caravaca (Fig. 9b). However, within the ejecta layer, $[SO_3^-]$ is considerably enhanced above the "background" defined by the Ca proxy, and this enhancement grows monotonically larger from the base of the ejecta layer up to the fireball layer, where the product $[SO_3^-] \times f_{calcite}$ reaches a maximum. Thus, the SO_3^- profile in the ejecta and fireball layers is explained by the initial arrival of clastic ejecta from the Yucatán platform (type-I calcites) followed by a mixture of clastic ejecta with ever-increasing amounts of calcites enriched in SO_3^-. We propose that the SO_3^--enriched calcites are vapor-plume condensates (type-II calcites) with $[SO_3^-]$ values higher than those in the target rocks.

The data from Bass River (New Jersey) are more equivocal in the sense that the as-measured SO_3^- and Mn^{2+} concentrations presently fall outside of the fields currently taken to define type-I and type-II calcites. However, later correction for spectrometer desensitization by water molecules (see above) may move the Bass River data point to a position in the presently defined type-II field (i.e., to some point in the gray-gradient area of Fig. 12 lying along the dashed arrow). Thus, there is a possibility that this sample too may eventually prove to contain a substantial component of vapor-plume material.

8
Conclusions

Electron-spin-resonance analyses of calcites from KT-boundary deposits from both proximal and distal locations demonstrate that this technique is very promising for identifying various types of Chicxulub impact debris. The present analyses have isolated three types of calcites distinguishable on the basis of their ESR-measured concentrations of SO_3^- and Mn^{2+} ions substitutional for CO_3^{2-} and Ca^{2+}, respectively, in the calcite lattice. Type I corresponds with limestone ejecta clasts, type II with ejecta that includes a significant component of calcites formed in the vapor plume, and type III with several forms of Chicxulub ejecta characterized by SO_3^- depletion. Calcite types I and II are promising tracers for clastic ejecta from the Yucatán platform and impact-vapor-plume condensates, respectively. Analyses of these tracers in samples from four distal KT ejecta layers (Bass River, Blake Nose, Caravaca, and Sopelana) suggest that debris from Chicxulub is distributed globally as clastic limestone ejecta and/or vapor plume condensates. These findings are consistent with sequestering of CO_2 and SO_3 in the rapidly condensing vapor plume, a process that may have important ramifications for future studies of the climatic effects of the Chicxulub impact.

Acknowledgements

This project was jump-started by the early availability of Spanish KT stratigraphic sequences carefully collected and kindly offered by E. Robin (LSCE UMR CEA/CNRS 1572, Gif-sur-Yvette, France). The authors are also indebted to J. Urrutia-Fucugauchi (Instuto de Geofisica, UNAM, México) and R.K. Olsson (Dept. Geological Sciences, Rutgers U., USA) for provision of the Guayal spheroid and Bass River spherules, respectively, and to the Ocean Drilling Project for samples from the Leg 171B/1049/A core. P.I. Premovič (Lab. for Geochemistry and Cosmochemisty, Dept. Chem., U. Niš, Yugoslavia) kindly recorded some of the early data for the ODP samples. Funding for this research was provided by the Fulbright Foundation, Universités de Paris 6 & 7, and the U.S. National Aeronautics and Space Administration Exobiology Program, in part through a contract with the Jet Propulsion Laboratory (California Institute of Technology).

References

Alvarez LW, Alvarez W, Asaro F, Michel HV (1980) Extraterrestrial cause for the Cretaceous-Tertiary extinction. Science 208: 1095-108

Deer WA, Howie RA, Zussman J (1975) An Introduction to Rock-Forming Minerals, Longman, London, 528 pp

Dunlop DJ (1972) Magnetite: Behavior near the single-domain threshold. Science 176: 41-43

Feigl FJ, Anderson JH (1970) Defects in crystalline quartz: electron paramagnetic resonance of E' vacancy centers associated with germanium impurities. Journal of Physics and Chemistry of Solids 31: 575-589

Fouke BW, Zerkle AL, Alvarez W, Pope KO, Ocampo AC, Wachtman RJ, Grajales Nishimura JM, Claeys P, Fischer AG (2002) Cathodoluminescence petrography and isotope geochemistry of KT impact ejecta deposited 360 km from the Chicxulub crater, at Albion Island, Belize. Sedimentology 49: 117-138

Gerasimov MV, Dikov YP, Yakovlev OI, Wlotzka F (1994) High-temperature vaporization of gypsum and anhydrite: Experimental results (abstract). Lunar and Planetary Science 25, 413-414

Grajales JM, Moran DJ, Padilla P, Sanchez MA, Cedillo E, Alvarez W (1996) The Loma Tristes Breccia: a K/T impact-related breccia from southern México [abs.]. Geological Society of America Abstract with Program 28: A-183, p. 93

Griscom DL (1984) Ferromagnetic resonance of precipitated phases in natural glasses. Journal of Non-Crystalline Solids 67: 81-118

Griscom DL (1990) Electron Spin Resonance. In: Uhlmann DR, Kreidl NJ (eds), Glass Science and Technology, Vol. 4B, Advances in Structural Analysis. Academic Press, New York, pp 151-261

Griscom DL (2001) Amorphous Materials: Electron Spin Resonance. In: Encyclopedia of Materials: Science and Technology, Elsevier Science, The Netherlands, pp 179-186

Griscom DL, Beltrán-López V (2002) ESR Spectra of Limestones from the Cretaceous-Tertiary Boundary: Traces of a Catastrophe. Proceedings of the International Symposium on Electron Spin Resonance Dating and Dosimetry, Osaka, Japan. (in press)

Griscom DL, Beltrán-López V, Merzbacher CI, Bolden E (1999) Electron spin resonance of 65-million-year-old glasses and rocks from the Cretaceous-Tertiary boundary. Journal of Non-Crystalline Solids 253: 1-22

Gupta SC, Ahrens, TJ, Yang, W (2001) Shock-induced vaporization and global cooling from the K/T impact. Earth and Planetary Science Letters 188: 399-412

Hyde JH (1961) EPR Standard sample data, 1 p. Varian Associates, Palo Alto, California

Ikeya M (1993) New Applications of Electron Spin Resonance - Dating, Dosimetry and Microscopy, World Scientific, Singapore, 500 pp

Izett GA (1990) The Cretaceous/Tertiary boundary interval, Raton Basin, Colorado and New Mexico, and its content of shock-metamorphosed minerals: evidence relevant to the K/T boundary extinction theory. Geological Society of America Special Paper 249: 1-100

Kai A, Miki T (1991) Sulfite radicals in irradiated calcite. Japanese Journal of Applied Physics 30: 1109-1110

Kai A, Miki T (1992) Electron spin resonance of sulfite radicals in irradiated calcite and aragonite. Radiation Physics and Chemistry 40: 469-476

Kastner M (1999) Oceanic minerals: Their origin, nature of their environment, and significance. In: Geology, Mineralogy, and Human Welfare. Proceedings of the National Academy of Sciences USA 96: 3380-3387

Martínez-Ruiz F, Ortega-Huertas M, Kroon D, Smit J, Palomo-Delgado I, Rocchia R (2001) Geochemistry of the Cretaceous-Tertiary boundary at Blake Nose (ODP Leg 171B). In: Kroon D, Norris RD, Klaus A (eds) Western North Atlantic Paleogene and Cretaceous Paleooceanography. Geological Society London Special Publications 183: 131-148

Melosh HJ (1989) Impact Cratering - A Geologic Process. Oxford Monographs on Geology and Geophysics #11, Oxford University Press, New York, 245 pp

Miura Y, Ohkura Y, Ikeya M, Miki T, Ruckledge J, Takaoka N, Neilsen TFD (1985) ESR data of Danish calcite at Cretaceous and Tertiary boundary. In: ESR Dating and Dosimetry, IONICS, Tokyo, pp 469-476

Norris RD, Kroon D, Klaus A et al. (1998) Proceedings of the Ocean Drilling Program, Initial Reports 171B. Ocean Drilling Program, College Station, Texas, 47-91

Ocampo AC, Pope KO, Fischer AG (1996) Ejecta blanket deposits of the Chicxulub crater from Albion Island, Belize. In: Ryder G, Fastovsky D, Gartner S (eds) The Cretaceous-Tertiary Event and Other Catastrophes in Earth History. Geological Society of America Special Paper 307: 75-88

Ocampo AC, Pope KO, Fischer AG (1997) Carbonate ejecta from the Chicxulub crater: Evidence for ablation and particle interactions under high temperatures and pressures. Lunar and Planetary Science 28, CD ROM abstract # 1861.

Odom AL, Rink WJ (1988) Natural accumulation of Schottky-Frenkel defects: implications for a quartz geochronometer. Geology 17: 55-58

Olsson RK, Miller KG, Browning JV, Habib D, Sugarman PJ (1997) Ejecta layer at the Cretaceous-Tertiary boundary, Bass River, New Jersey (Ocean Drilling Program Leg 174AX). Geology 25: 588-580

Ohno S, Sugita S, Kadono T, Hasegawa S (2002) Mass spectroscopic observation of sulfur chemistry in laser-simulated impact vapor clouds [abs.]. Lunar and Planetary Science 33, CD ROM abstract # 1634

Pierazzo E, Kring DA, Melosh HJ (1998) Hydrocode simulation of the Chicxulub impact event and the production of climatically active gases. Journal of Geophysical Research 103: 28,607-28,625

Poole CP Jr. (1983) Electron Spin Resonance, Wiley-Interscience, New York, 780 pp

Pope KO, Ocampo AC, Fischer AG, Morrison J (1996) Carbonate condensates in the Chicxulub ejecta from Belize [abs.]. Lunar and Planetary Science 27: 1045-1046

Pope KO, Baines KH, Ocampo AC, Ivanov BA (1997) Energy, volatile production, and climatic effects of the Chicxulub Cretaceous/Tertiary impact. Journal of Geophysical Research 102: 21,645-21,664

Pope KO, Ocampo AC, Fischer AG, Alvarez W, Fouke BW, Webster CL, Vega FJ, Smit J, Fritsche AE, Claeys P (1999) Chicxulub impact ejecta from Albion Island, Belize. Earth and Planetary Science Letters 170: 351-364

Pope KO, Ocampo AC, Fischer AG, Fouke BW, Wachtman RJ (2000) Anatomy of the Chicxulub ejecta blanket. Geol. Soc. Amer. Abstracts Program 32: 163

Premovič PI, Nikolič DN, Tonsa IR, Pavlovič MS, Premovič MP, Dulanovič DT (2000) Copper and copper porphyrins of the Cretaceous-Tertiary boundary at Stevns Klint (Denmark). Earth and Planetary Science Letters 177: 105-118

Premovič PI, Nikolič DN, Pavlovič MS, Bratislav ŽT (2002) Geochemistry of the Cretaceous-Tertiary boundary (Fiskeler) at Stevns Klint (Denmark): Trace metals (Cu, Ni, Cr, Ir, Au) of kerogen and biogenic calcite. Earth and Planetary Science Letters (accepted for publication).

Rink WJ, Odom AL (1991) Natural alpha recoil particle radiation and ionizing radiation sensitivities in quartz detected with EPR: implications for geochronometry. Nuclear Tracks and Radiation Measurements 18, Nos. 1 & 2: 163-173

Robin E, Boclet D, Bonté P, Froget L, Jéhanno C, Rocchia R (1991) The stratigraphic distribution of Ni-rich spinels in Cretaceous-Tertiary boundary rocks at El Kef (Tunisia), Caravaca (Spain), and Hole 761C (Leg 122). Earth and Planetary Science Letters 107: 715-721

Silsbee RH (1961) Electron spin resonance in neutron-irradiated quartz. Journal of Applied Physics 32: 1459-1462

Serway RA, Marshall SA (1967) Electron spin resonance absorption spectra of CO_3^- and CO_3^{3-} molecule ions in single crystal calcite. Journal Chemical Physics 46: 1949-1952

Shepherd RA, Graham, WRM (1984) EPR of Mn^{2+} in polycrystalline dolomite. Journal of Chemical Physics 81: 6080-6084

Smit J (1999) The global stratigraphy of the Cretaceous-Tertiary boundary impact ejecta. Annual Reviews of Earth Planetary Science 27: 75-113

Weeks RA (1956) Paramagnetic resonance of lattice defects in irradiated quartz. Physical Review 130: 570-576

Weil JA, Bolton JR, Wertz JE (1996) Electron Paramagnetic Resonance: Elemental Theory and Practical Applications, Wiley, New York, 568 pp

Petrography and Geochemistry of a Deep Drill Core from the Edge of the Morokweng Impact Structure, South Africa

Wolf Uwe Reimold[1] and Christian Koeberl[2]

[1]Impact Cratering Research Group, School of Geosciences, University of the Witwatersrand, Private Bag 3, P.O. Wits 2050, Johannesburg, South Africa. (reimoldw@geosciences.wits.ac.za)
[2]Institute of Geochemistry, University of Vienna, Althanstrasse 14, A-1090 Vienna, Austria. (christian.koeberl@univie.ac.at)

Abstract. A large impact structure has in recent years been described from an area about 80 km northwest of the town of Vryburg in the North West Province of South Africa. The Morokweng impact structure was formed 145 Ma ago, around the time of a minor mass extinction marking the Jurassic-Cretaceous boundary. Early size estimates for this structure ranged from 140 to 340 km, whereas recent geophysical modeling favors a mere 70-80 km diameter. If a >200 km diameter would be indicated, this impact event likely would have had globally catastrophic effects.

A more than 3.4 km long drill core obtained ca. 40 km to the southwest of the center of the Morokweng Structure has been analyzed petrographically, and the major lithologies were analyzed for their major and trace element compositions. Only a single, 10-cm-wide, bedding-parallel injection of impact breccia was recovered in this drillcore. All other lithological sections are unshocked and generally undeformed (with exception of several narrow cataclastic zones that were probably formed prior to the impact event). Several thick sections of felsic granophyric rocks of granitic mineralogical and chemical composition are texturally very similar to the impact melt rock from Morokweng, but have distinct mineralogical (lack of a substantial pyroxene component) and chemical (less FeO, different alkali element contents) differences to the impact melt rock from Morokweng, the so-called Morokweng Granophyre. U-Pb SHRIMP dating of representative samples of felsic granophyre established beyond doubt that these rocks were formed in Archean times and are unrelated to the 145 Ma old impact melt rock. The general lack of deformation in the drill core rocks strongly suggests that this borehole was sunk outside or, at best, close to the edge, of the Morokweng impact structure – providing a strong case for a maximum diameter of about 70 km for this structure.

1
Introduction

A large impact structure was identified in the northwestern parts of South Africa (Andreoli et al. 1995; Corner et al. 1997; Hart et al. 1997; Koeberl et al. 1997; Reimold et al. 1999; Andreoli et al. 1999; Koeberl and Reimold 2002; Reimold et al. 2002). Known as the Morokweng Structure after a township located just off its center, this impact structure was first identified on the basis of a strong, positive, near-central aeromagnetic anomaly over the so-called Ganyesa Dome, centered at 23°32'E/26°20'S ca. 80 km northwest of the town of Vryburg in North West Province. Outcrop in this sand- and calcrete-covered region is extremely scarce, but drill core from the area of this anomaly, as well as a small number of surface samples, provided evidence of shock metamorphic deformation in basement rocks (Andreoli et al. 1995; Corner et al. 1997; Hart et al. 1997). The source of the magnetic anomaly was soon recognized to be a thick body of impact melt rock, because of the presence of a very significant meteoritic component (Koeberl et al. 1997, 2000; Hart et al. 1997; McDonald et al. 2000) and some evidence in the form of shock metamorphosed clasts in this melt rock. The then available shallow boreholes provided evidence of an at least 150 m thick impact melt layer, above brecciated and stongly shock metamorphosed basement granitoids (Reimold et al. 1999; Andreoli et al. 1999). More recently, McDonald et al. (2000) noted that a melt layer with a minimum thickness of 410 m was encountered in a new borehole [M3], and Dutta et al. (2001) and Hart et al. (2002) reported impact melt rock thicknesses of ≥ 500 and 870 m, respectively, for this new drill core. M3 allegedly (Hart et al. 2002) entered granitoids to a final depth in excess of 1000 m where drilling was terminated in brecciated basement. They concluded that the impact melt body in the center of the Morokweng structure had an overall thickness of > 800 m and that the granitoids described by earlier workers from the shallow WF boreholes represented "granite boulders. In contrast, Reimold et al. (1999) had suggested that these granitoids represented mega-breccia below the impact melt body. It is suggested here that the currently known variation in thicknesses of impact melt rock from Morokweng, as illustrated in Fig. 2 of Hart et al. (2002), does not clarify the issue whether the average thickness of Morokweng impact melt is > 800 m or whether, alternatively, the sub-impact melt body geometry of the crater floor (in the central uplift region) is complex – as one would expect after collapse of a central uplift structure and further post-impact modification.

Early Morokweng workers used regional geophysical and geomorphological evidence to speculate on the size of this impact structure. Corner et al. (1997) proposed that Morokweng could be as large as 340 km, whereas Andreoli et al. (1995; 1999) favored values between more than 300 and 140 km. Bootsman et al. (1999) and Reimold et al. (1999) allowed a maximum size of 200 km, based on geomorphological constraints. Just recently, Dutta et al. (2001) reported a diameter for the Morokweng structure of ≥ 150 km. In contrast, detailed gravity and magnetic modeling of the Morokweng structure by Henkel et al. (2002) strongly suggests that this impact structure can not be larger than 70-80 km in diameter, consistent with the disruption of regional dike swarms clearly evident on the regional aeromagnetic anomaly map.

This greatly reduced maximum diameter for Morokweng poses a severe constraint for speculations that have, in the past, linked the existence of this large impact structure with the minor mass extinction at the Jurassic/Cretaceous boundary. The age of the Morokweng Structure has been dated conclusively by two groups (Koeberl et al. 1997; Hart et al. 1997) applying single zircon U-Pb isotopic techniques at 145 ± 1 Ma. This age coincides with the presently preferred age for the Jurassic/Cretaceous boundary and the associated faunal extinction event. Some evidence for a possible involvement of an impact event with the J/K mass extinction has been presented, but remains unconfirmed (Kudielka et al. 2001).

Here we report stratigraphic, petrographic and geochemical detail for a 3420 m long drill core obtained about 40 km from the center (in accordance with the width of the central aeromagnetic anomaly) of the Morokweng impact structure. The purpose of this investigation was to pursue further evidence to constrain the maximum size of the Morokweng impact structure and, should it be found that this borehole would extend into the interior of the impact structure, obtain more information on structure and impact breccia distribution within the crater. First results of this study were recently published by Reimold et al. (2002).

2
The Morokweng Impact Structure

The Morokweng impact structure formed in a composite target composed of Phanerozoic sedimentary and igneous rocks of the ca. 350-180 Ma Karoo Supergroup, overlying a thick sequence of Proterozoic (Griqualand West Sequence and Olifantshoek Supergroup of approximately 1.8-2.25 Ga age, as well as Transvaal Supergroup (ca. 2.15-2.6 Ga) metasedimentary rocks above a crystalline basement of a variety of Archean granitoids. The region is also intruded by numerous mafic dikes that probably represent lavas of the Ventersdorp Supergroup (ca. 2.7 Ga) and/or of the Ongeluk Formation (ca. 2.2 Ga), as well as Karoo dolerite dike swarms. The entire region is covered by Tertiary and Quaternary sand and calcrete of the Kalahari Group deposited since the Tertiary. The local geology around the center of the impact structure is shown in Fig. 1, largely as it was determined from shallow hydrogeologic drilling (Geological Survey of South Africa 1974). The basement granitoids in this area are widely covered by a thick succession of carbonates and other supracrustals of the Transvaal Supergroup, and the Griqualand West and Olifantshoek sequences are represented by thick layers of banded ironstones/jaspilites and cherts. The latter lithologies are well exposed in subhorizontal formations to the west and northwest of the structure. Locally, Dwyka tillite of the Karoo Supergroup has been encountered.

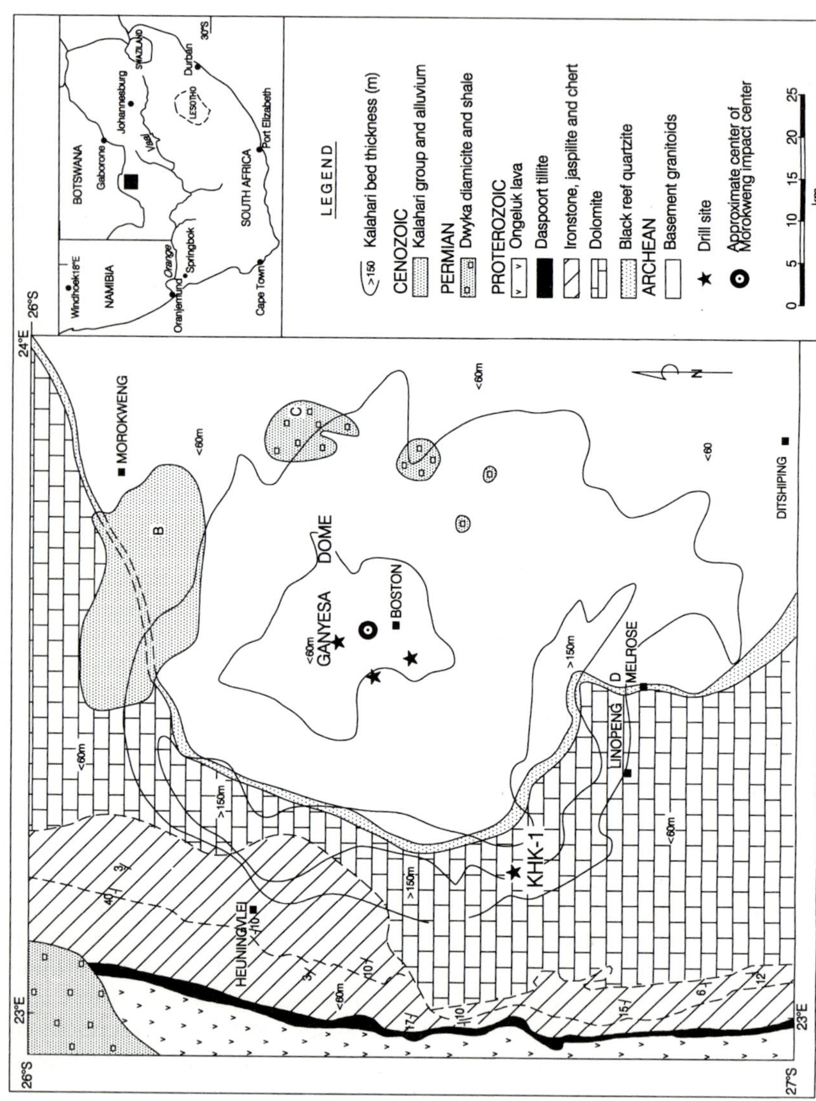

Fig. 1. Simplified geological map of the area of the Morokweng impact structure, based on hydrogeological drilling by the Geological Survey of South Africa (1974). Note the locations of the three boreholes into the central impact melt body and the position of deep borehole KHK-1.

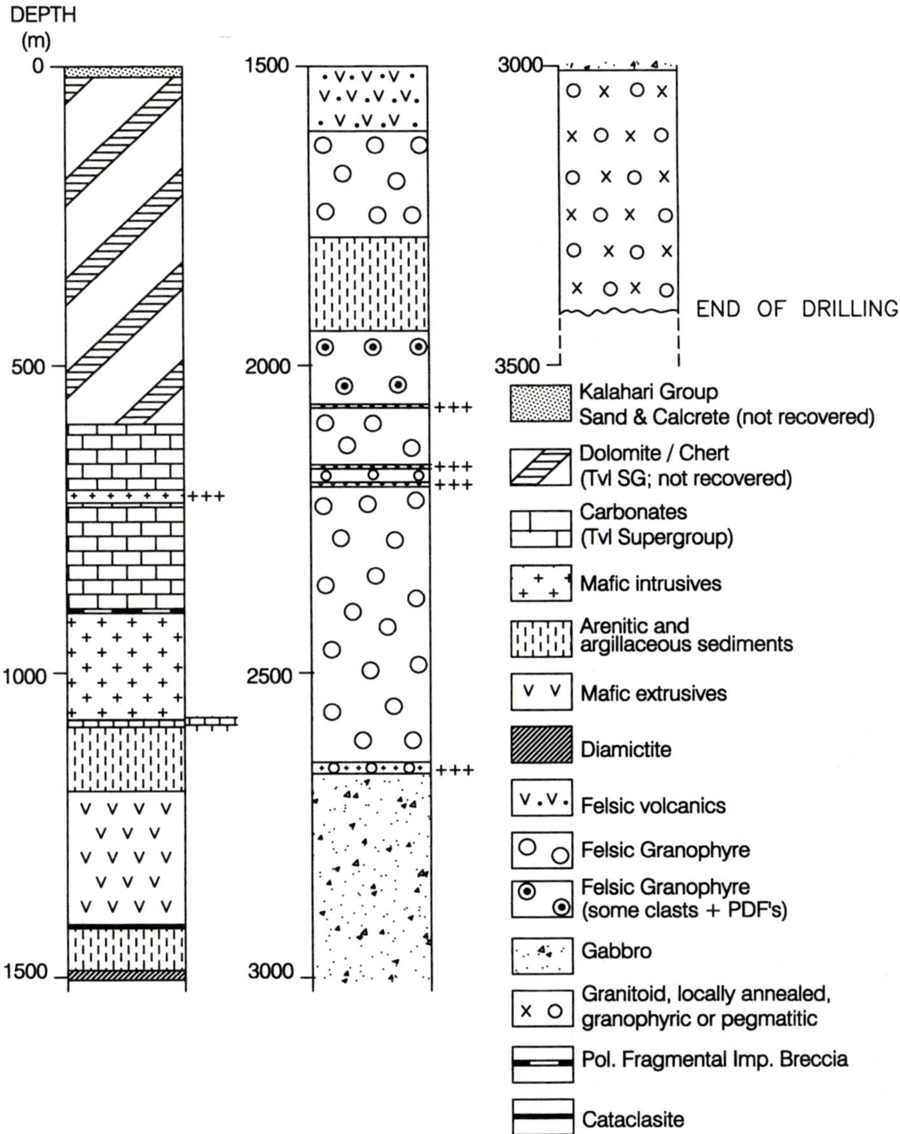

Fig. 2. Schematic stratigraphic column for the KHK-1 drill core. Abbreviations: Tvl = Transvaal; SG = Supergroup; Pol = polymict; Imp = impact.

Figure 1 also illustrates the locations of the three shallow boreholes in the central part of the structure, which – as discussed - provided the critical evidence for an impact origin of the Morokweng structure. Also shown is the position of borehole KHK-1, a 3420-m-deep exploration borehole drilled by Anglogold Limited on the farm Kelso 351 (~23°12'E/26°40'S; Fig. 1), approximately 40 km southwest of the center of the aeromagnetic anomaly. On the presumption that a borehole at this distance from the center would have either intersected impactites or shock metamorphosed strata of the crater fill or floor to the large Morokweng impact structure, we investigated this core in detail. Representative samples of all lithologies were studied petrographically and were analyzed by X-ray fluorescence spectrometry and instrumental neutron activation analysis for their major and trace element abundances. Details on the analytical procedures, accuracies and precisions can be obtained in Koeberl (1993) and Reimold et al. (1994), as well as Kudielka et al. (2001), and references therein.

2
Stratigraphy of the KHK-1 Borehole and some Petrographic Detail

Figure 2 provides a schematic drill core log of the KHK-1 borehole. The top 599.15 m were unfortunately not recovered by the company, but information was obtained that it consisted exclusively of dolomite and chert, presumably of the 2.25-2.6 Ga Transvaal Supergroup. We were also informed that no particular deformation features had been recorded and that these strata were oriented approximately subhorizontal. There is, of course, a minor chance that some thin layers of possibly impact-related formation might have been missed in this unrecovered top of the core. The recovered drill core commences at a depth of 599.15 m. The Transvaal meta-sedimentary sequence continues with near-horizontal stratigraphic contacts (bedding), to a depth of 889.1 m. Rock deformation is extremely scarce. Only a few, up to several centimeter wide, cataclastic zones were logged. At the depth of 889.1 m, a wide gabbroic intrusion was intersected that extends to the depth of 1083 m. The contact at 889.1 m is characterized by a 10-cm-wide breccia layer consisting of a polymict mixture of angular fragments of granitoids and metasedimentary rocks (carbonate and arenitic lithologies), besides a small but significant component of angular to rounded glass fragments (Figs. 3a and b). Shock deformation in the form of single sets of planar deformation features was encountered in two quartz clasts in this breccia. Consequently, this narrow layer has been classified as impact breccia, specifically of the type of polymict, suevitic impact breccia.

Between 1083 and 1093.3 m depth, a zone comprising dolomite, presumably still of Transvaal Supergroup association, was transected, before a sequence of arenitic metasedimentary rocks follows to 1200 m depth. This is succeeded by a

Fig. 3. Photomicrographs of salient features in lithologies of drill core KHK-1 (all images correspond to widths of field of view of 3.4 mm and were taken with crossed polarizers). (a) Angular, in part annealed, felsic mineral clasts and an elongated melt fragment (upper central area, above a biotite-rich clast) in polymict suevitic impact breccia from 889.1 m depth in drill core KHK-1. (b) Another micro-image of polymict suevitic breccia from 889.1 m depth: a large, medium-grained gabbro clast on left side and a number of carbonate clasts of generally very angular shapes. (c) Plagioclase crystal in felsic granophyre from 1950.5 m depth. The crystal has been partially melted resulting in a myrmekitic pattern of microscopic melt veinlets – somewhat reminiscent of the appearance of checkerboard feldspar texture well known from clasts in impact melt rock from many impact structures. (d) Similar to (c), but more pronounced partial melting of an alkali feldspar grain in felsic granophyre from 1996 m depth. (e) Partially melted granitic clast in felsic granophyre from 1998.8 m depth. (f) An example of the micropegmatitic texture (granophyric intergrowth) of quartz and feldspar minerals that is characteristic for the textures of the so-called felsic granophyres in drill core KHK-1 (sample from ca. 3300 m depth). This texture is very similar to that of felsic portions of the impact melt rock from this impact structure (the so-called Morokweng Granophyre, named in analogy to the Vredefort Granophyre).

complex series of volcanic flows to a depth of 1417.8 m. This sequence will be studied in future. The volcanic sequence is directly underlain by a 1.2-m-wide clastic breccia (characterized by highly angular clasts and thought to be the product of local cataclasis) that completely lacks any evidence of shock deformation. This breccia overlies more metasedimentary rocks, including a layer logged as diamictite, a polymict breccia completely devoid of shock deformation. It should be noted that a number of diamictite horizons occur in various stratigraphic positions throughout the Transvaal and Griqualand West sequences.

This section of the drillcore is followed by further felsic volcanics and a felsic granophyre to a depth of 1787.3 m, where more metasedimentary rock, including two more thin bands thought to represent diamictite, also devoid of shock deformation, is reached. This metasedimentary rock extends to a depth of 1948.2 m, where a thick layer of felsic granophyre commences. It consists of micropegmatitic intergrowths of quartz and various feldspar minerals, with minor biotite and amphibole (Figs. 3c,d). This package extends to the depth of 2669.4 m, where a thick gabbroic intrusion is entered. The felsic granophyre is texturally similar to the equally granophyric impact melt rock from Morokweng that is known as the Morokweng Granophyre (e.g., Koeberl and Reimold 2002). Detailed descriptions of this rock type have been given, for example, by Reimold et al. (1999) and Andreoli et al. (1999). Whereas the felsic granophyres of KHK-1 are predominantly quartz-feldspar rocks that contain only minor amounts or accessory components of mafic minerals, the Morokweng Granophyre contains a large pyroxene component. The gabbro intrusion continues to a depth of 3012 m, where locally pegmatoidal (mostly micropegmatoidal) granitoids are encountered, which continue, with the exception of a narrow dioritic intrusion at around 3408 m depth, to the final drilling depth of 3420 m. Absolutely no characteristic petrographic evidence of shock metamorphism was detected in the minerals of any of the numerous felsic granophyre samples studied by us, although a number of feldspar crystals have been encountered that have internal textures reminiscent of beginning checkerboard feldspar formation (Figs. 3d,e). These observations are interpreted to indicate that at least some of these granophyric materials are derived from melting of older granitic phases. Three sequences of granophyric materials were encountered in this drill core and were individually sampled for chemical analysis (cf. below). These granophyres have been termed I-III and comprise samples from 1775.4 m depth, and the zones between 1950.5 and 1998 m, and 2075.6 and 2617.8 m depth, respectively (compare Fig. 2).

All lithological contacts encountered in drill core KHK-1, as well as the internal bedding of metasedimentary rock sequences, are consistently (sub)horizontal. The only deformation zones observed involve the thin impact breccia layer at 889.1 m depth and several thin cataclasite occurrences. These cataclasites are generally composed of highly angular clasts that are cemented by silica or carbonate infills. There are no indications at what time(s) they might have formed – any time between Transvaal Supergroup deposition at > 2.15 Ga or since. Chert and carbonate cemented cataclasites are well-known to occur throughout the Transvaal Supergroup in various parts of the Kaapvaal Craton.

3
Geochemistry

Major and trace element data for a large number of KHK-1 samples of all important lithologies in this drill core are listed in Table 1. The samples analyzed include various mafic intrusives encountered in the drill core, the assumed diamictite layers, a number of mafic intrusives, the felsic volcanics and felsic granophyres, and the basement granitoids. Two ternary diagrams (FeO-MgO-total alkali elements and CIPW normative Orthoclase-Albite-Anorthite contents) are shown in Figs. 4a and b, and present a comparison of the compositions of the various lithologies of KHK-1 with those of the impact melt rock (Morokweng Granophyre – Koeberl and Reimold 2002) and Archean granitoids of the basement underneath the central part of the impact structure (data from Koeberl and Reimold 2002). First-order observations (Fig. 4a) show that the Archean basement granitoids are somewhat enriched in the alkali elements in comparison to all other analyzed lithologies. Compositions of felsic volcanics are very diverse but overlap with some of the felsic granophyre samples. The impact melt rock field shows rather limited variation (see also Koeberl and Reimold 2002), in comparison to felsic granophyre and volcanic compositions. Only a small number of felsic granophyre samples has compositions that straddle the impact melt field, whereas the majority of felsic granophyre samples is much enriched in alkali elements. In terms of normative feldspar compositions (Fig. 4b), felsic volcanics and, especially, felsic granophyre samples have compositions that are generally distinct from the Morokweng Granophyre composition. Basement granitoids are generally different in composition from Morokweng Granophyre, but overlap the compositional fields of felsic volcanics and granophyres. Whether felsic volcanics and granophyres are genetically related or not is a topic that deserves further investigation.

Felsic granophyres are generally similar with regard to major element abundances, but the sample from 1775.4 m depth is distinct by lower Na and higher K contents. Felsic volcanics are characterized by generally similar compositions to those of felsic granophyres and show internal variation similar to that of the felsic granophyres (compare Table 1, Reimold et al. 2002). Particular differences between felsic granophyre and Morokweng Granophyre compositions include much higher Fe and Ca contents in the impact melt rock, with MgO values being rather similar. In comparison to basement granitoids, felsic granophyres and felsic volcanics have somewhat higher Fe and MgO contents. Granite basement from KHK-1 compares compositionally very well with granitic basement from the center of the impact structure (see Koeberl and Reimold 2002).

In Table 2, average compositions (and standard deviations) have been compiled for the most important KHK-1 lithologies, in comparison to impact melt rock and basement granitoids from the drill cores of the central part of the impact structure. Fig. 5 compares the chondrite-normalized rare earth element (REE) patterns for average compositions of KHK-1 felsic volcanics and felsic granophyres, Morokweng impact melt rock, and basement granitoids from the KHK-1 drill core and from borehole WF5 (described by, e.g., Reimold et al. 1999, and termed here

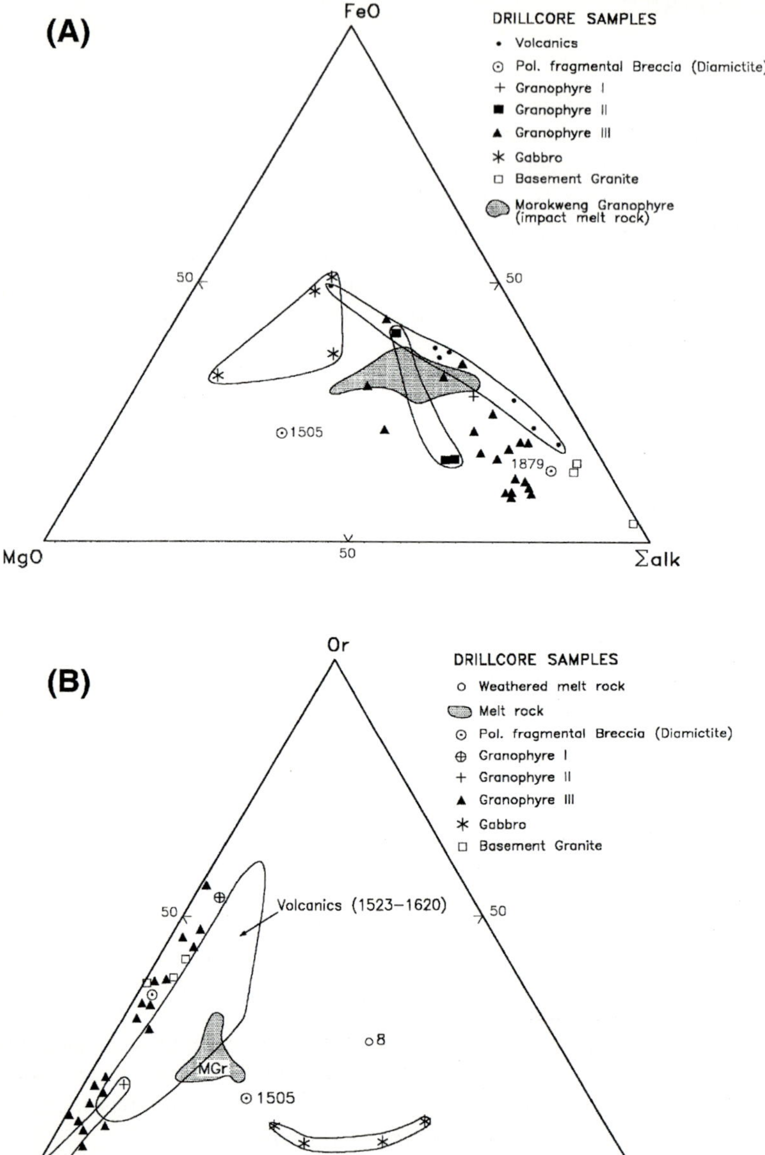

Fig. 4. Ternary diagrams comparing the chemical compositions of important KHK-1 lithologies with the compositional field of Morokweng Granophyre (impact melt rock). (a) FeO-MgO-total alkali element abundances; (b) CIPW normative orthoclase-albite-anorthite contents.

Table 1. Chemical composition of representative samples from KHK-1 drill core lithologies.

	Gabbroic Intrusions		Shale	Diamictite	Felsic Volcanics								
	1006.0	1022.9	1391.2	1505.0	1523.0	1528.4	1536.7	1545.6	1555.0	1581.0	1589.1	1600.51	1620.3
SiO_2	47.23	47.51	68.78	57.72	72.54	67.75	62.04	74.32	69.47	71.31	71.68	69.89	64.61
TiO_2	2.41	2.55	0.50	0.33	0.71	0.83	0.92	0.69	0.82	0.71	0.84	0.84	1.00
Al_2O_3	12.21	12.77	9.00	12.33	11.58	13.14	13.42	11.07	13.47	11.10	12.16	12.57	13.22
Fe_2O_3	18.65	18.84	7.43	4.78	4.93	6.43	9.91	2.04	4.06	4.70	3.29	5.95	5.71
MnO	0.30	0.28	0.14	0.13	0.09	0.09	0.11	0.09	0.09	0.10	0.09	0.09	0.12
MgO	5.77	5.12	5.69	9.20	2.39	3.27	5.02	0.69	1.22	1.89	1.16	2.16	2.41
CaO	9.41	9.77	2.89	3.29	1.08	1.08	1.39	2.11	1.68	1.97	1.10	0.65	2.41
Na_2O	2.74	2.65	2.71	3.15	4.27	2.39	2.18	2.83	5.80	2.25	1.84	4.98	3.06
K_2O	0.76	0.33	0.01	1.67	0.75	2.64	1.48	4.15	1.92	2.46	6.58	1.41	3.02
P_2O_5	0.16	0.23	0.28	0.14	0.17	0.19	0.38	0.16	0.20	0.15	0.15	0.19	0.18
LOI	0.27	-0.07	1.63	6.07	1.31	2.62	3.15	1.86	1.55	3.32	1.28	1.35	4.21
Total	99.91	99.98	99.06	98.81	99.82	100.43	100.00	100.01	100.28	99.96	100.17	100.08	99.95
Sc	39.4	37.8	n.d.	9.44	8.26	19.3	14.6	7.31	10.2	6.98	7.18	10.5	10.1
V	564	526	115	61	44	101	98	35	51	32	39	62	68
Cr	93.4	73.2	1010	647	9.26	12.2	5.41	46.2	8.5	10.1	12.1	9.31	8.06
Co	54.1	46.7	23	24.8	5.03	3.54	5.91	1.45	2.99	24.6	2.43	3.74	4.09
Ni	74	64	217	339	15	16	23	10	14	12	6	10	10
Cu	290	312	67	3	2	<2	12	<2	<2	10	<2	<2	<2
Zn	127	134	53	227	77	108	152	27	56	80	35	83	90
Ga	43	60	n.d.	11	8	43	10	12	22	18	16	38	26
As	0.4	0.41	n.d.	13.4	1.01	2.19	0.65	1.31	0.32	1.11	1.07	0.71	0.25
Se	0.09	0.12	n.d.	0.22	0.28	0.31	0.19	0.21	0.29	0.21	0.11	0.38	0.27
Br	0.2	0.5	n.d.	0.9	0.6	0.9	1.1	0.4	0.5	0.2	0.9	0.4	0.8
Rb	26.1	13.1	7	47.5	23.1	155	82.5	75.8	41.2	125	115	40.5	116
Sr	132	138	48	89	70	60	75	70	41	67	32	35	44
Y	34	40	32	12	76	76	81	57	86	79	72	75	57
Zr	126	146	90	144	467	535	580	395	572	477	525	558	505
Nb	8	9	8	11	23	24	27	21	25	26	28	27	28
Sb	0.14	0.35	n.d.	0.71	0.95	1.87	0.32	1.05	0.85	0.95	0.75	0.68	0.75
Cs	0.25	0.93	n.d.	3.92	0.59	2.63	1.36	0.98	0.54	2.01	1.05	0.41	2.12
Ba	89	55	21	305	210	565	285	670	356	350	1090	304	505
La	6.99	8.29	n.d.	24.5	97.3	85.3	48.6	26.8	13.5	19.5	19.9	21.5	11.5
Ce	17.6	22.4	n.d.	46.5	183	168	101	62.8	35.2	43.1	47.5	48.5	28.5
Nd	13.2	16.7	n.d.	19.8	75.1	82.2	49.9	33.6	25.5	23.8	27.3	30.1	17.1
Sm	4.19	5.01	n.d.	3.15	12.3	14.7	11.5	7.61	9.61	7.17	6.75	8.31	5.01
Eu	1.65	1.85	n.d.	0.91	2.71	3.48	3.21	1.42	2.47	1.74	1.27	1.64	0.98
Gd	5.5	6.78	n.d.	2.83	14.5	12.4	12.3	8.99	13.1	9.79	9.05	9.28	6.7
Tb	1.11	1.19	n.d.	0.43	2.08	2.23	2.03	1.45	2.32	1.87	1.65	1.93	1.41
Tm	0.49	0.62	n.d.	0.23	1.07	1.46	1.29	0.84	1.41	1.04	1.18	1.36	1.12
Yb	2.97	3.78	n.d.	1.15	7.31	9.92	9.18	5.91	1.01	7.43	8.63	9.59	8.05
Lu	0.44	0.59	n.d.	0.19	1.08	1.44	1.28	0.89	1.49	1.15	1.38	1.44	1.28
Hf	3.23	4.16	n.d.	4.05	12.1	15.1	14.1	10.1	16.3	12.1	14.3	15.2	14.6
Ta	0.21	0.29	n.d.	1.03	1.44	2.07	1.48	1.24	1.95	1.55	1.85	2.15	1.87
W	1.5	0.9	n.d.	1.8	0.9	2.3	1.2	1.2	3.1	2.5	6.5	5.5	2.5
Ir (ppb)	<2	<1	n.d.	0.5	<1	<1	<0.6	0.3	<0.5	<2	<1	<1.5	<0.8
Au (ppb)	3	9	n.d.	4	<3	<5	1.5	0.2	<2	1.8	1.4	1.4	3.2
Th	0.69	0.91	n.d.	6.25	10.2	13.1	14.5	9.03	14.1	11.3	12.9	14.3	14.1
U	0.84	0.24	n.d.	1.29	2.51	4.16	3.62	2.44	9.76	3.69	4.32	5.26	4.56
K/U	7540	11458	n.d.	10788	2490	5288	3407	14173	1639	5556	12693	2234	5519
Zr/Hf	39.0	35.1	n.d.	35.6	38.6	35.4	41.1	39.1	35.1	39.4	36.7	36.7	34.6
La/Th	10.1	9.1	n.d.	3.92	9.54	6.51	3.35	2.97	0.96	1.73	1.54	1.50	0.82
Th/U	0.82	3.79	n.d.	4.84	4.06	3.15	4.01	3.70	1.44	3.06	2.99	2.72	3.09
La_N/Yb_N	1.59	1.48	n.d.	14.4	8.99	5.81	3.58	3.06	9.03	1.77	1.56	1.51	0.97
Eu/Eu*	1.05	0.97	n.d.	0.93	0.62	0.79	0.82	0.52	0.67	0.63	0.50	0.57	0.52

Major elements in wt%, trace elements in ppm, except as noted. All Fe as Fe_2O_3.

Table 1. (cont.)

	Felsic Granophyre	Felsic Volcanics			Felsic Granophyre								Clast
	1775.4	1879.6	1902.7	1950.5	1996.0	1998.0	1998.1	2075.6	2085.5	2100.6	2108	2148.8	2148.8
SiO_2	70.39	67.61	69.17	66.23	72.86	71.69	68.52	70.72	n.d.	67.25	72.69	71.06	50.08
TiO_2	0.79	0.34	0.38	0.71	0.60	0.41	0.63	0.26	n.d.	0.36	0.82	0.23	2.25
Al_2O_3	12.13	13.32	14.42	13.44	12.64	13.65	13.94	15.15	n.d.	16.03	12.69	15.04	11.77
Fe_2O_3	4.75	2.06	3.32	7.44	2.33	2.40	3.39	1.71	n.d.	2.81	2.66	1.29	16.97
MnO	0.09	0.11	0.11	0.10	0.10	0.09	0.10	0.08	n.d.	0.09	0.08	0.09	0.26
MgO	2.17	1.30	1.75	3.60	2.94	3.14	4.79	1.92	n.d.	2.74	1.49	2.14	5.27
CaO	0.67	1.40	0.52	0.51	0.36	0.28	0.37	0.47	n.d.	0.49	0.59	0.48	9.22
Na_2O	2.11	4.69	3.56	4.38	6.82	7.16	5.94	6.84	n.d.	5.98	6.91	6.49	2.39
K_2O	5.51	4.24	5.27	1.37	0.26	0.01	0.15	2.14	n.d.	2.40	1.25	2.00	0.80
P_2O_5	0.18	0.11	0.13	0.18	0.15	0.11	0.17	0.11	n.d.	0.10	0.19	0.10	0.21
LOI	1.33	4.67	1.39	2.15	1.39	1.49	2.32	1.09	n.d.	1.65	1.08	1.17	0.03
Total	100.12	99.85	100.02	100.11	100.45	100.43	100.32	100.49			99.90	100.45	100.09	99.25
Sc	10.1	4.79	3.98	11.1	n.d.	n.d.	n.d.	n.d.	n.d.	3.82	n.d.	n.d.	n.d.
V	47	34	37	106	55	61	85	24	30	66	30	28	453
Cr	7.35	28.1	29.4	17.5	23	23	23	27	20	13.1	25	24	86
Co	5.46	3.01	8.04	10.9	11	10	13	9	10	2.56	9	<9	46
Ni	12	24	25	16	13	17	28	10	<9.	12	<9	<9	82
Cu	<2	<2	<2	<2	<2	<2	<2	<2	<2	<2	<2	<2	246
Zn	70	47	118	69	45	42	67	29	28	36	27	27	134
Ga	30	23	32	37	n.d.	n.d.	n.d.	n.d.	n.d.	12	n.d.	n.d.	n.d.
As	0.14	0.4	0.95	0.48	n.d.	n.d.	n.d.	n.d.	n.d.	0.35	n.d.	n.d.	n.d.
Se	0.24	0.15	0.15	0.16	n.d.	n.d.	n.d.	n.d.	n.d.	0.09	n.d.	n.d.	n.d.
Br	0.9	0.9	0.9	0.7	n.d.	n.d.	n.d.	n.d.	n.d.	0.9	n.d.	n.d.	n.d.
Rb	101	162	139	63.8	<3	<3	<3	40	60	97.4	37	45	47
Sr	33	92	51	56	48	62	52	84	103	91	101	99	138
Y	87	20	20	40	22	22	20	7	4	6	4	8	39
Zr	525	162	162	266	190	158	201	134	78	150	109	137	161
Nb	26	14	12	15	13	11	13	9	8	10	8	9	9
Sb	0.92	0.25	0.95	1.52	n.d.	n.d.	n.d.	n.d.	n.d.	0.19	n.d.	n.d.	n.d.
Cs	0.81	2.24	2.81	1.64	n.d.	n.d.	n.d.	n.d.	n.d.	1.29	n.d.	n.d.	n.d.
Ba	685	776	1130	295	34	28	31	471	565	744	655	536	233
La	22.8	15.8	45.4	10.6	n.d.	n.d.	n.d.	n.d.	n.d.	8.98	n.d.	n.d.	n.d.
Ce	50.7	33.7	79.1	25.3	n.d.	n.d.	n.d.	n.d.	n.d.	21.6	n.d.	n.d.	n.d.
Nd	34.4	17.3	33.9	19.9	n.d.	n.d.	n.d.	n.d.	n.d.	10.4	n.d.	n.d.	n.d.
Sm	11.2	3.79	5.77	6.97	n.d.	n.d.	n.d.	n.d.	n.d.	1.75	n.d.	n.d.	n.d.
Eu	3.18	1.15	1.49	2.08	n.d.	n.d.	n.d.	n.d.	n.d.	0.49	n.d.	n.d.	n.d.
Gd	13.1	4.1	5.52	7.69	n.d.	n.d.	n.d.	n.d.	n.d.	1.39	n.d.	n.d.	n.d.
Tb	2.35	0.56	0.76	1.11	n.d.	n.d.	n.d.	n.d.	n.d.	0.17	n.d.	n.d.	n.d.
Tm	1.32	0.34	0.35	0.57	n.d.	n.d.	n.d.	n.d.	n.d.	0.088	n.d.	n.d.	n.d.
Yb	8.64	1.62	1.29	3.98	n.d.	n.d.	n.d.	n.d.	n.d.	0.58	n.d.	n.d.	n.d.
Lu	1.25	0.22	0.16	0.58	n.d.	n.d.	n.d.	n.d.	n.d.	0.063	n.d.	n.d.	n.d.
Hf	13.9	4.21	4.35	6.98	n.d.	n.d.	n.d.	n.d.	n.d.	4.25	n.d.	n.d.	n.d.
Ta	1.81	0.95	0.91	0.85	n.d.	n.d.	n.d.	n.d.	n.d.	0.24	n.d.	n.d.	n.d.
W	2.5	0.7	0.95	2.5	n.d.	n.d.	n.d.	n.d.	n.d.	0.7	n.d.	n.d.	n.d.
Ir (ppb)	0.2	0.1	0.4	<0.8	n.d.	n.d.	n.d.	n.d.	n.d.	0.1	n.d.	n.d.	n.d.
Au (ppb)	1.5	4.1	1.5	41	n.d.	n.d.	n.d.	n.d.	n.d.	1.5	n.d.	n.d.	n.d.
Th	13.1	11.7	7.71	8.83	n.d.	n.d.	n.d.	n.d.	n.d.	5.52	n.d.	n.d.	n.d.
U	4.64	5.53	1.91	2.82	n.d.	n.d.	n.d.	n.d.	n.d.	0.53	n.d.	n.d.	n.d.
K/U	9896	6389	22993	4048	n.d.	n.d.	n.d.	n.d.	n.d.	37736	n.d.	n.d.	n.d.
Zr/Hf	37.8	38.5	37.2	38.1	n.d.	n.d.	n.d.	n.d.	n.d.	35.3	n.d.	n.d.	n.d.
La/Th	1.74	1.35	5.89	1.20	n.d.	n.d.	n.d.	n.d.	n.d.	1.63	n.d.	n.d.	n.d.
Th/U	2.82	2.12	4.04	3.13	n.d.	n.d.	n.d.	n.d.	n.d.	10.4	n.d.	n.d.	n.d.
La_N/Yb_N	1.78	6.59	23.8	1.80	n.d.	n.d.	n.d.	n.d.	n.d.	10.5	n.d.	n.d.	n.d.
Eu/Eu*	0.80	0.89	0.81	0.87	n.d.	n.d.	n.d.	n.d.	n.d.	0.96	n.d.	n.d.	n.d.

Table 1. (cont.).

	Felsic Granophyre												
	2169.6	2176.5	2194.0	2218.7a	2218.7b	2221.6	2231.1	2233.8	2238.8	2289.3	2292.15	2308.4	2343.9
SiO_2	67.97	66.88	75.38	68.61	67.22	71.51	69.81	71.12	66.80	73.11	73.09	72.86	72.42
TiO_2	0.19	0.36	0.58	0.41	0.40	0.64	0.64	0.69	0.70	0.64	0.68	0.73	0.75
Al_2O_3	13.96	17.10	11.73	16.10	16.94	14.09	13.75	12.47	12.78	11.71	11.84	11.57	12.74
Fe_2O_3	3.51	1.96	1.39	1.27	1.49	1.98	3.24	3.85	5.37	2.29	2.71	2.40	2.51
MnO	0.10	0.08	0.09	0.09	0.09	0.09	0.09	0.10	0.11	0.09	0.10	0.08	0.10
MgO	4.72	2.22	1.66	2.31	2.49	1.74	2.47	2.00	4.94	2.72	1.49	2.26	1.77
CaO	0.93	0.88	0.61	0.52	0.59	0.57	0.80	0.76	1.09	1.04	0.40	1.32	0.71
Na_2O	5.45	7.02	7.44	4.58	4.81	6.54	6.74	4.82	4.74	4.26	3.52	2.85	3.50
K_2O	0.95	2.29	0.16	4.25	4.77	1.34	1.01	3.22	0.48	2.99	4.79	4.91	4.53
P_2O_5	0.10	0.13	0.16	0.11	0.10	0.05	0.18	0.36	0.40	0.13	0.12	0.12	0.17
LOI	2.50	1.36	0.93	1.38	1.52	0.97	1.33	1.02	2.39	1.23	0.98	1.03	1.11
Total	100.38	100.28	100.13	99.63	100.42	99.52	100.06	100.41	99.80	100.21	99.72	100.13	100.31
Sc	n.d.	n.d.	n.d.	n.d.	n.d.	7.81	n.d.	6.98	12.4	n.d.	n.d.	8.91	n.d.
V	37	49	36	63	61	56	79	71	94	30	38	39	32
Cr	25	21	10	20	18	7.65	<9	8.8	7.54	<9	18	8.78	<9.
Co	10	<9	9	9	10	2.66	9	2.21	5.81	10	<9	3.36	10
Ni	21	<9	9	12	11	16	9	14	16	16	<9	13	<9.
Cu	<2	<2	<2	<2	<2	<2	<2	<2	<2	<2	<2	<2	<2
Zn	57	28	29	29	27	24	44	28	76	38	24	34	30
Ga	n.d.	n.d.	n.d.	n.d.	n.d.	23	n.d.	11	29	n.d.	n.d.	19	n.d.
As	n.d.	n.d.	n.d.	n.d.	n.d.	0.72	n.d.	0.35	0.62	n.d.	n.d.	0.57	n.d.
Se	n.d.	n.d.	n.d.	n.d.	n.d.	0.31	n.d.	0.23	0.26	n.d.	n.d.	0.32	n.d.
Br	n.d.	n.d.	n.d.	n.d.	n.d.	1.8	n.d.	1.4	0.7	n.d.	n.d.	1.7	n.d.
Rb	24	53	4	122	115	21.5	4	43.5	17.1	54	84	83.5	72
Sr	83	98	50	153	149	36	38	40	20	51	49	40	60
Y	7	8	47	7	7	101	56	85	114	55	74	61	85
Zr	109	157	464	142	136	483	447	469	460	463	484	473	529
Nb	7	10	27	8	8	25	27	26	28	24	22	22	25
Sb	n.d.	n.d.	n.d.	n.d.	n.d.	0.11	n.d.	0.16	0.35	n.d.	n.d.	0.48	n.d.
Cs	n.d.	n.d.	n.d.	n.d.	n.d.	0.39	n.d.	0.39	1.44	n.d.	n.d.	0.42	n.d.
Ba	344	759	47	2250	2397	298	52	491	75	749	1005	880	899
La	n.d.	n.d.	n.d.	n.d.	n.d.	18.5	n.d.	82.6	103	n.d.	n.d.	56.8	n.d.
Ce	n.d.	n.d.	n.d.	n.d.	n.d.	45.3	n.d.	179	248	n.d.	n.d.	120	n.d.
Nd	n.d.	n.d.	n.d.	n.d.	n.d.	27.3	n.d.	93.4	129	n.d.	n.d.	57.6	n.d.
Sm	n.d.	n.d.	n.d.	n.d.	n.d.	8.03	n.d.	17.8	26.8	n.d.	n.d.	10.5	n.d.
Eu	n.d.	n.d.	n.d.	n.d.	n.d.	1.33	n.d.	2.01	2.95	n.d.	n.d.	1.98	n.d.
Gd	n.d.	n.d.	n.d.	n.d.	n.d.	12.2	n.d.	17.9	23.6	n.d.	n.d.	12.7	n.d.
Tb	n.d.	n.d.	n.d.	n.d.	n.d.	2.33	n.d.	2.33	3.25	n.d.	n.d.	1.68	n.d.
Tm	n.d.	n.d.	n.d.	n.d.	n.d.	1.42	n.d.	1.07	1.74	n.d.	n.d.	1.05	n.d.
Yb	n.d.	n.d.	n.d.	n.d.	n.d.	9.44	n.d.	7.34	11.5	n.d.	n.d.	7.85	n.d.
Lu	n.d.	n.d.	n.d.	n.d.	n.d.	1.33	n.d.	0.99	1.62	n.d.	n.d.	1.17	n.d.
Hf	n.d.	n.d.	n.d.	n.d.	n.d.	15.6	n.d.	12.7	13.9	n.d.	n.d.	14.1	n.d.
Ta	n.d.	n.d.	n.d.	n.d.	n.d.	1.86	n.d.	1.55	1.85	n.d.	n.d.	1.62	n.d.
W	n.d.	n.d.	n.d.	n.d.	n.d.	0.8	n.d.	1.7	0.9	n.d.	n.d.	1.5	n.d.
Ir (ppb)	n.d.	n.d.	n.d.	n.d.	n.d.	0.1	n.d.	0.1	<0.8	n.d.	n.d.	<0.8	n.d.
Au (ppb)	n.d.	n.d.	n.d.	n.d.	n.d.	2	n.d.	1	1.5	n.d.	n.d.	2	n.d.
Th	n.d.	n.d.	n.d.	n.d.	n.d.	4.53	n.d.	13.9	13.5	n.d.	n.d.	11.9	n.d.
U	n.d.	n.d.	n.d.	n.d.	n.d.	1.99	n.d.	1.85	1.98	n.d.	n.d.	3.73	n.d.
K/U	n.d.	n.d.	n.d.	n.d.	n.d.	5611	n.d.	14505	2020	n.d.	n.d.	10970	n.d.
Zr/Hf	n.d.	n.d.	n.d.	n.d.	n.d.	31.0	n.d.	36.9	33.1	n.d.	n.d.	33.5	n.d.
La/Th	n.d.	n.d.	n.d.	n.d.	n.d.	4.08	n.d.	5.94	7.63	n.d.	n.d.	4.77	n.d.
Th/U	n.d.	n.d.	n.d.	n.d.	n.d.	2.28	n.d.	7.51	6.82	n.d.	n.d.	3.19	n.d.
La_N/Yb_N	n.d.	n.d.	n.d.	n.d.	n.d.	1.32	n.d.	7.60	6.05	n.d.	n.d.	4.89	n.d.
Eu/Eu*	n.d.	n.d.	n.d.	n.d.	n.d.	0.41	n.d.	0.34	0.36	n.d.	n.d.	0.52	n.d.

Table 1. (cont.).

	Felsic Granophyre			Gabbro				Granitic Basement			Diorite	Granite
	2395.7	2541.25	2617.8	2722.0	2725.2	2974.3	2998.1	3357.6	3371.6	3396.0	3408.7	3400.0
SiO_2	70.33	70.34	66.76	53.02	54.49	53.25	57.20	70.69	n.d.	71.78	58.19	70.20
TiO_2	0.82	0.74	0.82	1.63	0.89	0.34	0.27	0.15	n.d.	0.25	1.03	0.25
Al_2O_3	11.64	11.94	12.22	13.58	15.09	14.26	18.91	15.82	n.d.	14.81	16.74	14.45
Fe_2O_3	5.42	5.50	8.13	12.80	11.80	8.14	5.62	0.49	1.68	2.11	10.16	2.21
MnO	0.10	0.11	0.12	0.22	0.20	0.20	0.13	0.08	n.d.	0.10	0.19	0.09
MgO	2.73	1.98	3.62	5.76	6.72	11.69	4.47	0.21	n.d.	0.54	2.37	0.78
CaO	0.32	0.44	0.49	7.67	5.63	7.32	8.30	0.37	n.d.	0.72	1.75	1.12
Na_2O	2.26	3.90	3.25	3.65	3.32	1.68	2.93	6.61	4.01	4.63	4.64	4.49
K_2O	4.74	3.08	2.38	0.86	0.60	0.88	0.74	5.32	n.d.	4.59	2.30	5.23
P_2O_5	0.18	0.19	0.23	0.20	0.14	0.09	0.10	0.06	n.d.	0.08	0.33	0.08
LOI	1.30	1.33	1.86	1.26	1.55	2.47	1.23	0.41	n.d.	0.86	2.20	0.83
Total	99.84	99.55	99.88	100.65	100.43	100.32	99.90	100.21		100.47	99.90	99.73
Sc	12.7	8.93	11.6	n.d.	n.d.	n.d.	n.d.	1.46	1.96	n.d.	n.d.	n.d.
V	54	49	80	350	264	103	50	<15	n.d.	20	63	18
Cr	9.15	8.31	10.4	60	18	881	42	10.2	9.53	10	11	13
Co	7.31	6.81	12.3	55	46	43	23	0.95	3.68	<9	15	10
Ni	5	10	12	57	40	160	110	5	13	<9	<9	<9
Cu	<2	26	<2	163	29	<2	75	30	28	<2	2	5
Zn	60	36	93	116	111	74	36	9	37	37	114	38
Ga	13	17	31	n.d.	n.d.	n.d.	n.d.	44	15	n.d.	n.d.	n.d.
As	0.51	0.79	0.92	n.d.	n.d.	n.d.	n.d.	0.74	0.81	n.d.	n.d.	n.d.
Se	0.25	0.21	0.22	n.d.	n.d.	n.d.	n.d.	0.07	0.08	n.d.	n.d.	n.d.
Br	0.8	0.8	0.1	n.d.	n.d.	n.d.	n.d.	0.2	2.8	n.d.	n.d.	n.d.
Rb	102	59.6	77.9	36	29	39	13	151	171	149	159	170
Sr	48	50	33	172	185	274	230	30	127	124	218	148
Y	89	64	73	33	24	10	5	8	9	10	32	9
Zr	470	450	499	126	104	48	33	117	150	155	495	151
Nb	24	24	23	9	8	5	6	8	8	9	23	9
Sb	0.17	0.14	0.25	n.d.	n.d.	n.d.	n.d.	0.25	0.55	n.d.	n.d.	n.d.
Cs	1.59	0.78	1.63	n.d.	n.d.	n.d.	n.d.	0.26	0.41	n.d.	n.d.	n.d.
Ba	919	587	540	250	197	333	130	1030	1180	1111	801	1193
La	48.3	43.9	20.5	n.d.	n.d.	n.d.	n.d.	13.1	26.1	n.d.	n.d.	n.d.
Ce	88.8	90.9	46.5	n.d.	n.d.	n.d.	n.d.	24.3	44.5	n.d.	n.d.	n.d.
Nd	39.7	46.3	29.5	n.d.	n.d.	n.d.	n.d.	8.59	19.8	n.d.	n.d.	n.d.
Sm	8.09	9.04	8.08	n.d.	n.d.	n.d.	n.d.	1.24	2.39	n.d.	n.d.	n.d.
Eu	1.25	2.57	3.09	n.d.	n.d.	n.d.	n.d.	0.35	0.69	n.d.	n.d.	n.d.
Gd	10.1	10.8	11.6	n.d.	n.d.	n.d.	n.d.	1.5	2.3	n.d.	n.d.	n.d.
Tb	1.76	1.53	2.17	n.d.	n.d.	n.d.	n.d.	0.17	0.28	n.d.	n.d.	n.d.
Tm	1.74	0.93	1.28	n.d.	n.d.	n.d.	n.d.	0.11	0.17	n.d.	n.d.	n.d.
Yb	16.2	6.65	8.32	n.d.	n.d.	n.d.	n.d.	0.58	0.68	n.d.	n.d.	n.d.
Lu	2.78	0.99	1.27	n.d.	n.d.	n.d.	n.d.	0.088	0.11	n.d.	n.d.	n.d.
Hf	13.5	12.5	14.7	n.d.	n.d.	n.d.	n.d.	3.98	4.07	n.d.	n.d.	n.d.
Ta	1.59	1.39	1.79	n.d.	n.d.	n.d.	n.d.	0.28	0.39	n.d.	n.d.	n.d.
W	2.2	1.3	0.9	n.d.	n.d.	n.d.	n.d.	3.4	3.5	n.d.	n.d.	n.d.
Ir (ppb)	0.3	0.1	<1	n.d.	n.d.	n.d.	n.d.	<0.5	<0.7	n.d.	n.d.	n.d.
Au (ppb)	2.2	2.5	3	n.d.	n.d.	n.d.	n.d.	0.7	1.2	n.d.	n.d.	n.d.
Th	12.2	11.1	12.9	n.d.	n.d.	n.d.	n.d.	9.52	8.53	n.d.	n.d.	n.d.
U	14.7	2.78	3.94	n.d.	n.d.	n.d.	n.d.	0.85	1.11	n.d.	n.d.	n.d.
K/U	2687	9233	5034	n.d.	n.d.	n.d.	n.d.	52157	n.d.	n.d.	n.d.	n.d.
Zr/Hf	34.8	36.0	33.9	n.d.	n.d.	n.d.	n.d.	29.4	36.9	n.d.	n.d.	n.d.
La/Th	3.96	3.95	1.59	n.d.	n.d.	n.d.	n.d.	1.38	3.06	n.d.	n.d.	n.d.
Th/U	0.83	3.99	3.27	n.d.	n.d.	n.d.	n.d.	11.2	7.68	n.d.	n.d.	n.d.
La_N/Yb_N	2.01	4.46	1.67	n.d.	n.d.	n.d.	n.d.	15.3	25.9	n.d.	n.d.	n.d.
Eu/Eu*	0.42	0.79	0.98	n.d.	n.d.	n.d.	n.d.	0.78	0.90	n.d.	n.d.	n.d.

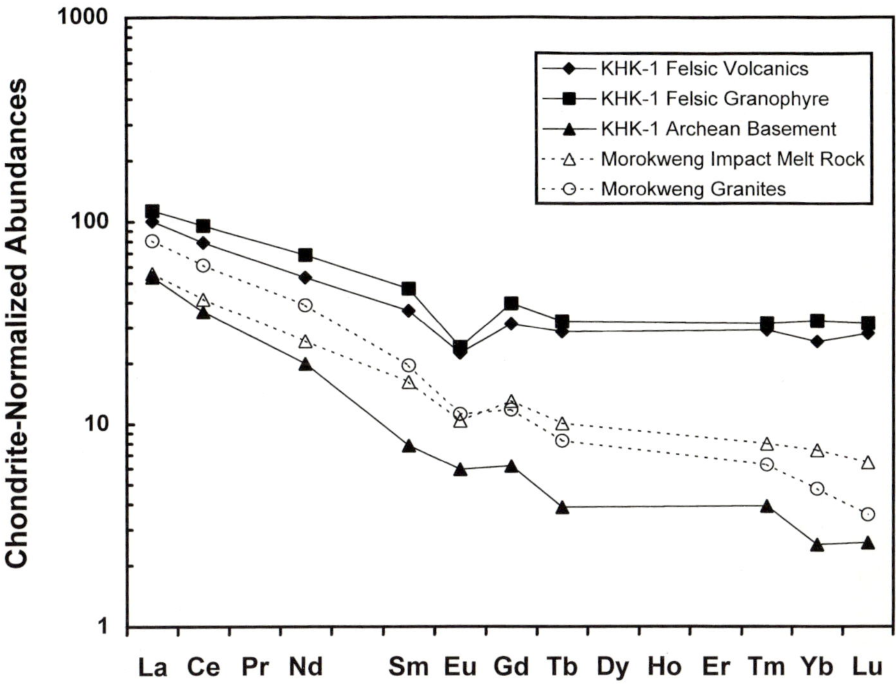

Fig. 5. Chondrite-normalized REE patterns for major lithologies in KHK-1 drill core in comparison to average compositions of Morokweng Granophyre (impact melt rock) and Archean basement granitoids from drill cores into the central part of the structure (compare Table 2). Normalization factors from Taylor and McLennan (1985).

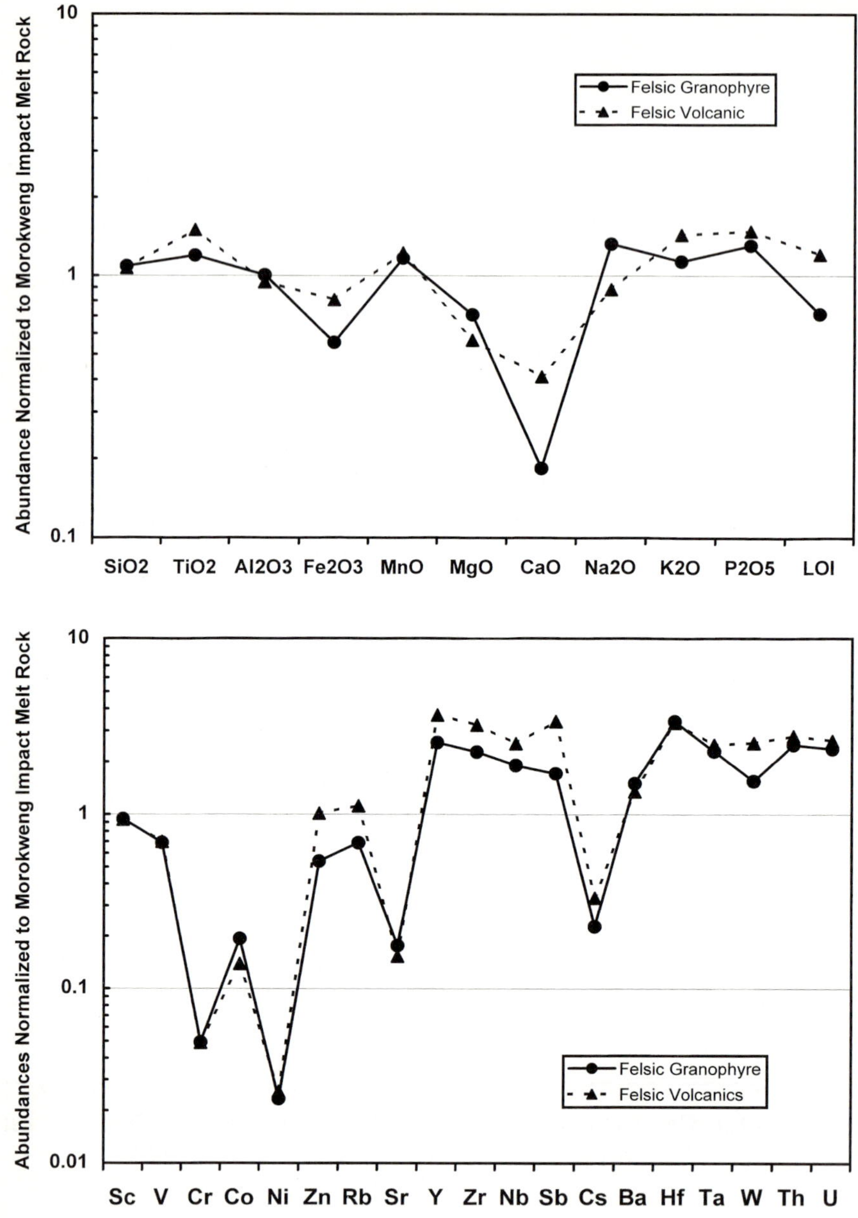

Fig. 6. Comparison of average compositions, in terms of major elements (a) and selected trace elements (b), of Felsic Volcanics and Felsic Granophyre with that of impact melt rock (Morokweng Granophyre). Data for the average compositions are listed in Table 2.

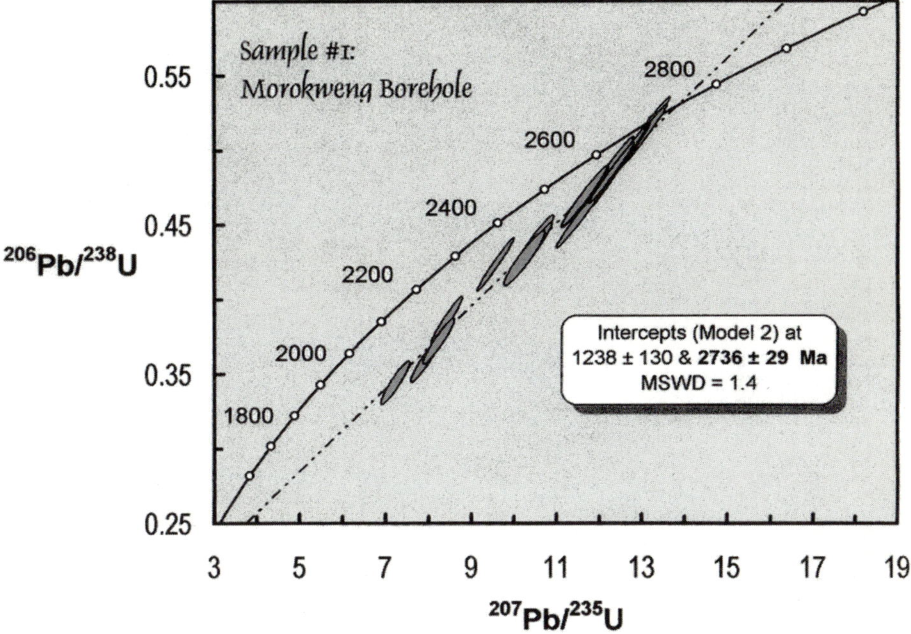

Fig. 7. Concordia diagram for SHRIMP U-Pb data for zircon grains from a felsic granophyre sample from 2289.5 m depth in drill core KHK-1 (data in Reimold et al. 2002, where also details about experimental procedures and data reduction are presented). The upper intercept is interpreted to represent the formation age of this rock at 2739 ± 29 Ma. A lower intercept of 1238 ± 130 Ma was also obtained for this sample and interpreted as an indication for Namaquan-Kibaran metamorphic overprint at that time, which is documented for rocks from the northwestern parts of South Africa (e.g., Robb et al. 1999).

Morokweng Granites). It is obvious that felsic volcanics and felsic granophyres have near-identical patterns that are flatter and enriched with regard to abundances of all elements in comparison to the other lithologies. Basement granitoids from KHK-1 and the other boreholes in the center of the structure also have very similar patterns but cover a significant range of abundances. This range also includes the pattern for average impact melt rock. In Figs. 6a and b, major and trace element contents in average felsic volcanics and felsic granophyre from the KHK-1 drill core are compared against the respective abundances in Morokweng impact melt rock (Morokweng Granophyre; data after Koeberl and Reimold 2002). Both diagrams demonstrate, once again, the chemical similarity between felsic volcanic

Table 2. Average compositions (with standard deviations) of major rock types from the KHK-1 deep borehole compared to rocks from the three central Morokweng drill holes.

	KHK-1 Felsic Volcanic		KHK-1 Felsic Granophyre		KHK-1 Gabbro		KHK-1 Granitic Basement		Morokweng Impact Melt Rock		Morokweng Granites	
SiO_2	69.13	3.54	70.22	2.53	53.59	0.79	63.95	9.54	64.68	1.74	73.47	0.79
TiO_2	0.73	0.21	0.58	0.20	0.95	0.65	0.21	0.08	0.49	0.13	0.23	0.05
Al_2O_3	12.68	1.08	13.49	1.70	14.31	0.76	17.37	2.18	13.44	0.48	13.67	0.58
Fe_2O_3	4.76	2.26	3.27	1.86	10.91	2.45	2.60	2.69	5.90	0.91	1.70	0.23
MnO	0.10	0.01	0.09	0.01	0.21	0.01	0.11	0.04	0.081	0.021	0.02	0.01
MgO	2.11	1.21	2.64	1.00	8.06	3.18	2.34	3.01	3.73	0.70	0.41	0.09
CaO	1.40	0.60	0.63	0.26	6.87	1.09	4.34	5.61	3.41	0.64	1.28	0.29
Na_2O	3.44	1.32	5.13	1.64	2.88	1.06	4.52	1.89	3.88	0.32	4.64	0.40
K_2O	3.08	1.80	2.44	1.76	0.78	0.16	3.03	3.24	2.15	0.24	3.81	0.68
P_2O_5	0.18	0.07	0.16	0.08	0.14	0.06	0.08	0.03	0.12	0.03	0.06	0.01
LOI	2.43	1.24	1.44	0.46	1.76	0.63	0.82	0.58	2.01	0.50	0.46	0.16
Total	100.05		100.11		100.47		99.34		99.90		99.75	
Sc	9.38	4.41	9.44	2.75	n.d.		1.71	0.35	10.0	2.7	2.11	0.65
V	54.6	25.0	53.8	22.1	239	125	50.0		78.1	26.7	20.4	7.8
Cr	16.2	12.7	16.3	6.9	320	487	20.6	18.6	331	81	13.0	3.2
Co	5.89	6.46	8.19	3.1	48	6	9.2	12.0	42.3	4.0	4.56	3.0
Ni	15.0	6.4	13.6	4.9	86	65	42.7	58.4	583	112	14.0	10.6
Cu	8	5.29	<2		96	95	44.3	26.6	33	16	12.2	15.7
Zn	79	37	42.2	19.2	100	23	27.3	15.9	78	10	46.6	25.8
Ga	23	11	22	9.2	n.d.		29.5	20.5	17.1	5.7	14.1	6.8
As	0.91	0.55	0.55	0.23	n.d.		0.78	0.05	0.83	0.38	0.89	0.71
Se	0.23	0.08	0.23	0.07	n.d.		0.08	0.01	0.40	0.13	0.20	0.10
Br	0.7	0.3	1.0	0.5	n.d.		1.5	1.8	0.3	0.1	0.2	0.1
Rb	98	48	60	34	35	5.1	112	86	87.4	11.4	162	24
Sr	58	19	66	34	210	56	129	100	377	31	424	131
Y	64	23	44	36	22	12	7.3	2.1	17	3	14	3
Zr	449	151	315	171	93	40	100	60.3	139	14	168	27
Nb	23.2	5.5	17.4	7.9	7.3	2.1	7.3	1.2	9	1	10	1
Sb	0.85	0.42	0.43	0.45	n.d.		0.40	0.21	0.25	0.11	0.20	0.09
Cs	1.52	0.87	1.04	0.54	n.d.		0.34	0.11	4.56	1.84	2.95	1.31
Ba	567	319	628	586	260	69	780	568	419	40.6	841	287
La	36.8	29.6	41.6	31.8	n.d.		19.6	9.2	20.4	2.2	29.7	8.9
Ce	75.5	53.9	91.6	73.1	n.d.		34.4	14.3	39.8	3.9	58.4	19.3
Nd	37.8	22.2	48.8	36.4	n.d.		14.2	7.9	18.3	1.71	27.6	7.6
Sm	8.41	3.32	10.8	6.89	n.d.		1.82	0.81	3.74	0.37	4.51	1.11
Eu	1.96	0.86	2.09	0.88	n.d.		0.52	0.24	0.91	0.10	0.98	0.37
Gd	9.61	3.28	12.1	5.84	n.d.		1.90	0.57	3.98	0.51	3.61	0.54
Tb	1.66	0.58	1.87	0.84	n.d.		0.23	0.08	0.58	0.07	0.48	0.08
Tm	1.04	0.39	1.12	0.51	n.d.		0.14	0.04	0.28	0.04	0.22	0.04
Yb	6.36	3.44	8.05	4.16	n.d.		0.63	0.07	1.84	0.26	1.18	0.16
Lu	1.07	0.47	1.20	0.70	n.d.		0.10	0.02	0.25	0.04	0.14	0.02
Hf	12.0	4.21	12.2	3.65	n.d.		4.03	0.06	3.60	0.34	3.99	0.84
Ta	1.59	0.43	1.46	0.52	n.d.		0.34	0.08	0.63	0.1	0.79	0.25
W	2.49	1.92	1.50	0.70	n.d.		3.45	0.07	0.97	0.4	6.46	13.6
Ir	<0.5		<0.5		n.d.		<0.7		22.5	5.52	0.6	0.3
Au	1.9	1.2	5.8	12.4	n.d.		1.0	0.35	5.5	1.36	1.6	1.3
Th	12.1	2.31	10.7	3.35	n.d.		9.03	0.70	4.31	1	7.23	0.99
U	4.34	2.13	3.90	3.98	n.d.		0.98	0.18	1.65	0.6	1.45	0.87
K/U	5918		5217		n.d.		25765		10901		21908	
Zr/Hf	37.3		25.8		n.d.		24.8		38.5		42.1	
La/Th	3.05		3.87		n.d.		2.17		4.74		4.10	
Th/U	2.8		2.8		n.d.		9.2		2.6		5.0	
La_N/Yb_N	3.9		3.5		n.d.		21.0		7.5		16.9	
Eu/Eu*	0.67		0.56		n.d.		0.86		0.72		0.74	

Major elements in wt%, trace elements in ppm, except Ir and Au in ppb. All Fe as Fe_2O_3. Data for the Morokweng lithologies from Koeberl and Reimold (2002).

and felsic granophyre, and emphasize the significant differences between these two lithologies and the Morokweng Granophyre not only with respect to major elements Ti, Fe, and CaO, but also to most trace elements.

A further difference between these rock types has been discussed by Reimold et al. (2002): felsic granophyre samples do not have pronounced contents of siderophile elements that could be taken as evidence for the presence of a possible meteoritic component. These rocks are in this regard totally different from the Morokweng Granophyre that contains a significant (2-5%) meteoritic component (Koeberl et al. 1997; Hart et al. 1997; Koeberl and Reimold 2002).

4
U-Pb SHRIMP Dating Results

Reimold et al. (2002) reported results of single zircon U-Pb SHRIMP dating performed on seven samples from all three packages of felsic granophyre, basement granitoid, and a gabbro specimen from drill core KHK-1. The concordant results obtained by these workers prove beyond doubt that the felsic granophyres, as well as the basement granitoids, represent Archean lithologies. Reliable ages obtained scatter between 2690 and 2880 Ma. Lower intercept ages, defined by trends of strongly discordant data and intercepts with the concordia line, indicate later overprint at about 1300-900 Ma, probably related to the Namaqua-Kibaran regional tectono-magmatic phase (e.g., Robb et al. 1999). Clearly, these felsic granophyres are not coeval with the Morokweng impact melt rock of 145 Ma age. An example of a Concordia diagram for SHRIMP U-Pb data for zircon grains from a felsic granophyre sample (from 2289.5 m depth) is shown in Fig. 7 (for analytical data, see Reimold et al. 2002).

5
Discussion and Conclusions

Whilst the felsic granophyres encountered in the KHK-1 borehole are texturally similar to the Morokweng impact melt rock (the Morokweng Granophyre), they are very different with regard to both mineralogical and chemical compositions. No evidence of meteoritic contamination of felsic granophyre has been detected. Whether or not the felsic granophyres are related to the felsic volcanics intersected in this borehole needs to be further investigated. The fact that Reimold et al. (2002) obtained Archean ages for zircon grains of felsic granophyre is evidence that the formation of these rocks is not related to the impact event. The granophyres represent an integral part of the regional geological evolution, prior to the Namaquan-Kibaran magmato-tectonic phase of about 1300-900 Ma ago.

The lithologies in the KHK-1 drill core are generally undeformed. With the exception of a small number of narrow cataclastic zones that are either of tectonic or impact origin, only a single, very narrow occurrence of impact breccia has been detected. This thin, bedding-parallel zone is interpreted as an injection of polymict, suevitic impact breccia. Its uniqueness in this drill core and scarcity over a length of 3400 m is interpreted as a strong indication that this borehole was sunk outside of the actual crater structure. That impact breccia has been intersected at all could suggest that the position of this borehole is not too far from the crater rim that should be expected to be transected by a host of impact breccia veins. Consequently, we propose that this result further constrains the crater size of the Morokweng Structure to about 70 to, at maximum, 80 km diameter. This finding is in agreement with the geophysical interpretations of Henkel et al. (2002).

In conclusion, the chemical data show that there is no relationship between Morokweng impact melt rock and granophyric rocks encountered in the KHK-1 core. A maximum crater diameter of 70-80 km for the Morokweng impact structure is supported by these new results from a deep borehole. This finding does not exclude that this impact event had a major environmental effect on the fauna in a large part of the world, but raises the question whether the Morokweng impact event could indeed have caused a global mass extinction at the time corresponding to the Jurassic-Cretaceous boundary.

Acknowledgments

We are grateful to Anglogold Limited and their former Consulting Geologist, Gerald Cantello, for permission to study and sample the KHK-1 core and to publish these results. Our research is supported by grants from the National Research Foundation of South Africa and the Research Council of the University of the Witwatersrand (to W.U.R.), as well as the Austrian Fonds zur Förderung der wissenschaftlichen Forschung (grant Y58-GEO to C.K.). This is University of the Witwatersrand Impact Cratering Research Group Contribution No. 47.

References

Andreoli MAG, Ashwal LD, Hart RJ, Smith CB, Webb SJ, Tredoux M, Gabrielli F, Cox R, Hambledon-Jones BB (1995) The impact origin of the Morokweng ring structure, southern Kalahari, South Africa [abs.]. Centennial Geocongress of the Geological Society of South Africa, RAU, Johannesburg, Extended Abstracts I: 541-544

Andreoli MAG, Ashwal LD, Hart RJ, Huizenga JM (1999) A Ni- and PGE-enriched quartz norite impact melt complex in the Late Jurassic Morokweng impact structure, South Africa. In: Dressler BO, Sharpton VL (eds) Large Meteorite Impacts and Planetary Evolution II. Geological Society of America Special Paper 339: 91-108

Bice DM, Newton CR, McCauley S, Reiners PW, McRoberts CA (1992) Shocked quartz at the Triassic-Jurassic boundary in Italy. Science 255: 443-446

Bootsman CS, Reimold WU, Brandt D (1999) Evolution of the Molopo drainage and its possible disruption by the Morokweng impact event at the Jurassic-Cretaceous boundary. Journal of African Earth Sciences 29: 669-678

Corner B, Reimold WU, Brandt D, Koeberl C (1997) Morokweng impact structure, Northwest Province, South Africa: Geophysical imaging and some preliminary shock petrographic studies. Earth and Planetary Science Letters 146: 351-364

Dutta RK, Sidras-Haddad E, Andreoli MAG, Hart RJ, Cloete T (2001) Proton microprobe characterization of a unique siderophile element mineral assemblage in the melt sheet of the giant 145 Ma Morokweng impact structure, Southern Kalahari, South Africa. Nuclear Instruments and Methods in Physics Research B 181: 551-556

Geological Survey of South Africa (1974) Geological map, sheet 2622 Morokweng (Scale 1:250 000). Council for Geoscience, Pretoria, South Africa

Hart RJ, Andreoli MAG, Tredoux M, Moser D, Ashwal LD, Eide EA, Webb SJ, Brandt D (1997) Late Jurassic age for the Morokweng impact structure, southern Africa. Earth and Planetary Science Letters 147: 25-35

Hart RJ, Cloete M, McDonald I, Carlson RW, Andreoli MAG (2002) Siderophile-rich inclusions from the Morokweng impact melt sheeet, South Africa: possible fragments of a chondritic meteorite. Earth and Planetary Science Letters 198: 49-62

Henkel H, Reimold WU, Koeberl C (2002) Magnetic and gravity model of the Morokweng impact structure, South Africa. Journal of Applied Geophysics: 49: 129-147

Koeberl C (1993) Instrumental neutron activation analysis of geochemical and cosmochemical samples: A fast and proven method for small sample analysis. Journal of Radioanalytical and Nuclear Chemistry 168: 47-60

Koeberl C, Reimold WU (2002) Geochemistry and petrography of impact breccias and target rocks from the 145 Ma Morokweng impact structure, South Africa. Geochimica et Cosmochimica Acta 66: in press

Koeberl C, Armstrong RA, Reimold WU (1997) Morokweng, South Africa: a large impact structure of Jurassic-Cretaceous boundary age. Geology 25: 731-734

Koeberl C, Peucker-Ehrenbrink B, Reimold WU (2000) Meteoritic component in impact melt rocks from the Morokweng, South Africa, impact structure: An Os isotopic study [abs.]. Lunar and Planetary Science 30: abs.#1595 (CD-ROM)

Koeberl C, Peucker-Ehrenbrinck B, Reimold WU, Shukolyukov A, Lugmair G (2002) A comparison of the osmium and chromium isotopic methods for the detection of meteoritic components in impactites: Examples from the Morokweng and Vredefort impact structures, South Africa. In: Koeberl C, MacLeod K (eds) Catastrophic Events and Mass Extinctions: Impact and Beyond. Geological Society of America Special Paper 356: 607-617

Kudielka G, Koeberl C, Montanari A, Newton J, Reimold WU (2001) Stable-isotope and trace element stratigraphy of the Jurassic/Cretaceous boundary, Bosso River Gorge, Italy. In: Buffetaut E, Koeberl C (eds) Geological and Biological Effects of Impact Events. Impact Studies Series 1, Springer-Verlag, Heidelberg-Berlin, pp 25-68

McDonald I, Andreoli MAG, Hart RJ, Tredoux M (2000) Platinum-group elements in the Morokweng impact structure, South Africa: Evidence for the impact of a large ordinary chondrite projectile at the Jurassic-Cretaceous boundary. Geochimica et Cosmochimica Acta 65: 299-309

Reimold WU, Koeberl C, Bishop J (1994) Roter Kamm impact crater, Namibia: Geochemistry of basement rocks and breccias. Geochimica et Cosmochimica Acta 58: 2689-2710

Reimold WU, Koeberl C, Brandstätter F, Kruger FJ, Armstrong RA, Bootsman C (1999) Morokweng impact structure, South Africa: Geologic, petrographic, and isotopic results, and implications for the size of the structure. In: Dressler BO, Sharpton VL (eds) Large Meteorite Impacts and Planetary Evolution II. Geological Society of America Special Paper 339: 61-90

Reimold WU, Armstrong RA, Koeberl C (2002) A deep drillcore from the Morokweng impact structure, South Africa: petrography, geochemistry and constraints on the crater size. Earth and Planetary Science Letters 201: 221-232

Robb LJ, Armstrong RA, Waters DJ (1999) The history of granulite-facies metamorphism and crustal growth from single zircon U-Pb geochronology: Namaqualand, South Africa. Journal of Petrology 40: 1747-1770

Taylor SR, McLennan SM (1985) The Continental Crust: Its Composition and Evolution. Blackwell, Oxford, 312 pp

Stratigraphy, Paleomagnetic Results, and Preliminary Palynology across the Permian-Triassic (P-Tr) Boundary at Carlton Heights, Southern Karoo Basin (South Africa)

Dylan M. Schwindt[1], Michael R. Rampino[1,2], Maureen B. Steiner[3], and Yoram Eshet[4]

[1]Earth and Environmental Science Program, New York University, 100 Washington Square East, New York, NY 10003, USA. (mrr1@nyu.edu)
[2]NASA, Goddard Institute for Space Studies, 2880 Broadway, New York, NY 10025, USA.
[3]Department of Geology and Geophysics, University of Wyoming Laramie, WY 82071, USA.
[4]Geological Survey of Israel, Jerusalem, 95501, Israel.
Tel Hai Academic College, Tel Hai, 12210, Israel.

Abstract The most severe mass extinction of marine species and terrestrial vertebrates and plants is associated with the Permian-Triassic (P-Tr) boundary (~253 Ma). In order to investigate the relative timing of the biotic crises in terrestrial and marine environments, we studied the stratigraphy, paleomagnetism, and palynology of the Carlton Heights section in the southern Karoo Basin, South Africa. Stratigraphic and palynological study at Carlton Heights revealed the abrupt disappearance of Late Permian gymnosperm taxa and replacement by Triassic palynomorphs just below the boundary between the Balfour Formation and the Katberg sandstone, at a layer characterized by abundant remains of fungi. This "fungal spike" occurs globally in marine and terrestrial P-Tr boundary sections, and thus can be used to correlate the extinctions of marine fauna and terrestrial land plants and vertebrates in the P-Tr boundary interval. Our results suggest that the extinction of mammal-like reptiles at the end of the Permian may have preceded the fungal event and land-plant extinction within an interval of less than ~100,000 years, and possibly less than ~25,000 years.

Paleomagnetic data confirm multiple magnetizations of Karoo Basin P-Tr strata during Jurassic intrusive events. Mildly altered sedimentary strata, more altered sediments, and an intrusive dike all showed normal polarity magnetization stable to about 500-575°C and 60 mT; reversed polarity was exhibited primarily as trends at high demagnetization temperatures and field strengths. The normal polarity directions in the sedimentary and igneous rocks are statistically identical to the mean Jurassic direction during Karoo igneous activity. A Permian-Triassic magnetic signature is no longer identifiable.

1
Introduction

The end-Permian (~253 Ma; see Mundil et al. 2001) mass extinction eliminated more than 90% of marine species (Raup 1979; Jin et al. 2000). On land, an estimated 70% of terrestrial vertebrate families were eradicated (Maxwell 1992), insects suffered a major loss of taxa (Labandiera and Sepkoski 1993), and more than 90% of gymnosperm plant species died out (Retallack 1995; Visscher et al. 1996; Looy et al. 1999). The extinctions were accompanied by a dramatic negative shift of several per mil in the $\delta^{13}C$ values of ocean carbonate and organic matter (Magaritz et al. 1988, 1992), and atmospheric carbon dioxide (Morante 1996; MacLeod et al. 2000; Sephton et al. 2002).

Precise radiometric dating by Bowring et al. (1999) constrained the end-Permian extinction pulse in the oceans within an interval of less than 1 million years, and the negative carbon isotope shift to less than 165,000 years. Jin et al. (2000) further constrained the marine extinction to less than 100,000 years. Using the improved resolution provided by cycle analysis in the GK-1 core and nearby outcrops in the Carnic Alps (Austria), Rampino et al. (2000) estimated that the abrupt marine faunal change could have occurred in less than 8,000 years. The initial sharp negative carbon-isotope shift was estimated to have occurred rapidly within an interval of less than 30,000 years coincident with the marine extinctions, and the longer-term negative carbon-isotope excursion lasted about 500,000 years into the Early Triassic (Rampino et al. 2000).

The plant extinction is evidenced by the disappearance of almost all Late Permian gymnosperm pollen at an horizon containing almost exclusively fungal remains and woody debris (Visscher et al. 1996). This fungal event was followed by the appearance of an Early Triassic palynoflora dominated by lycopod spores and bisaccate gymnosperm pollen (Eshet et al. 1995; Visscher et al. 1996). The global proliferation of fungi has been interpreted as indicating massive destruction of terrestrial vegetation and accumulation of decayed organic debris (Ouyang and Utting 1990; Eshet et al. 1995; Visscher et al. 1996; Poort et al. 1997).

The fungal event has been observed globally both in terrestrial and shallow marine sequences. In marine sections, the brief increase in fungal remains has been correlated closely with the mass extinction of marine organisms (Visscher and Brugman 1986; Visscher et al. 1996; Twitchett et al. 2001). An abrupt negative shift in carbon isotopes in the oceans is also closely associated with the reduction in gymnosperms (Looy et al. 2001), and the zone of abundant fungal spores (Visscher et al. 1996; Wignall et al. 1996). Twitchett et al. (2001) proposed that the collapse of marine and terrestrial ecosystems began at the same stratigraphic level, and preceded the sharp negative excursion in $\delta^{13}C$.

One important remaining question is the correlation of the marine extinctions and terrestrial plant turnover with the extinction of non-marine vertebrates. The Upper Permian and Lower Triassic Beaufort Group of the Karoo Supergroup of South Africa is well known for its record of the succession of mammal-like reptiles across the P-Tr boundary (Smith and Ward 2001). The demise of the dicynodonts of the *Dicynodon* assemblage zone, and their replacement by the *Lystrosaurus* assemblage fauna has served in the past as the approximate definition of the Permian-Triassic boundary in the Karoo, although the first occurrence of *Lystrosaurus* is now known to predate the last occurrence of *Dicynodon* (Smith 1990, 1995; Smith and Ward 2001).

The Beaufort Group consists of an apparently uninterrupted succession of alluvial sedimentation (Catuneanu and Elango 2001). Sediments shed from the Cape Fold Belt region produced a fluvial network which prograded across the Karoo Basin during Late Permian time. Lacustrine mudstones, and fluvial overbank mudstones and channel sandstones make up a sedimentary sequence up to 6 km thick in the Karoo Basin of the southern part of South Africa. A change from predominantly green mudstone and sandstone deposited by high sinuosity river systems to multistoried channel and sheet sandstones intercalated with maroon mudstones, typical of deposition by braided streams, occurs close to the Permian-Triassic boundary as defined by the vertebrate assemblages (Smith 1990, 1995; Ward and Smith 2000).

The sequence of polarity changes of the geomagnetic field across the P-Tr boundary has been uncertain until recently (Steiner et al. 1993). The uncertainty was the result of poor agreement among the numerous magnetostratigraphic sequences of the Permian marine realm, and poor or non-existent biostratigraphy in many P-Tr terrestrial sequences. However, recent work (e.g., Steiner et al. 1989, 1993; Gialanella et al. 1997; Scholger et al. 2000) has led to a better understanding of the pattern of geomagnetic field polarity changes within this time interval. Compilation of global magnetostratigraphic studies across the P-Tr boundary now indicates a change from reversed polarity to normal polarity approximately at the base of the Triassic (Steiner 2001). In Italy, the change from reversed to normal polarity occurs at the base of the Tesero Horizon of the Werfen Formation (Scholger et al. 2000). Thus, it is closely correlated with the negative shift in $\delta^{13}C$ identified by Magaritz et al. (1988) and the fungal spike identified by Visscher and Brugman (1986) in the lower Tesero Horizon. In China, the polarity change (Steiner et al. 1989) is also known to occur close to the negative shift in $\delta^{13}C$ (Bowring et al. 1998).

Previous paleomagnetic studies of the Permian-Triassic boundary interval in the Karoo Basin found widespread remagnetization (Ballard et al. 1986) due to heating by Jurassic igneous activity (Hargraves et al. 1997). However, Kirschvink and Ward (1998) and Pehl et al. (2001) recently reported the successful recovery of Late Permian and Early Triassic magnetic directions from the Lootsberg Pass section in the Karoo. They suggested that the latest Permian and earliest Triassic, as defined by vertebrate remains, were reversed in polarity followed by a period of normal polarity in the early Triassic. Thus, the potential exists to use magnetostratigraphy in the Karoo basin as a correlation tool within the basin, as well as between marine and terrestrial sequences in general.

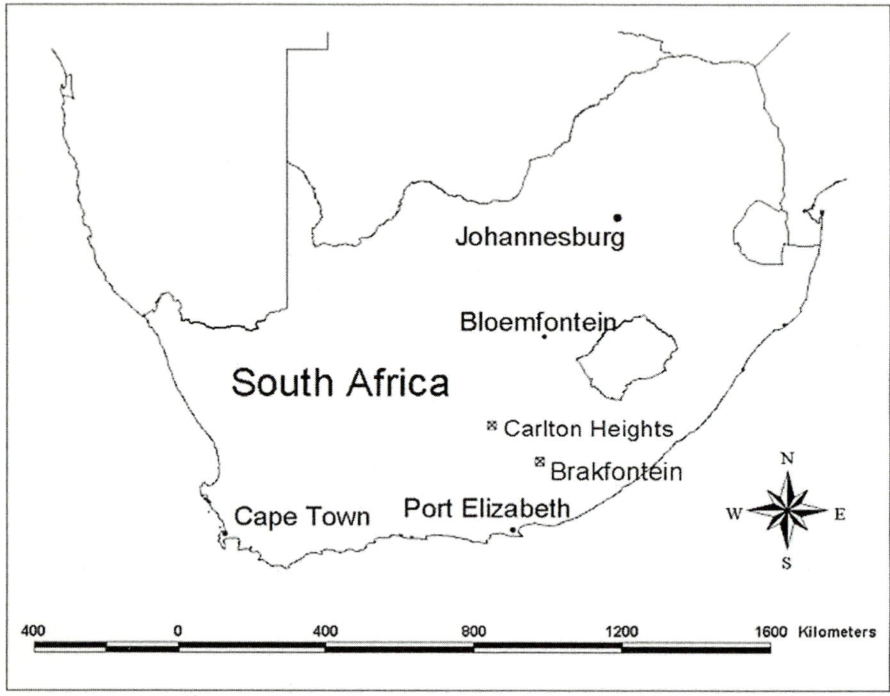

Fig. 1. Map of South Africa showing locations of the Carlton Heights and Brakfontein sections (southern Karoo Basin).

In an attempt to correlate the terrestrial sequence in the Karoo with the global stratigraphic indicators of the P-Tr boundary interval, we studied the stratigraphy, magnetostratigraphy, and palynology of the well-exposed Carlton Heights section, located along the Graaff-Reinet - Colesburg highway between Middleburg and Neupoort, SA (Fig. 1). Carbon isotopes of carbonate nodules in the Carlton Heights section were also studied and the results will be reported elsewhere.

Samples were collected in gullies to the SE and below the main highway (on the B.P. Erasmus farm, "Beskuitfontein") and along the main highway itself about 500 meters north of the Carlton Heights railroad stop (31°13.03'S latitude, 24°56.96'E longitude). To evaluate the reliability of magnetostratigraphy within the Karoo Basin, samples were also collected from another section 150 km away at Brakfontein (Fig. 1) (32°11.5'S, 26°08'E; Groenewald, 1989), located 23 km southwest of Tarkastad (on the Hartebeest Laagt and Brakfontein farms of L. Stein and P. Vandertavor, respectively).

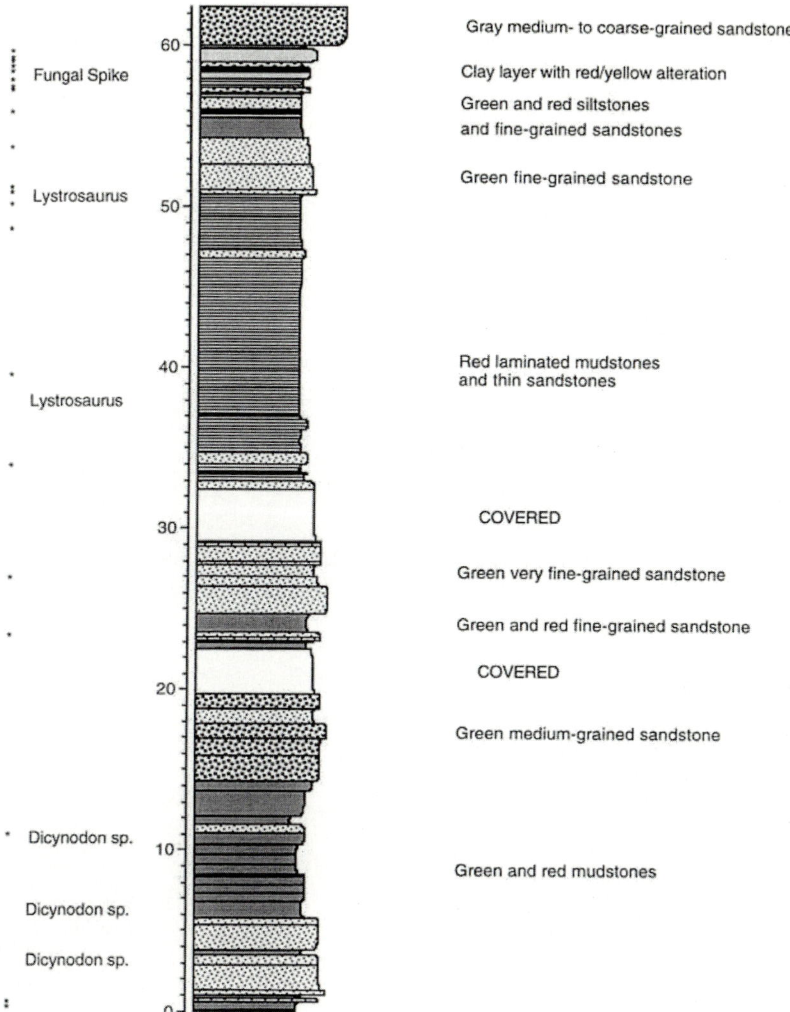

Fig. 2. Stratigraphic section across the P-Tr boundary in the Karoo Supergroup strata at Carlton Heights, South Africa. Vertical scale in meters. Asterisks show positions of palynological samples. The uppermost 11 meters in the Katberg Formation sandstones studied are not shown.

2
Stratigraphy of the Carlton Heights Section

At the Carlton Heights locality (Fig. 1), flat lying, predominantly greenish mudstone, siltstone and thin tabular sandstones dominate the lower part of the section (Fig. 2). These were previously mapped as Upper Permian (Balfour Formation) (Keyser 1977) based on lithology and stratigraphic position relative to the subsequent widespread change from mudstones to sandstones. Our discovery of a hip bone and other skeletal elements of *Dicynodon sp.* in the lower part of the section (between 2.8 and 10.5 m above the base of the section) supports a late Permian age for these sediments (Fig. 2).

Fig. 3. Laminated to massive maroon mudstone with thin siltstone interbeds along the railway cut northeast of the Carlton Heights railway stop (person for scale). This unit is correlative with the P-Tr event beds of Smith and Ward (2001).

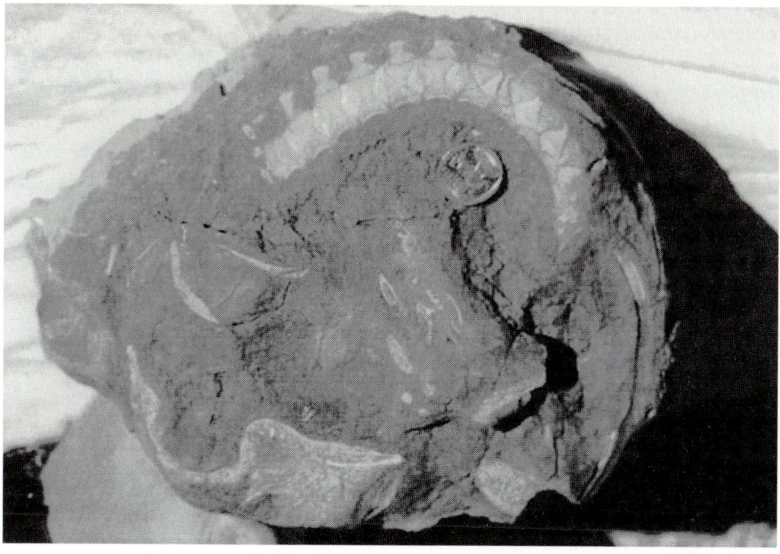

Fig. 4. *Lystrosaurus* fossil from ~38 m (Balfour Formation) above the base of the Carlton Heights section.

At about 33.5 m above the base of the section, the green mudstone/sandstone sequence is overlain by ~17.5 m of laminated to massive maroon mudstone with occasional thin sands (Fig. 3), (well-exposed in the railroad cut just to the south of our section) capped by a 4 to 5 m thick grayish-green sand unit. *Lystrosaurus* first occurs in our section at 38 m above the base (Fig. 4).

Based on lithology and stratigraphic position, we correlate this unit with a similar unit identified by Smith and Ward (2001) at the Bethulie and Lootsberg Pass sections in the Karoo. At these localities, Smith and Ward (2001) identified a 3 to 5 m thick distinctly laminated maroon mud rock made up of thinly bedded dark reddish-brown and olive-gray siltstone-mudstone couplets. This unit is stratigraphically unique in all three sections, and has been identified as composed of lifeless "event beds" that were deposited after the extinction of most of the Late Permian vertebrate taxa (Smith and Ward 2001). An increased number of *Lystrosaurus* fossils were noted near the top of this unit at Carlton Heights ~51 m above the base of the section. (Fig. 2).

At ~56 m above the base of the Carlton Heights section, the maroon mudstone unit grades upward into thin-bedded alternating green and red siltstones and fine sandstones showing abundant sub-horizontal cylindrical burrows. The thin-bedded burrowed unit is overlain (at 58.5 m above the base of the section) by a thin (~5 cm thick) very fine-grained layer (Fig. 5). This laterally continuous layer is heavily burrowed, and is marked by red and yellow-brown alteration. Clay-mineral analysis by qualitative x-ray diffraction methods shows that the layer is composed predominantly of illite and illite-smectite. The layer also contains quartz, low albite, gypsum, chlorite, mica, and jarosite, and thus has a heavily weathered detrital signature.

Fig. 5. Roadcut on main highway about 500 meters north of the Carlton Heights RR stop (31° 13.03" S latitude, 24° 56.96" E longitude), showing the location of the fungal spike zone (hand for scale). The ~1-m-thick fungal spike is overlain by a thin (~5-cm-thick) layer of heavily burrowed, fine-grained (predominantly illite and illiite/smectite) material showing red to yellow-brown alteration (white dashed line). The base of the first multistoried sandstone in the section (about 1 m above the fine-grained layer) is interpreted as the base of the Katberg Formation.

About 0.5 m above this layer, the first laterally widespread multistoried sandstone with an erosional base containing lenses of mud-pebble and carbonate-nodule conglomerates (identified with the base of the Katberg Formation) is encountered at 59 m above the base of the section, just above the main highway (Fig. 2).

The Katberg Formation at Carlton Heights (>270 m in total thickness) is represented by a facies dominated by gray and white fine- to coarse-grained sandstone that is typically multistoried and laterally extensive (Fig. 6). The sands commonly show scoured bases with lenses of intra-formational mud-pebble and pedogenic carbonate-nodule conglomerates, horizontal stratification, and large-scale trough cross stratification (Smith 1995). The thick multistoried sands are interbedded with thin red mudstone layers showing desiccation features, such as sand-filled mudcracks.

Fig. 6. Katberg Formation at Carlton Heights, showing multistoried sandstone with large-scale cross-bedding (45 cm hammer for scale).

The stratigraphy of the Brakfontein sections is similar to that of Carlton Heights (Groenewald 1989). Green overbank mudstones and sandstones typical of meandering stream deposits dominate the Upper Permian portion of the section. The green-colored strata are overlain by laminated-to-massive red mudstone interbedded with a few thin tabular sands. These beds are overlain by the multistoried, laterally widespread sandstones of the Katberg Formation.

3
Paleomagnetic Studies

We sampled a total of 73 meters of sedimentary strata of the Carlton Heights section at 0.5 to 3 m intervals, and 70 meters of sedimentary strata of the Brakfontein section at 1 to 12 m intervals. At Carlton Heights, a narrow (0.7 m) vertical dolerite dike was also sampled in order to obtain the magnetic directions during the period of Jurassic igneous activity in the area. Six samples were taken across the width of the dike, and one sample from the baked contact. Paleomagnetic samples were drilled using a gasoline-powered coring machine and oriented with a magnetic compass. Analyses were performed with a three-axis 2G cryogenic magnetometer at the University of Wyoming paleomagnetic laboratory.

In previous work, Ballard et al. (1986) performed both alternating field (AF) and thermal demagnetization on samples of Karoo sediment from two localities. They found AF demagnetization to be ineffective in separating remanence components. For that reason, and the known thermal history of the area, we employed thermal demagnetization on our sediment samples. Twenty two or more heating steps from $150°$ through $680°$ C were used for the sediments. The dolerite samples were subjected to both AF demagnetization (20 field-strength steps from 20 mT through 120 mT) and thermal demagnetization (the same 22 steps used for sediment samples).

Table 1. Paleomagnetic data: Locality main directions measured initially (NRM) and after thermal demagnetization (Below and Above 575°C)

Locality	Declination	Inclination	n	a_{95}	k
NRM_0					
Carlton Hts. Strata	338.5	-56.7	119	3.4	16
Brakfontein Strata	341.2	-63.4	17	9.5	15
Carlton Hts. Dolerite	347.1	-51.4	12	4.1	113
$\leq 575°$ C					
Carlton Hts. Strata	334.5	-55.3	57	2.3	71
Brakfontein Strata	335.8	-61.4	16	4.3	73
Carlton Hts. Dolerite	345.6	-56.2	12	2.7	261
$\geq 575°$ C					
Carlton Hts. Strata	159	54	33*		
Brakfontein Strata	161	54	12*		
Carlton Hts. Dolerite	172.5	54	11*		

n = number of samples
a_{95} = half-angle of 95% confidence circle about the mean
k = precision parameter for sample population
*maximum intersection directions read from stereonet.

Sediments at Carlton Heights and Brakfontein show similar demagnetization patterns. At Carlton Heights, the mean of the initially measured remanence directions (NRM_0) has a declination of $338.5°$ and inclination of $-56.7°$ (Table 1), compared with $341.2°$, $-63.4°$ from Brakfontein. The directions for an axial-dipole field is $0°$, $50.6°$ for Carlton Heights, $0°$, $51.6°$ for Brakfontein, and $339.0°$, $-65.7°$ (Carlton Heights) and $337.7°$, $-65.4°$ (Brakfontein) for the currently observed geomagnetic field. Remnant intensities ranged from 1×10^{-3} to 5×10^{-2} mA/m at Carlton Heights and generally were between 1 to 2×10^{-1} mA/m at Brakfontein although several sandstone samples containing heavy mineral layers had intensities of 6×10^{-2} to 1×10^{-1} mA/m .

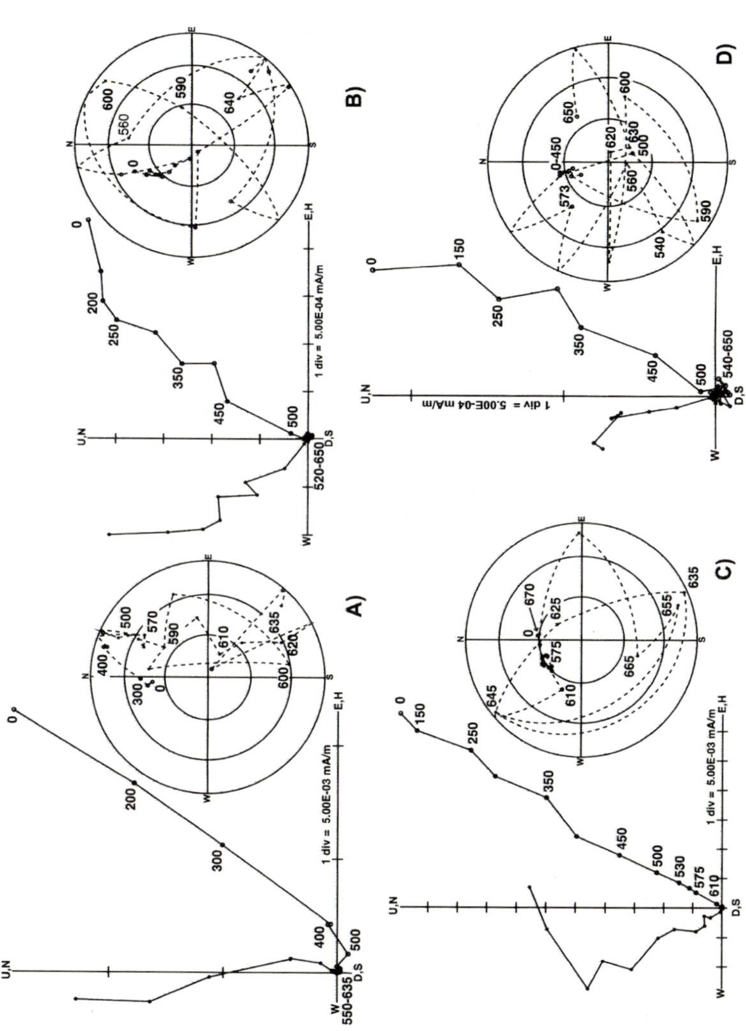

Fig. 7. Typical responses to thermal demagnetization displayed by the sediment samples from Carlton Heights and Brakfontein. Equal-area stereographic plots show lower (upper) hemisphere directions as solid (open) circles. Orthogonal-axes plots show the horizontal projection of the magnetic vector (declination) as solid symbols and the vertical projection (inclination) as open symbols.

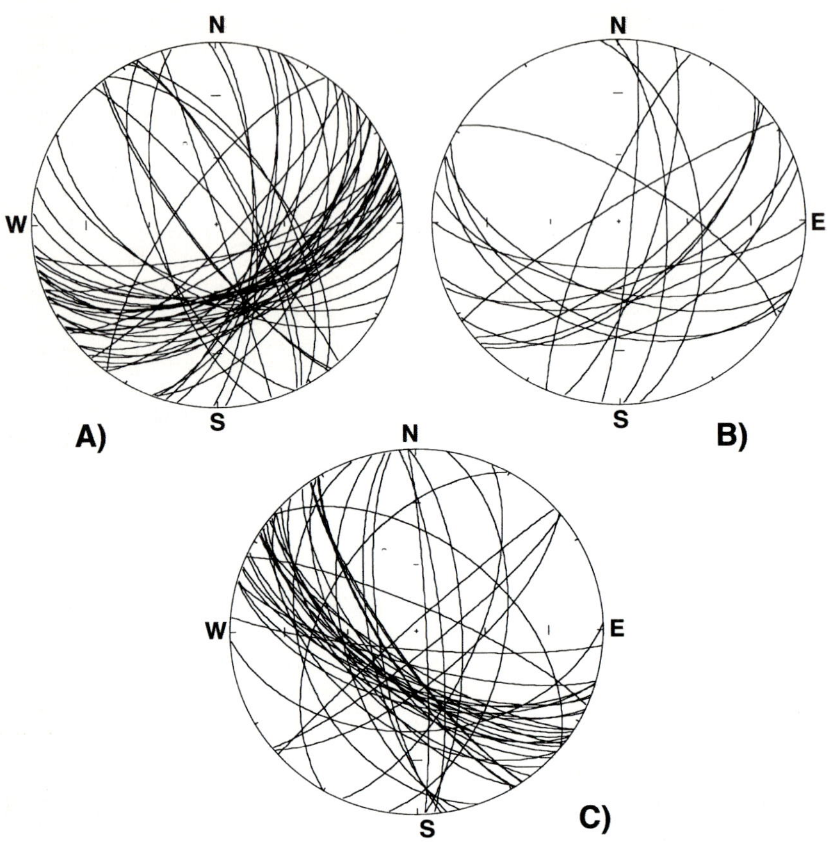

Fig. 8. Half-planes representing the great-circle paths observed and their intersection directions for: (a) Carlton Heights sediment samples; (b) Brakfontein sediment samples; and (c) Carlton Heights dolerite and baked contact samples.

Thermal demagnetization of Carlton Heights samples through 575°C gave sample directions that generally remained in the vicinity of the NRM_0 direction (Fig. 7). Above 500-575°C, directions began to move away and to, or towards, the direction of the opposite polarity (Fig. 7). The directional response was commonly erratic, in some samples clearly moving towards and away from the pre-575°C direction. The opposite polarity direction commonly appears several times within the demagnetization sequence. The same behavior was observed in the samples from Brakfontein.

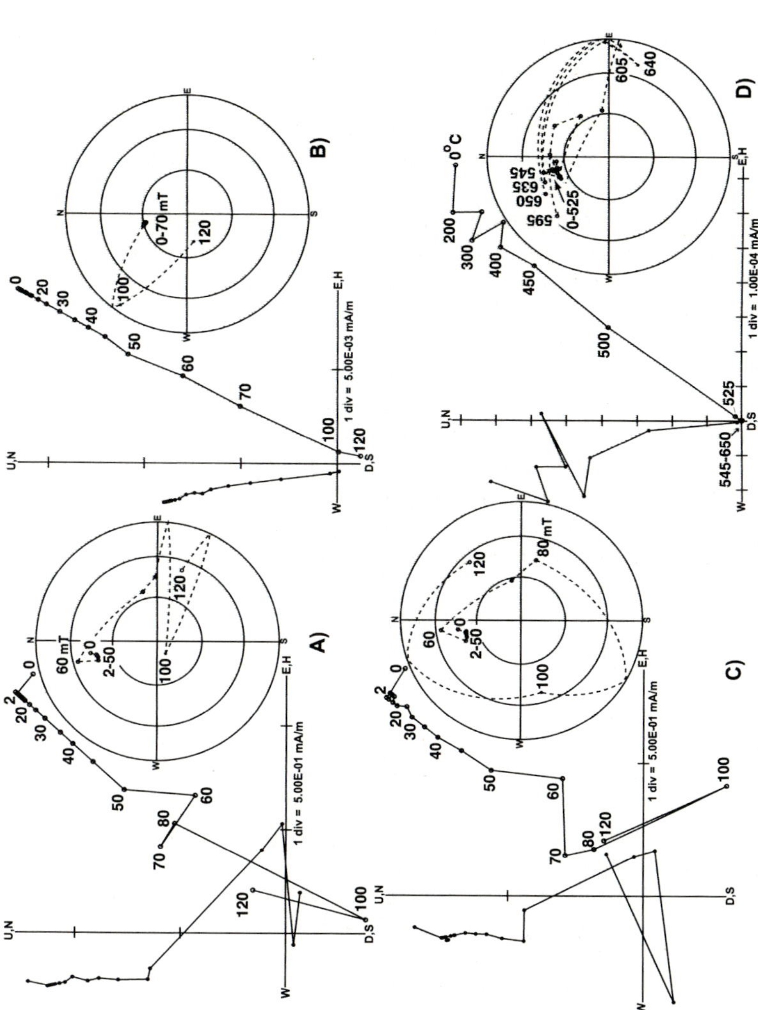

Fig. 9. Typical AF demagnetization (a - c) and thermal demagnetization (d) behavior for the dolerite and baked-contact samples at Carlton Heights. Equal-area stereographic plots show lower (upper) hemisphere directions as solid (open) circles. Orthogonal-axes plots show the horizontal projection of the magnetic vector (declination) as solid symbols and the vertical projection (inclination) as open symbols.

The component that is stable to thermal demagnetization at $\geq$575°C was calculated for each sample using least-squares analysis of the linear sequences of directions in that temperature interval. Mean directions for the >575°C steps were then calculated for each locality using standard Fisherian statistics. The mean directions are well defined at both localities, and are quite similar (Table 1).

The trends toward a reversed polarity direction exhibited at >575°C demagnetization never display stable end points. However, great-circle paths are well delineated by these trends; great-circle paths calculated for each locality are shown in Figs. 8A and B. Many of the paths are similar, but some are distinctly different, allowing determination of an intersection point. The great circles observed at Carlton Heights and Brakfontein intersect at the same point, suggesting that the same reversed polarity remanence was imprinted at both sites some 150 km apart.

The dolerite samples from Carlton Heights show NRM_0 directions and demagnetization responses similar to those of the sedimentary rocks. Dolerite sample responses to AF demagnetization were similar through 50 mT field strengths for three samples, 60 mT for two samples, and 70 mT for the sample closest to the dike margin. The samples exhibited a very stable direction near 371.1°, -51.4° through these demagnetization levels (Table 1). Above these demagnetization levels, samples responded with trends away from that direction and into a direction of opposite polarity (Figs. 9a-c). The baked contact sediment showed the same paleomagnetic direction through treatments to 25 mT, after which polarity directions moved slightly away from that direction (less than 10°).

Thermal treatment of dolerite samples gave results identical to those observed in the sedimentary samples. Generally, the NRM_0 direction was maintained between 545-585°C, after which directions trended irregularly towards opposite polarity (Fig. 9d). Again, the dolerite sample closest to the contact differed slightly in that its NRM_0 direction is stable only to 500°C, after which directions moved to the opposite hemisphere. The NRM_0 direction of the baked contact sediment was not stable to thermal demagnetization, and successive heating steps produced direction that trend to the opposite polarity and oscillate back and forth through continued heating. A mean was calculated from the combined directions held at the lower AF and thermal demagnetization levels in the dolerite samples (Table 1). Great circles were calculated for the dolerite sample from both AF and thermal demagnetization responses (Fig. 8c).

Essentially the same intersection point among great circles was obtained from the Carlton Heights dolerite as was found in the Carlton Heights sediments, and a very similar direction occurs in the Brakfontein sediments (Fig. 8c, Table 1). Thus, it seems that a series of remagnetization events affected both the dolerite and sediments at Carlton Heights, and the sediments Brakfontein by inference from the similarity of directions. The results for Carlton Heights demonstrate that these remagnetization events occurred after the emplacement of the dolerite dike.

These results suggest that the magnetic signatures in the sediment samples at Carlton Heights and Brakfontien are the result of remagnetization. The normal polarity direction observed in igneous rocks throughout the Karoo Basin was projected to the Carlton Heights (335.5°, -54.4°) and Brakfontein (335.2°, -55.6°)

sites. The direction during the Jurassic igneous activity at Carlton Heights and Brakfontein is statistically identical to that of the dolerite dike at Carlton Heights and to the less than 575°C direction in the sediments at both sites. The Jurassic intrusive activity in the Karoo Basin has been dated by ^{40}Ar-^{39}Ar techniques at 183±1 Ma, but it spanned at least one, and perhaps as many as seven reversals of the geomagnetic field (Duncan et al. 1997; Hargraves et al. 1997; Pálfy and Smith 2000), suggesting that the igneous activity may have had a duration of several million years.

The stability of the sedimentary rocks and dolerite samples suggests that the lower stability normal polarity magnetization was the second magnetization imprinted into these rocks. Moreover, although samples are few in number, the dolerite and baked-contact suggest better preservation of the reversed polarity direction in the middle of the dike, and better expression of the normal polarity direction on the dike edges and in the baked-contact sample. This sequence is also in accord with the reversed polarity imprint having been first and the normal second, and agrees with the general conclusion of Hargraves et al. (1997) that reversed geomagnetic field polarity prevailed during the initiation of igneous activity, followed by a later period of normal polarity.

Regardless of the timing, the results of our paleomagnetic study clearly indicate that both sedimentary strata and the dolerite intruding them were magnetized twice in the geomagnetic field directions that prevailed during Jurassic intrusive activity. No non-Jurassic directions were obtained, i.e., no recognizable Permian or Triassic paleomagnetic directions. Our study at Carlton Heights was motivated by the fact that the sedimentary strata were the least altered that we could locate in the southern Karoo Basin. The similarity of results from the widely spaced sites of Carlton Heights and Brakfontein indicate that multiple remagnetization of sedimentary strata was widespread throughout the Karoo Basin during the episode of Jurassic intrusions.

4
Palynological Results

We analyzed twenty-nine samples from the Balfour Formation and the lower Katberg sandstones at Carlton Heights for palynomorphs, of which seven were barren (Fig. 2). Three palynological assemblage zones were identified (Fig. 10): 1) the Late Permian *Kausipollenites schaubergeri* Zone, dominated by taxa of the form genera *Protohaploxypinus* and *Falcisporites*; 2) an interval composed almost entirely of fungal cell remains (*Reduviasporonites* or its junior synonyms *Chordecystia* or *Tympanicysta*) (Visscher et al. 1996) and abundant recycled woody material—the fungal spike zone (Fig. 10). The fungal spike zone is only ~1 m thick (from 57.6 to 58.6 m above the base of the section) (Figs. 2 and 5). 3) The Early Triassic *Kaeuselisporites-Lunatisporites* Zone, dominated by species of the lycopod *Kaeuselisporites* and the bisaccate pollen *Lunatisporites* and *Platysaccus*.

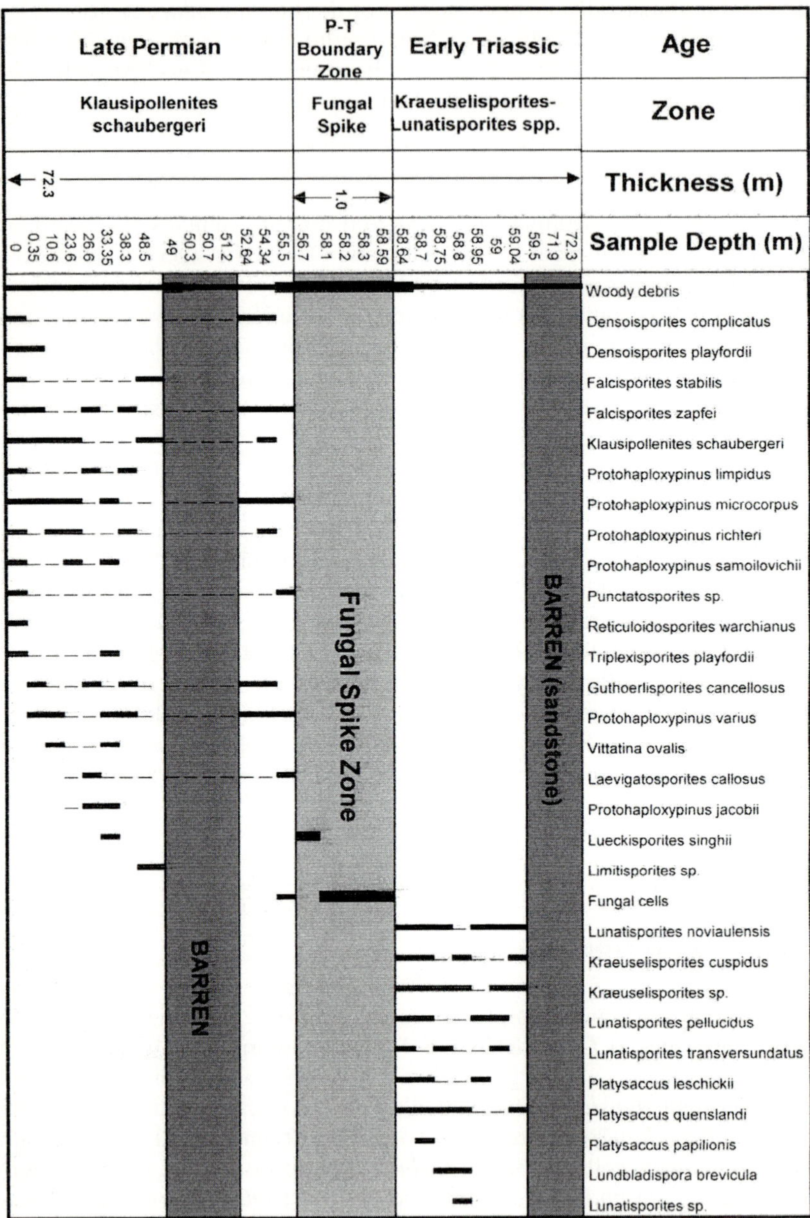

Fig. 10. Distribution chart of palynomorphs identified in the Carlton Heights stratigraphic section (compare with Fig. 2). Note that the chart is not to scale vertically. The total section is 72.3 m thick and the fungal spike interval is only ~1 m in thickness.

The interval from ~49 to 51.2 m in the section is barren of palynomorphs, and 8 of the 17 Late Permian palynomorph taxa that we identified last occur at or below this barren zone. The other nine Late Permian taxa last occur at or just below the base of the fungal spike zone (Fig. 10). The last occurrences just prior to the barren zone could represent a slightly earlier decrease in plant diversity, or poor preservation of pollen and spores in this zone.

Previous palynological studies of Karoo Supergroup rocks showed evidence for a major turnover or extinction of palynomorphs at or near the P-Tr boundary (as defined by the vertebrate assemblages) (Anderson 1977; Stapleton 1978; Utting 1979; Nyambe and Utting 1997). In sub-Equatorial Africa, a zone rich in fungal spores has been reported in sections spanning the P-Tr boundary from Kenya (Hankel 1992) and Madagascar (Wright and Askin 1987).

5
Discussion

In several sections in the Karoo Basin, Ward and Smith (2001) identified a zone of laminated maroon mudrock that seems to be devoid of vertebrate fossils just after the last appearance of *Dicynodon* zone fauna. They propose that these "event beds" mark the extinction of Late Permian vertebrates and thus the P-Tr boundary in the Karoo. At Carlton Heights, this unit occurs some 10 to 20 meters below the base of the fungal spike and the level at which most Late Permian palynomorphs disappear. This correlation suggests that the extinction of Late Permian mammal-like reptiles may have occurred prior to the land plant extinction event.

The worldwide fungal proliferation near the P-Tr boundary has been interpreted as reflecting global destruction of arboreous vegetation, a major loss of standing biomass, and the build-up of decaying vegetation on land (Visscher and Brugman, 1986; Visscher et al. 1996). Recovery from the extinction and renewed diversification in land plants were relatively slow, taking about 4 Ma (Eshet et al. 1995; Looy et al. 1999).

In the Tesero section in the southern Alps (Italy), the fungal spike occurs near the base of the Tesero Horizon of the Werfen Formation (Visscher and Brugman 1986). The disappearance of Late Permian marine fauna and an abrupt negative shift in carbon-isotope ratios in marine carbonates and organic carbon have been recognized at about the same level (Broglio-Loriga and Cassinis 1992; Margaritz et al. 1992; Rampino and Adler 1998; Rampino et al. 2000). The abrupt $\delta^{13}C$ shift can be explained by the rapid loss of primary productivity in the oceans and on land (Broecker and Peacock 1999). A similar negative $\delta^{13}C$ isotope shift has been reported from some terrestrial sections (Morante 1996; Krull and Retallack 2000), and from molecular fossils in land-plant leaf cuticles deposited in marine sediments, reflecting the coeval shift in atmospheric carbon isotopes (Sephton et al. 2002).

Recently, a similar negative excursion in carbon-isotope ratios has been reported from carbonate soil nodules and bone material from the Bethulie section in the Karoo Basin (MacLeod et al. 2000). At Bethulie, the initiation of the δ^{13}C excursion coincides with the local first appearance of Lystrosaurus, and δ^{13}C values begin to return to their former levels after the local last appearance of Dicynodon (Macleod et al. 2000).

In the oceans, the carbon-isotope shift, major marine-extinction level and the fungal spike have been estimated to be have occurred within an interval of less than 30,000 years (Twitchett et al. 2001). At Carlton Heights, the last recorded occurrence of Dicynodon is 10.8 m above the base of the section, or about 48 m below the base of the fungal spike zone (Fig. 2). At average sedimentation rates for Karoo rocks (~50 cm/1,000 years; Catuneanu and Elango 2001), the local last occurrence datum (LOD) of Dicynodon at Carlton Heights would have been about 100,000 years prior to the fungal spike.

The true last occurrence of Dicynodon, however, could lie considerably closer to the fungal spike level. Our calculation, using Marshall's (1990) formula for estimating the sampling error, for the top of the range of Dicynodon at Carlton Heights puts the 95% confidence level of last occurrence about 35 m above the current LOD of Dicynodon, or only ~13 m below the fungal spike zone. This would reduce the interval between the predicted last occurrence of Dicynodon and the fungal spike to roughly 25,000 years at average rates of Karoo sedimentation. The presence of Lystrosaurus fossils ~20 m below the fungal spike at Carlton Heights suggests that the first occurrence of Lystrosaurus predates the earliest Triassic (as defined in marine sections) by at least 40,000 years.

The fungal spike itself was apparently a very short-lived event. In the Carlton Heights sequence, the zone of maximum fungal spore abundance spans only ~1 meter of the sedimentary record (Figs. 5 and 10). At minimum estimated accumulation rates for the sediments of the Balfour Formation, this would mean about 2,000 years duration for the episode, with burrowing and the mixed depositional regimes making this an upper limit.

The heavily weathered layer corresponding to the top of the fungal spike zone may indicate a change in the climate or environment that promoted a brief period of intense weathering at the time of the land-plant extinction. Sedimentologic evidence for the sudden regional loss of terrestrial vegetation is indicated by the marked lithologic change to the first thick, multistoried channel and sheet sandstones typical of braided streams (Katberg Formation) (Ward et al. 2000). This change occurs less than 1 m above the fungal spike zone at Carlton Heights (Figs. 2 and 5).

6
Conclusions

Stratigraphic and palynological study at Carlton Heights identified a zone at the top of the Balfour Formation containing solely fungal spores coincident with the disappearance of typical Late Permian gymnosperm palynomorphs. This fungal spike occurs just below the first Katberg Sandstone, which represents a basin-wide change to braided stream patterns, probably related to the loss of vegetation (Ward et al. 2000). The fungal spike apparently represents a proliferation of fungi upon large volumes of decaying plant matter.

The fungal spike has a global distribution, and in marine sections it seems to be coeval with the abrupt negative shift in carbon isotopes and extinction of marine fauna that marks the end of the Permian (Twitchett et al. 2001). The local stratigraphy at Carlton Heights, when combined with the recent work of Smith and Ward (2001) on other Karoo sections, suggests that the disappearance of typical Late Permian vertebrates may have preceded the fungal spike zone and related extinction of gymnosperms by a time interval that we estimate from average sedimentation rates to be about 25,000 to 100,000 years.

Paleomagnetic analysis of sediment samples from the Carlton Heights and Brakfontein sections of the Karoo suggests that Permian–Triassic sediments in the Karoo are recording a widespread double overprint of Jurassic age. Many of the sediment samples show a reversed trend at high demagnetization temperatures, and similar demagnetization patterns are found in a Jurassic intrusion at Carlton Heights. Our results indicate that the Jurassic igneous events in the Karoo Basin resulted in multiple magnetic overprints in sediments and dolerites, and the widespread loss of Permian-Triassic magnetic directions.

Acknowledgements

We thank P. J. Hancox and N. Tabor for help in the field, R. C. Reynolds for x-ray mineralogical analyses, P. D. Ward and R. Smith for helpful discussions, and C. Koeberl and M. Sephton for helpful reviews. We are grateful to the B.P. Erasmus, G. Van Zyl, L. Stein, and P. Vandertavor families for permissions and assistance. Great-circle analysis was performed using the PALEOMAG software package (Version 2.3) developed by C. Jones.

References

Anderson JM (1977) The biostratigraphy of the Permian and Triassic: Part 3, A review of Gondwana Permian palynology with particular reference to the northern Karoo Basin, South Africa. Memoirs of the Botanical Survey of South Africa 41: 1-300

Ballard MM, Van der Voo R, Hälbich IW (1986) Remagnetizations in late Permian and early Triassic rocks from southern Africa and their implications for Pangea reconstructions. Earth and Planetary Science Letters 79: 412-418

Bowring SA, Erwin DH, Jin YG, Martin MW, Davidek K, Wang W (1998) U/Pb Zircon geochronology and tempo of the end-Permian mass extinction. Science 280: 1039-1045

Broecker WS, Peacock S (1999) An ecologic explanation for the Permo-Triassic carbon and sulfur isotope shifts. Global Biogeochemical Cycles 13: 1167-1172

Broglio-Loriga C, Cassinis G (1992) The Permo-Triassic boundary in the Southern Alps (Italy) and in adjacent Periadriatic regions. In: Sweet WC, Zunyi Y, Dickins JM, Hongfu (eds) Permo-Triassic events in the eastern Tethys. Cambridge University Press, Cambridge, pp 78-97

Catuneanu O, Elango HN (2001) Tectonic control on fluvial styles: The Balfour Formation of the Karoo Basin, South Africa. Sedimentary Geology 140: 291-313.

Cirilli S, Radrizzani CP, Ponton M, Radrizzani S (1998) Stratigraphical and palaeoenvironmental analysis of the Permian-Triassic transition in the Badia Valley (Southern Alps, Italy). Palaeogeography, Palaeoclimatology, Palaeoecology 138: 85-113

Duncan RA, Hooper PR, Rehacek J, Marsh JS, Duncan AR (1997) The timing and duration of the Karoo igneous event, southern Gondwana. Journal of Geophysical Research 12: 18,127-18,138

Eshet Y, Rampino MR, Visscher H (1995) Fungal event and palynological record of ecological crisis and recovery across the Permian-Triassic boundary. Geology 23: 967-970

Gialanella PR, Heller F, Haag M, Nurgaliev D, Borisov A, Burov B, Jasonov P, Khasanov D, Ibragimov S, Zharkov I (1997) Late Permian magnetostratigraphy on the eastern Russian platform. Geologie en Mijnbouw 76: 145-154

Groenewald GH (1989) Stratigrafie en sedimentologie van die Groep Beaufort in die Nooordoos Vrystaat (Stratigraphy and sedimentology of the Beaufort Group in the Northeast Freestate). Bulletin of the Geological Survey of South Africa 96: 1-62

Hankel O (1992) Late Permian to Early Triassic microfloral assemblages from the Maji Ya Chumvi Formation, Kenya. Review of Palaeobotany and Palynology 72: 129-147

Hargraves RB, Rehacek J, Hooper PR (1997) Paleomagnetism of the Karoo igneous rocks in southern Africa. South African Journal of Geology 100: 195-212

Jin YG, Wang Y, Wang W, Shang QH, Cao CQ, Erwin DH (2000) Pattern of marine mass extinction near the Permian-Triassic boundary in South China. Science 289: 432-436

Keyser N (1977) Geological Map of the Republic of South Africa and the Kingdoms of Lesotho and Swaziland. South African Council for Geoscience, Johannesburg

Kirschvink JL, Ward PD (1998) Magnetostratigraphy of Permian/Triassic boundary sediments in the Karoo of southern Africa [abs]. Journal of African Earth Sciences 27: 124

Krull ES, Retallack GJ (2000) Delta C-13 depth profiles from paleosols across the Permian-Triassic boundary: Evidence for methane release. Geological Society of America Bulletin 112: 1459-1472

Labandiera CC, Sepkoski JJ, Jr (1993) Insect diversity in the fossil record. Science 261: 310-315

Looy CV, Brugman WA, Dilcher DL, Visscher H (1999) The delayed resurgence of equatorial forests after the Permian-Triassic ecologic crisis. Proceedings of the National Academy of Sciences of the USA 96: 13857-13862

Looy CV, Twitchett RJ, Dilcher DL, Van Konijnenburg-Van Cittert JHA, Visscher H (2000) Life in the end-Permian dead zone. Proceedings of the National Academy of Sciences of the USA 98: 7879-7883

MacLeod KG, Smith RMH, Koch PL, Ward PD (2000) Timing of mammal-like reptile extinctions across the Permian-Triassic boundary in South Africa. Geology 28: 227-230

Margaritz M, Bar R, Baud, A, Holser WT (1988) The carbon-isotope shift at the Permian/Triassic boundary in the southern Alps is gradual. Nature 331: 337-339

Margaritz M, Krishnamurthy RV, Holser WT (1992) Parallel trends in organic and inorganic carbon isotopes across the Permian-Triassic boundary. American Journal of Science 292: 727-739

Marshall CR (1990) Confidence intervals on stratigraphic ranges. Paleobiology 16: 1-10

Maxwell WD (1992) Permian and Early Triassic extinction of nonmarine tetrapods. Paleontology 35: 571-583

Morante R (1996) Permian and early Triassic isotopic records of carbon and strontium in Australia and a scenario of events about the Permian-Triassic boundary. Historical Biology 11: 289-310

Mundil R, Metcalfe I, Ludwig KR, Renne PR, Oberli F, Nicoll RS (2001) Timing of the Permian-Triassic biotic crisis: implications from new zircon U/Pb age data (and their limitations). Earth and Planetary Science Letters 187: 131-145

Nyambe IA, Utting J (1997) Stratigraphy and palynostratigraphy, Karoo Supergroup (Permian and Triassic), mid-Zambezi Valley, southern Zambia. Journal of African Earth Sciences 24: 563-583

Ouyang S, Utting J (1990) Palynology of Upper Permian and Lower Triassic rocks, Meishan, Changxing County, Zhejiang Province, China. Review of Palaeobotany and Palynology 66: 65-103

Pálfy J, Smith P (2000) Synchrony between Early Jurassic extinction, oceanic anoxic event, and the Karoo-Ferrar flood basalt volcanism. Geology 28: 747-750

Pehl CW, Kirschvink JL, Ward PD (2001) Towards a magnetostratigraphy of the terrestrial Permian-Triassic boundary [abs.]. Geological Society of America, Abstracts with Programs 33 (6): A-390

Poort R J, Clement-Westerhof JA, Looy CV, Visscher H (1997) Aspects of Permian palaeobotany and palynology. 17. Conifer extinction in Europe at the Permian-Triassic junction: Morphology, ultrastructure and geographic/stratigraphic distribution of *Nuskoisporites dulhuntyi* (prepollen of *Ortiseia, Walchiaceae*). Review of Palaeobotany and Palynology 97: 9-39

Rampino MR, Adler AC (1998) Evidence for abrupt latest Permian mass extinction of Foraminifera: Results of tests for the Signor-Lipps effect. Geology 26: 415-418

Rampino MR, Prokoph A, Adler A (2000) Tempo of the end-Permian event: High-resolution cyclostratigraphy at the Permian-Triassic boundary. Geology 28: 643-646.

Raup DM (1979) Size of the Permo-Triassic bottleneck and its evolutionary implications. Science 206: 217-218

Retallack GJ (1995) Permian-Triassic life crisis on land. Science 267: 77-80

Scholger R, Mauritsch HJ, Brandner R (2000) Permian-Triassic boundary magnetostratigraphy from the Southern Alps (Italy). Earth and Planetary Science Letters 176: 495-508

Sephton MA, Looy CV, Veefkind RJ, Brinkhuis H, De Leeuw JW, Visscher H (2002) A synchronous record of $\delta^{13}C$ shifts in the oceans and atmosphere at the end of the Permian. In: Koeberl C, MacLeod KG (eds) Catastrophic Events and Mass Extinctions: Impacts and Beyond, Geological Society of America Special Paper 356: 455-462

Smith RMH (1990) A review of stratigraphy and sedimentary environments of the Karoo Basin of South-Africa. Journal of African Earth Sciences 10: 117-137

Smith RMH (1995) Changing fluvial environments across the Permian-Triassic boundary in the Karoo Basin, South-Africa and possible causes of tetrapod extinctions. Palaeogeography, Palaeoclimatology, Palaeoecology 117: 81-104

Smith RMH, Ward PD (2001) Pattern of vertebrate extinctions across an event bed at the Permian-Triassic boundary in the Karoo Basin of South Africa. Geology 29: 1147-1150

Stapleton RP (1978) Microflora from a possible Permo-Triassic transition in South Africa. Review of Palaeobotany and Palynology 25: 253-258

Steiner MB (2001) Magnetostratigraphic correlation and dating of West Texas and New Mexico Late Permian strata, in Geology of the Llano Estacado. In: Lucas SG, Ulmer-Schoolle DS (eds) Guidebook, Guidebook for 52nd Field Conference, New Mexico Geological Society, pp 59-68

Steiner MB, Ogg J, Zhang Z, Sun S (1989) The Late Permian/Early Triassic magnetic polarity time scale and plate motions of South China. Journal of Geophysical Research 94: 7343-7363

Steiner MB, Morales M, Shoemaker EM (1993) Magnetostratigraphic, biostratigraphic and lithologic correlation in Triassic strata of the western U.S. In: Aissioui D, McNeill D, Hurley N (eds) Application of Paleomagnetism to Sedimentary Geology, Society of Economic Paleontologists and Mineralogists, Special Publication 49: 41-577

Twitchett RJ, Looy CJ, Morante R, Visscher H, Wignall PB (2001) Rapid and synchronous collapse of marine and terrestrial ecosystems during the end-Permian biotic crisis. Geology 29: 51-354

Utting J (1979) Pollen and spore assemblages from the Upper Permian of the North Luangava Valley, Zambia. Proceedings of the 4th International Palynology Conference, Moscow 2: 165-174

Visscher H, Brugman WA (1986) The Permian-Triassic boundary in the Southern Alps: A palynological approach. Memoire della Societá Geologia Italiana 34: 121-128

Visscher H, Brinkhuiss H, Dilcher DL, Elsik WC, Eshet Y, Looy CV, Rampino MR, Traverse A (1996) The terminal Paleozoic fungal event: Evidence of terrestrial ecosystem destabilization and collapse. Proceedings of the National Academy of Sciences USA 93: 2155-2158

Wang K, Geldsetzer HHJ, Krouse HR (1995) Permian-Triassic extinction: Organic $\delta^{13}C$ evidence from British Columbia, Canada. Geology 22: 580-584

Ward PD, Montgomery DR, Smith R (2000) Altered river morphology in South Africa related to the Permian-Triassic extinction. Science 289: 1740-1743

Wignall PB, Kozur H, Hallam A (1996) On the timing of palaeoenvironmental changes at the Permo-Triassic (P/Tr) boundary using conodont biostratigraphy. Historical Biology 12: 39-62

Wright RP, Askin RA (1987) The Permian-Triassic boundary in the Southern Morondava Basin of Madagascar as defined by plant microfossils. Geophysical Monographs 41: 157-166

Search for an Extraterrestrial Component in the Late Devonian Alamo Impact Breccia (Nevada): Results of Iridium Measurements

Christian Koeberl[1], Heinz Huber[1], Matthew Morgan[2] and John E. Warme[2]

[1] Institute of Geochemistry, University of Vienna, Althanstrasse 14, A-1090 Vienna, Austria.
(christian.koeberl@univie.ac.at)
[2] Colorado School of Mines, Golden, Colorado 80401-1887, USA.

Abstract. The Alamo Breccia is a Late Devonian sedimentary layer as much as 135 m in thickness that is widespread over southern Nevada, USA. Direct evidence for an impact origin of the Alamo Breccia includes shocked quartz grains within the Breccia matrix and broken fragments of distinctive carbonate lapilli beds within the heterolithic Breccia clast population. The Breccia was studied as a potential impactite because it was recognized as an anomalous stratigraphic unit compared to the contrasting and predictable carbonate platform strata of the Devonian platform. To date, only one possible crater locality has been identified within the complex geology of Nevada.

A search was made for the presence of a meteoritic component in 23 samples: 16 from the Breccia matrix, 6 from the lapilli, and 1 background sample from below the Breccia. The samples were analyzed by multiparameter γ-γ coincidence spectrometry after neutron activation. The results vary between 19 and 51 ppt iridium, and indicate only minor possible Ir enrichment in the matrix and lapilli clasts compared to the background sample. The lapilli samples averaged slightly higher Ir values than the matrix. Because these are carbonate samples, and highly diluted by the massive Breccia, as little as 0.04 ppb matrix Ir may indicate a small meteoritic component. The lapilli population should have higher concentrations of rock from the target zone than does the mixed proximal and distal Breccia matrix, but they appear to contain insignificantly higher Ir values than the matrix. This leaves a variety of possibilities: either the Breccia does not contain much of an extraterrestrial component, or the Alamo impact event was caused by an Ir-poor projectile – either an achondrite, or maybe even an cometary nucleus.

1
Introduction

The Alamo Breccia of southern Nevada, USA (Fig. 1), contains the signatures of an impact that occurred on or near the edge of the carbonate platform that formed the western edge of North America during Late Devonian time. Warme and Kuehner (1998) reviewed the history of discovery and interpretations of the Breccia through 1997. Several new Breccia localities have been discovered where the Breccia is within the platform facies (Fig. 1), and possible offshore equivalents are present (Sandberg et al. 2002). The Alamo Breccia has been designated as a Member of the Late Devonian Guilmette Formation. The Guilmette is situated between the Middle Devonian Simonson Dolomite and the Late Devonian West Range Limestone. Because of its unusual thickness and chaotic bedding, the Breccia is unique compared to the intercalated carbonate platform facies of the Guilmette and off-platform beds to the west. Evidence that links the Alamo Breccia with an impact event includes shocked quartz grains in the Breccia matrix (Warme and Sandberg 1995, 1996: Leroux et al. 1995), clasts of distinctive carbonate accretionary lapilli (Warme et al. 2002), and minor enrichment in iridium. Warme and Sandberg (1995) reported unpublished data by C. Orth and M. Attrep (Los Alamos National Lab), who measured a maximum of 133 ppt (parts per trillion, or 10^{-12} g/g) of Ir in two closely-spaced vertical profiles through the Breccia at the Worthington Mountains (Fig. 2). In this report we present results of new tests for Ir in 23 Breccia samples: 16 from the matrix, six from the lapilli-bearing clasts, and one from below the Breccia for background signal.

2
Alamo Breccia

To date the Alamo Breccia has been reported in 20 mountain ranges in the Basin and Range Province across the southern half of Nevada. Within the shallow-water carbonate-platform facies of the Guilmette formation the Breccia presently covers an area of approximately 16,000 km^2, occupies a volume of ~600 km^3, and ranges in thickness from ~1 to more than 100 meters (Warme and Kuehner 1998; Warme et al. 2002). The geology of the area is complex, comprised of Early to Late Paleozoic mainly carbonate rocks, Tertiary volcanic rocks, and Quaternary alluvial valley fills. However, recent fieldwork indicates that the area and volume given above are conservative; the surficial area of the Breccia has been expanded into deep-water facies to the west (Morrow et al. 1998; Sandberg et al. 2002), and post-Devonian crustal compression in southern Nevada may have exceeded subsequent extension so that the present map distribution of the Breccia is less than its original area.

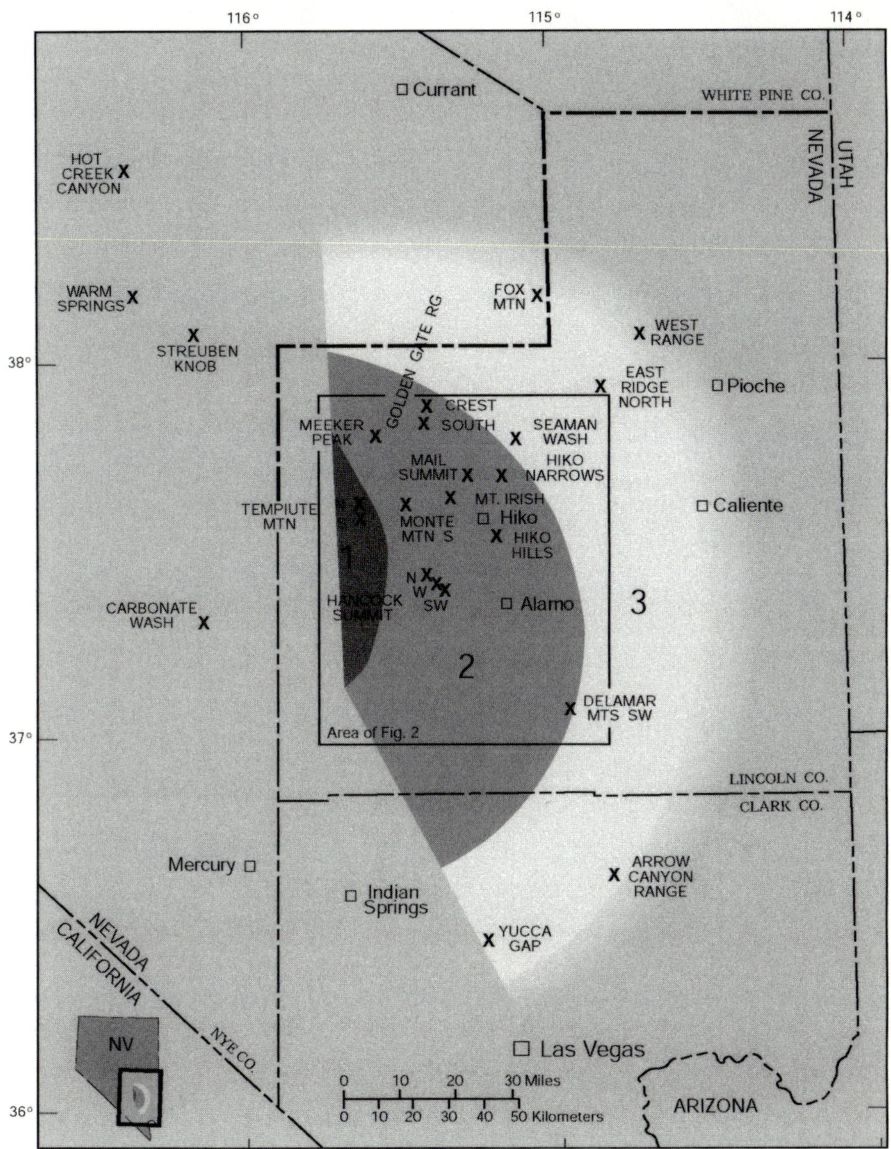

Fig. 1. Index map of southeastern Nevada showing the rough semi-circular distribution of the Alamo Breccia in Zones 1 (dark gray), 2 (light gray), and 3 (white). X's denote the location of Upper Devonian measured stratigraphic sections used as background for this report (from Warme and Sandberg 1996). Squares represent habitation centers.

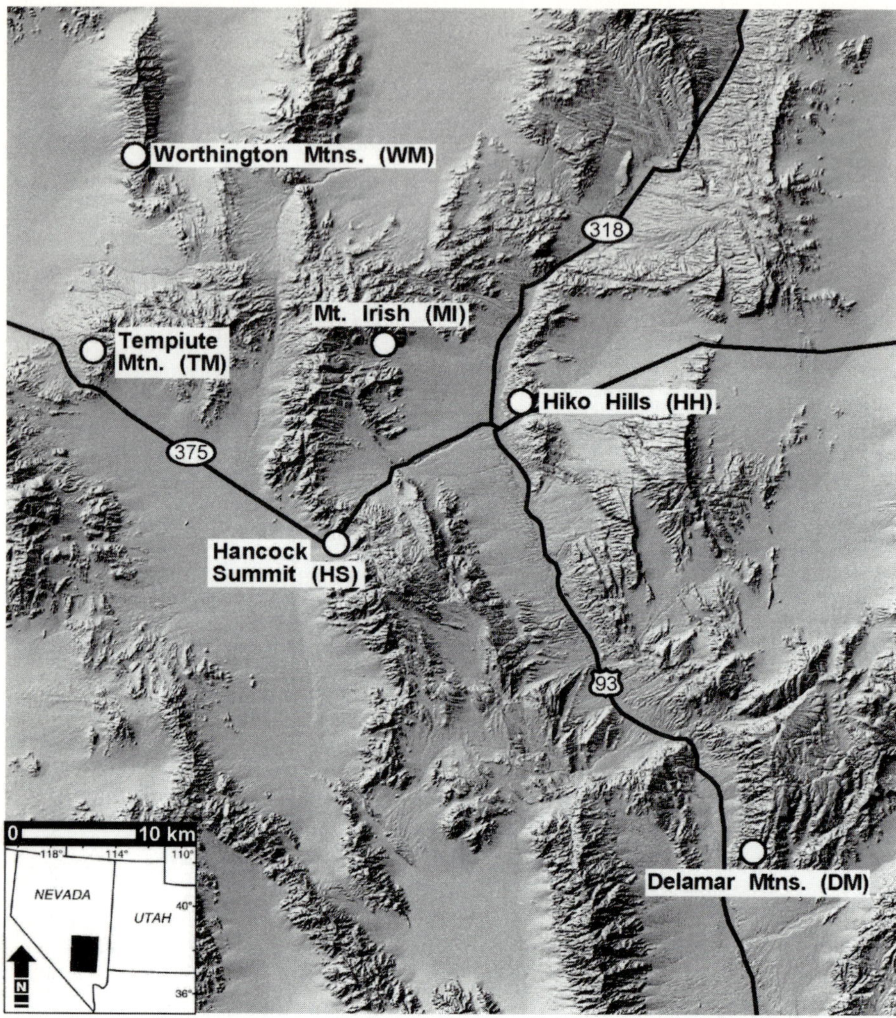

Fig. 2. Hillshaded digital elevation model (DEM) of the central portion of the Alamo Breccia distribution within Devonian carbonate platform facies, showing highways and sample locations discussed in the text.

Samples analyzed for this report were mainly from the Alamo Breccia where it is intercalated with the shallow-water platform facies of the Guilmette in a semi-circular distribution (Fig. 1). In outcrop exposure the massive Breccia commonly forms a shear cliff (Figs. 3, 4). Probable correlative beds in Upper Devonian deepwater environments to the west of the platform are X's on the western side of Figure 1 (see Sandberg et al. 2002).

Fig. 3. Outcrop photograph showing ~60 m of the cliff-forming Alamo Breccia, view to the northwest, Hiko Hills South locality (Fig. 2). The arrow indicates the top of the thin A-Unit, underlain by the chaotic B-Unit that extends to the base of the photo. Note tilted, broken clasts suspended in heterolithic matrix of the Breccia. Not shown are about 25 m of C-megaslab and D-interval beneath the B-Unit.

2.1
Lateral Zones

The Alamo Breccia is composed mainly of limestone and dolostone fragments derived from the disintegration during impact of the Devonian carbonate platform. The present-day exposures of the Breccia across the Guilmette platform were divided into three semi-concentric Zones by Warme and Sandberg (1995, 1996) (Fig. 1). Zone 1 contains the thickest Breccia (~130 m), and was believed by Warme and Kuehner (1998) to represent crater-fill based on the relative thickness (~135 m) compared to Zones 1 and 2 (<100 m to <1m), the highly deformed underlying beds with possible impact-generated dikes and sills, and the distinctive anoxic thin-bedded deepwater limestone over it that represents a marine stratified water column. In contrast, Sandberg et al. (2002) continued to believe the notion that Zone 1 represents a deepwater environment, requiring an abrupt change in the trend of the platform margin compared to their earlier papers, and that crustal compression and extension though several post-Devonian orogenies insignificant for interpretation of paleogeography at the time of the Alamo Event. This issue is still under investigation.

The thickness of the Breccia in Zone 2 averages approximately 60 m, and contains carbonate platform blocks as much as 500 m in length and 70 m in height as well as stacked graded beds in the upper portion that are interpreted as tsunamites. In peripheral Zone 3 the Breccia thins eastward from 10 to less than 1 m, and is currently regarded as a stranded tsunamite deposited by the uprush of impact-induced tsunamis onto the landward edge of the platform (Warme and Kuehner 1998).

Fig. 4. Vertical cliff of Alamo Breccia at Mount Irish locality (Fig. 2), looking eastward. Arrow shows contact between ~60 m of A- and B-Units of the Alamo Breccia, below, with massive stromatoporoid reef, above. The cliff is ~65 m high, contains the A- and B-Units, and rests on a slope of C-Megaslab and D-Interval.

Shocked quartz occurs in all three Zones of the Alamo Breccia. In Zone 3 the shocked grains are sparse. Transmission electron microscopic studies (Leroux et al. 1995) confirmed the presence of planar deformation features (PDFs) in the quartz grains from the Alamo Breccia, indicating their origin by shock (e. g., Stöffler and Langenhorst 1994; Grieve et al. 1996). The Ir anomalies (Warme and Sandberg 1995; this work) and accretionary carbonate lapilli (Warme et al. 2002) have been found as of 2002 only in Zones 1 and 2.

2.2
Vertical Facies

For descriptive purposes, the Alamo Breccia was divided into four vertical lithofacies A to D designated as units , facies or intervals (Warme and Sandberg 1995, 1996; Warme and Kuehner, 1998). Herein we use the following terms, from the base upward: D-interval, C-megaslabs, B-unit and A-unit (Warme and Kuehner 1998). The D-interval is a thin (<1-~3 m) monomict breccia, interpreted to be an interval created by abrasion from the overlying C-megaslabs and/or by seismic shock, fluidization, and detachment during impact. The C-megaslabs ("slab" = ~66 to 1050 m in length; Blair and MacPherson 1999), commonly tens of meters thick and hundreds of meters long, appear to be transported not far from their original locations on the carbonate platform. The A- and B-units of the Alamo Breccia are both polymict breccias composed of limestone and dolostone lithoclasts that include stromatoporoid (calcareous sponge) heads, coral fragments, and other fossil debris. Unit B is commonly massive and appears to be a single depositional event. It is chaotic (disorganized) throughout, and contains twisted clasts and large megaslabs (Kuehner 1997) suspended in the carbonate matrix composed of particles sized from boulders down to gravel, sand and mud. Unit A is divided into as many as five clast-supported and more organized graded beds that become progressively thinner and finer-grained upward. Scour surfaces, commonly loaded and deformed, separate these Units. Warme et al. (2002) proposed a model whereby the highly disorganized Unit represented impactite and/or debrite, and the better organized and graded Unit A represents a series of tsunamites.

2.3
Lapilli

Accretionary carbonate lapilli in the Alamo Breccia were first described by Warme and Kuehner (1998) and are the subject of the report by Warme et al. (2002). The lapilli provide the strongest evidence for an impact origin of the Breccia. They occur within rare, isolated clasts of lapillistone scattered in the Breccia of the A-Unit and perhaps as low as the B-Unit (Fig. 5). In most occurrences the clasts are plastically deformed, but rare segments of intact beds show roughly size-sorted layers, varying proportions of matrix, and roughly graded lapilli layers (Fig. 6). The thickest confirmed bed of lapillistone is ~1.3 m.

The lapilli resemble silicate volcanic lapilli in structure, but are composed entirely of carbonate except for sparse silt-sized shocked quartz grains incorporated into the accretionary mantle and small crystals of diagenetic metallic oxides. Lapilli are variable in size and detail, but are generally less than 1 cm in diameter. Many exhibit a nucleus believed to be altered target rock, enveloped by a mantle of fine-grained particles, and coated by a very fine-grained peripheral

Fig. 5. Photograph of lapilli clast within matrix of Alamo Breccia, Hiko Hills South locality (Fig. 2).

Fig. 6. Photograph of lapilli bed within the A-Unit of the Alamo Breccia at Mount Irish. In this example the distribution of the dark grains suggests inverse grading, but the bed was probably inverted during transport. The dark grains are nuclei of lapilli, or nuclei liberated when some lapilli disintegrated upon impact to form the bed. Natural shading results in better view of the lapilli boundaries in the finer-grained portion of the bed if the photo is viewed as inverted.

crust, as is shown in Figures 7 and 8. Their preserved form is spherical, deformed,

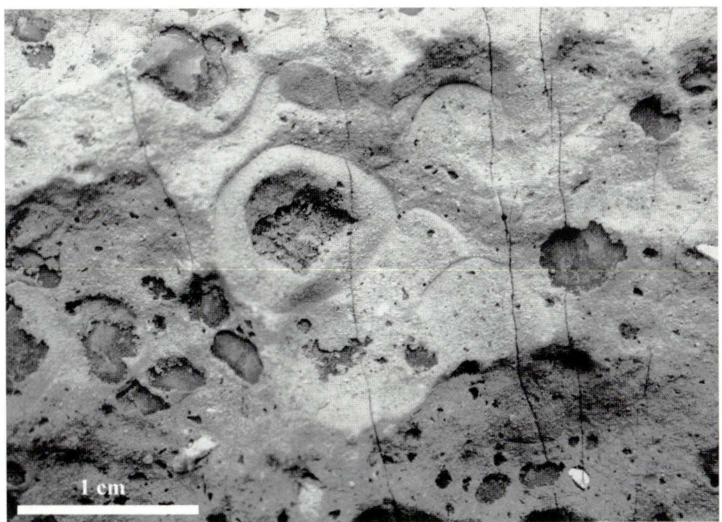

Fig. 7. Outcrop photograph of lapilli within a lapillistone bed at Mount Irish locality (Fig 2). Central lapillus is approximately 1 cm in diameter. Arcuate shapes are the boundaries of broken lapilli. Dark fragments are nuclei that were broken loose from lapilli upon impact with the accumulating bed.

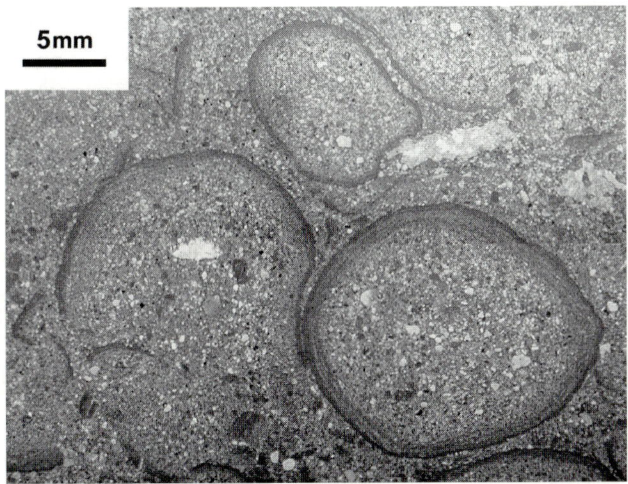

Fig. 8. Thin-section photomicrograph of lapilli comprised of central nuclei (specimen left of center), accretionary mantles and more brittle crusts. The lapilli are preserved as original spheres. as deformed, or as fragments. Segments of broken crusts are scattered in the matrix.

or broken. This condition implies that various degrees of damage occurred before or during deposition, and before bed hardening and penecontemporaneous rip-up and transportation by tsunamis, dewatering, and perhaps degassing of the Alamo Breccia soon after high-speed transportation, heating, and deposition.

Table 1. Sample list, sample locations, and iridium abundances in 23 samples from Alamo Breccia (AB) outcrops: 16 from Breccia matrix, six from lapilli clasts, and one from below the Breccia analyzed for background values.

Sample	Locality	Stratigraphic Location	Ir (ppt)	Error (ppt)
HH1	Hiko Hills South	Lapilli 6 m from AB top	45	13
HH2	Hiko Hills South	15 m from AB base	46	13
HH3	Hiko Hills South	21 m from AB base	35	14
HH4	Hiko Hills South	27 m from AB base	51	12
HH5	Hiko Hills South	Top of AB	41	12
HH6	Hiko Hills South	1.5 m from AB top	36	12
HH8	Hiko Hills South	17 m from AB top	28	12
HH9	Hiko Hills South	Lapilli 3 m from AB top	39	12
D1	Delamar Range	0.6 m from AB top	33	11
RR10	Mount Irish	Lapilli	34	11
MI1	Mount Irish	Lapilli (light-colored matrix)	35	12
MI2	Mount Irish	Lapilli (flame structures present)	43	13
HS1	W. Pahranagat Rng. (H. Sum.)	15 m from AB top	42	13
HS2	W. Pahranagat Rng. (H. Sum.)	Lapilli 21 m from AB top	24	10
HS3	W. Pahranagat Rng. (H. Sum.)	3 m from AB top	38	12
HS4	W. Pahranagat Rng. (H. Sum.)	3 m from AB top	30	11
HS5	W. Pahranagat Rng. (H. Sum.)	15 m from AB top	19	9
HS7	W. Pahranagat Rng. (H. Sum.)	69 m from AB top; D interval	38	12
TM1	Tempiute Mountain	1.5 m from AB base	27	10
TM2	Tempiute Mountain	38 m from AB base	20	9
TM5	Tempiute Mountain	45 m from AB base	30	11
TM6	Tempiute Mountain	9 m from AB top; dikes & sills	29	11
BASE	Mount Irish	70 m below AB	25	10

H. Sum. = Hancock Summit

Warme et al. (2002) proposed that the lapilli formed when some proportion of target rock was pulverized by impact pressure and calcined by impact heat, creating quicklime (CaO). The lapilli grew by adhesion of particles within the impact cloud, in the same fashion as silicate lapilli form during explosive volcanism, then precipitated as one or more beds over early ejecta and/or contemporaneous tsunamites. Rapid hydration and other processes initiated cementation of the lapilli in flight, and continued after precipitation so that portions of the bed survived tsunami reworking and Breccia settling. Coherent, isolated fragments of lapilli beds and deformed bed masses are preserved as much as 25 m beneath the top of the Breccia, indicating the thickness of reworked Breccia during the Alamo Event. Because the lapilli are interpreted to have formed in the impact cloud, they were expected to contain higher values of Ir than the Breccia matrix and non-lapilli clasts. However, average measured values of Ir in the lapilli are not much higher than in the Breccia matrix (Table 1), as discussed below.

3
Extraterrestrial Components

3.1
Background

As described in the reviews by Koeberl (1998) and Montanari and Koeberl (2000), the detection of meteoritic components in rocks can help confirm an impact origin for a suspected geologic structure, and also indicate the type of impactor. However, the detection of a meteoritic component in most impactites is difficult. Only very small amounts, usually less than 1 % by weight, of the meteoritic material are mixed with the vaporized, molten, or shocked terrestrial impact products. To differentiate these minute meteoritic components from the volumetrically overwhelming terrestrial signature if the target rocks is an analytical challenge. Identification of meteoritic components is commonly obtained by determination of the concentrations and interelement ratios of siderophile elements, especially the platinum group elements (PGEs), which are several orders of magnitude more abundant in meteorites than in terrestrial upper crustal rocks. Iridium is most often determined as a proxy for all PGEs, because it can be measured with the best detection limit of all PGEs by neutron activation analysis. Strong enrichment in siderophile elements in impact melts and other products usually indicates the presence of either a chondritic or an iron meteoritic component.

Due to analytical, mineralogical, and geological problems, such PGE analyses are not always easily obtained, or may yield ambiguous results. Meteoritic components have been identified for little more than 40 impact structures (see Koeberl, 1998, for a list), compared to more than 160 impact structures that have so far been identified on Earth.

3.2
Iridium Abundance Measurements

Iridium Coincidence Spectrometry (ICS) is a multiparameter γ-γ coincidence spectrometry method specifically suited for the determination of very low Ir concentrations, in the sub-parts per billion (ppb = 10^{-9} g/g) range, which is important for a variety of geological applications. Good sensitivity and precision can be obtained even for small sample sizes (less than 0.5 g). The ICS method is currently the only possibility for the non-destructive determination of Ir at ppt abundance levels in very small (less then 0.5g) geological samples (e.g., Koeberl and Huber 2000; Huber et al. 2001).

The main purpose of coincidence counting is background reduction and removal of spectral interferences. However, studies of impact-related rocks and deposits required detection limits that are at least one to two orders of magnitude lower then those obtainable with instrumental neutron activation analysis (INAA). Instrumental neutron activation analysis allows the determination of most trace elements without the need of radiochemical separation or complicated coincidence counting. The detection limit of Ir, using standard Ge-detectors with INAA, is a few ppb. This requirement led to the development of the ICS method, using high-purity germanium detectors, fast counting systems, and fast coincidence electronics (e.g., Koeberl and Huber 2000).

Compared to other methods, such as radiochemical neutron activation analysis (RNAA) or inductively coupled plasma source mass spectrometry (ICP-MS), using ICS for iridium measurement has several advantages (see discussion in Koeberl et al. 2000). Iridium has two naturally occurring isotopes, i.e., ^{191}Ir and ^{193}Ir, with high neutron capture cross sections, making Ir one of the most sensitive elements that can be determined by thermal neutron activation techniques. This method does not require dissolution or other chemical treatments of the samples and, therefore, avoids possible contamination.

Generally, the abundances of the platinum group elements (PGE) are several orders of magnitude higher in meteorites (chondrites and most iron meteorites) than in common terrestrial rocks. Ir abundances in chondrites range at 400-800 ppb, in iron meteorites between 0.1 ppb and 10 ppm, whereas continental rocks contain on the order of 0.02 ppb Ir. Nevertheless, the addition of only 0.1 wt% of a chondritic component (containing about 500 ppb Ir) to a crustal target rock would yield an elevated Ir content of about 0.5 ppb in the resulting impact breccia (e.g., Koeberl 1998). Thus, the detection of enrichments of Ir and other PGEs in impact-derived rocks provides good evidence for the presence of a meteoritic component.

3.3
Principles of Coincidence Counting

The method is based on the counting of coincidence intensities due to the cascade decay of ^{192}Ir after neutron bombardment. The ICS at the Institute of Geochemistry (University of Vienna) is designed as a fast-coincidence counting system and consists of two low energy HPGe-detectors, two (pre-) amplifiers, two analog-to-digital-converters (ADCs) and a multiparameter-analyzer (MPA) small bus box (Fast Com Tec MPA-SBB) with a PC based evaluation procedure containing specially adapted software to interpolate the resulting 3D-data. Only signals occurring within the coincidence time from both detectors are accepted by the MPA. The spectra (see Figs. 9, 10) are three-dimensional 1024*1024 matrices, where first the regions of 316.5 and 468.1 keV (the two main decay lines) and vice versa are isolated. The resulting two smaller matrices are used for the data evaluation procedure, which includes a 3D peak-fitting and background reduction to evaluate the peak volumes. The results are corrected for sample weight, live time, decay time and neutron flux and compared with the standards. The method as implemented in Vienna allows to measure Ir contents with detection limits in the lower parts per trillion (ppt = 10^{-12} g/g). For further details on precision and accuracy of this method, see Koeberl and Huber (2000).

3.4
Samples and Experimental Methods

Twenty three samples from different outcrop locations of the Alamo Breccia (Fig. 2) were analyzed in two measurement cycles. Table 1 gives the sample locations, vertical positions within the Breccia, and whether they were Breccia matrix (16 samples), lapilli (6 samples) or analyzed for non-Breccia Ir background (1 sample).

The rock samples were manually crushed in a plastic wrap and mechanically in an alumina jaw crusher, and then powdered using automatic agate mills. About 100 mg of the material were sealed in the quartz vials of 4 mm outer diameter. The samples and standards were packed and irradiated for 48 hours at the TRIGA MARK II reactor of the Institute of Isotopes, Budapest, Hungary, at a flux of ~ 7.10^{13} n.cm^{-2}s^{-1}. After a cooling period of three months the samples were gamma-counted for about one day each.

Geological reference materials were used as standards. An aliquot of powdered Allende standard meteorite (a carbonaceous chondrite, type CV3) from the Smithsonian Institution (containing 740 ppb Ir) was mixed intimately with high-purity quartz powder (Merck) and homogenized in a boron carbide mortar. To avoid errors due to the inhomogeneity of the Allende meteorite, dilution series of four standards were prepared directly in high-purity quartz vials ("Suprapur"), with contents between 48 and 325 ppt Ir (cf. Fig. 9). For the standard dilution series a regression analysis was done to get the interpolated values of their volumes. Some of our measurements are close to our detection limit for the current irradiation and measurement parameters, which are between 5 and 30 ppt, depending on the rock type analyzed. Further details of the procedure are given in Koeberl and Huber (2000).

In addition, seven samples were analyzed by X-ray fluorescence (XRF) spectrometry for their major and trace element composition. Details for this method, including information on precision and accuracy, are reported by Reimold et al. (1994).

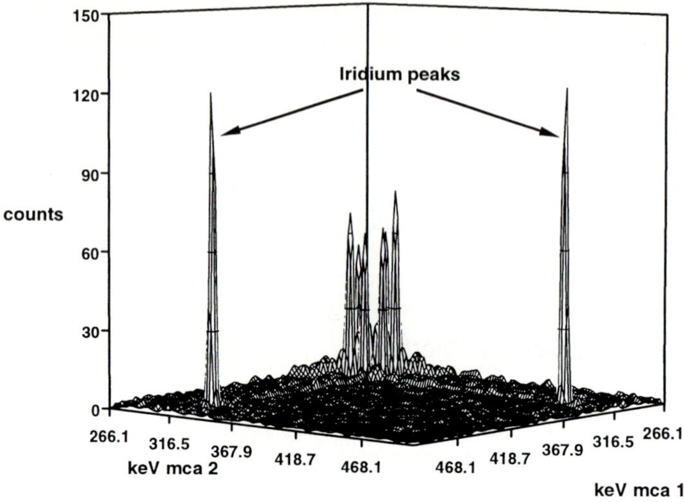

Fig. 9. 3D-multiparameter coincidence spectrum of a standard (Allende meteorite powder mixed with high purity quartz powder) containing 160 ppt iridium. The coincidence peaks at 316.5*468.1 keV and vice versa are used for the evaluation of the iridium content.

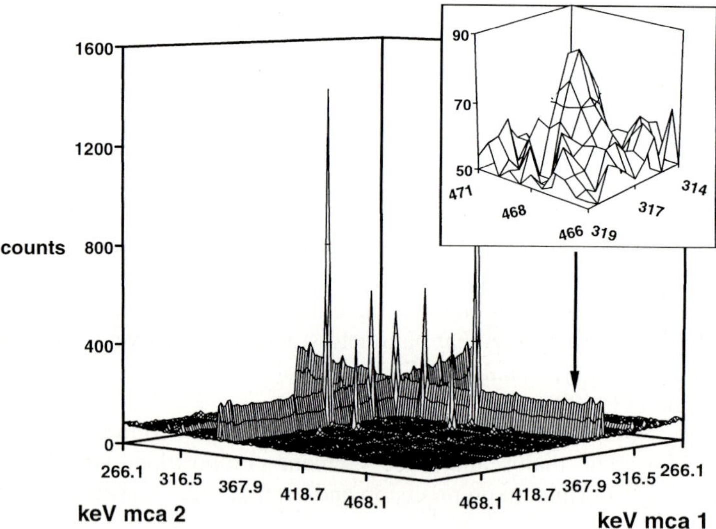

Fig. 10. Coincidence spectrum of sample MI 1 from the Alamo Breccia, containing 43 ±13 ppt iridium. The high ridges within the matrix at (e.g.) 344 keV are due to europium coincidences with Compton-scattered gamma rays.

4
Results and Discussion

The results of the coincidence spectrometry measurements for the 23 samples are shown in Table 1, and the results for the major and trace element XRF measurements of seven selected samples are given in Table 2. The major element contents confirm the carbonate-rich and silicate-poor nature of all samples. In fact, the silica and carbonate contents do not vary much between the individual samples; the only significant variation is in the MgO content and Sr content, depending on the amount of dolomite in the rock. No Cr, Co, or Ni enrichments are obvious, with the possible exception of sample HH4, which is the only one that has Ni and Cr above the detection limit (at 15 and 29 ppm, respectively). Interestingly, this is also the sample with the highest Ir value of all measured samples.

As shown in Table 1, the lowest Ir value obtained is 19 ± 9 ppt, and the highest value is 51 ± 14 ppt (sample HH4). All samples are silica-poor carbonates, including the lapilli, except for rare quartz grains, most of which are shocked, and diagenetic metallic oxides, which both occur in matrix and lapilli. Originally these were dominantly limestone. In some locations they remain limestone, whereas at others they are partially or entirely dolomitized. The results of our Ir abundance determinations are in broad agreement with earlier data for samples from Breccia Unit A in the Worthington Mountains (Fig. 2) by C. Orth and M. Attrep, as reported by Warme and Sandberg (1995). Orth and Attrep measured a range from 30 to 133 ppt and background values for the Guilmette Formation of <10 ppt. No details are available about their analytical precision. The high value of 133 ppt was obtained 1.5 m of the top of the Breccia. Even though our samples are from locations other than the Worthington Mountains, our results from similar stratigraphic levels are similar (e.g., HH5, HH6, D1, HS3). However, we do not observe values higher than 51 ppt Ir. For comparison between the two datasets it would have been interesting to re-analyze the same sample from the Worthington Mountains that gave the high Ir value. Differences may arise because Orth and Attrep used radiochemical neutron activation analysis, while we used a non-destructive technique. In addition, Ph. Claeys (personal communication, 2002) measured Ir contents for six other (mainly matrix-dominated) samples of Alamo Breccia: four samples from Hancock Summit and two from Mt. Irish. He obtained values ranging from 6 ± 20 to 75 ± 29 ppt, with an average of 47 ppt. These values are very similar to our data, confirming the rather Ir-poor nature of the Alamo Breccia.

Ignoring the errors introduced by the precision of the analyses, the range in Ir values for all 22 Breccia samples is 19-51 ppt. The average is 35 ppt. The samples that are closer to the top of the Breccia seem to have slightly higher Ir contents than lower ones, but this trend is not pronounced. One matrix sample from Hancock Summit (19 ppt) and one lapilli sample from the same locality (24 ppt) fall below the background sample result of 25 ppt. Thus, the data generated for this report only hint of an extraterrestrial component, and no strong, unambiguous signal. If the background sample of 25 ppt is representative of general background values, then the maximum content found in our Breccia dataset is higher than the background by only a factor of two.

Table 2. Major and trace element contents of the background sample and six Alamo Breccia samples (sample locations, see Table 1).

	BASE	HH4	HH5	HH6	MI1	MI2	HS5
SiO_2	2.54	2.63	2.54	2.51	0.99	1.20	0.72
TiO_2	0.02	0.04	0.03	0.03	0.03	0.03	0.02
Al_2O_3	0.58	0.82	0.60	0.59	0.37	0.50	0.11
Fe_2O_3	0.25	0.39	0.26	0.26	0.22	0.23	0.08
MnO	0.08	0.11	0.05	0.05	0.01	0.07	0.03
MgO	19.38	8.36	6.20	4.21	17.56	16.20	1.87
CaO	32.46	45.34	47.54	46.83	35.89	36.41	58.70
Na_2O	<0.01	0.30	<0.01	<0.01	<0.01	<0.01	0.21
K_2O	0.12	0.09	0.07	0.07	0.02	<0.01	<0.01
P_2O_5	0.02	0.02	0.02	0.02	0.02	0.02	0.02
LOI	43.58	40.95	41.71	44.67	43.91	44.47	37.56
Total	99.03	99.05	99.02	99.24	99.02	99.13	99.32
V	<15	<15	<15	<15	<15	<15	<15
Cr	<9	29	<9	<9	11	<9	<9
Co	<9	<9	<9	<9	<9	<9	<9
Ni	<9	15	<9	<9	<9	<9	<9
Cu	3	9	5	5	2	2	5
Zn	23	13	11	12	12	13	11
Rb	6	15	13	15	8	11	11
Sr	39	257	283	263	45	156	221
Y	<3	<3	<3	<3	<3	<3	<3
Zr	8	15	10	11	13	13	<6
Nb	<3	<3	<3	<3	<3	<3	<3
Ba	13	22	20	23	15	17	16

Major elements in wt%, trace elements in ppm. All Fe calculated as Fe_2O_3.

The 16 matrix samples contain the full range of values, and average 34 ppt. Values for the six matrix samples from Hiko Hills South average 40 ppt, the five from Hancock Summit average 34 ppt, and the four from Tempiute Mountain average 27 ppt. These averages diminish from higher to lower toward the west, which is presumably toward the crater. This trend may or may not be significant.

The six lapilli samples average 37 ppt, or 39 ppt if the lowest value for all samples (24 ppt) is discarded. Although the lapilli samples yield somewhat higher Ir values than the matrix, they are bracketed between the full range of matrix samples and do not contain statistically significant higher values. The lapilli population is expected to have higher contributions of rock from the target zone than does the mixed proximal and distal Breccia matrix, but the lapilli do not appear to contain insignificantly higher values than the matrix. However, it should be noted that the sample with the highest Ir content (51 ppt) also has the highest Cr and Ni values (in fact, the only such values above the detection limit). If we take these abundances at face value and calculate a meteoritic component based on average chondritic abundances (500 ppb Ir, 3600 ppm Cr, 1.4 wt% Ni; cf. Koeberl 1998), we arrive at chondritic contributions to the Breccia of 0.01, 0.8, and 0.1 wt%, respectively. These values are not corrected for any indigenous component, which is very difficult to do because the real background values are below the detection limit for the present measurements. Based on literature data for carbonates (cf. Koeberl 1998), though, it is likely that the Cr abundance would be the one with the most significant correction. Thus, a calculated chondritic contribution of 0.01 to 0.1 wt% is possible, but the data are simply not precise enough to arrive at an unambiguous conclusion.

A next step would be to try and determine the osmium isotopic composition for two or three samples with the highest Ir contents (and a background sample) to try top confirm the presence of an extraterrestrial component. Until such measurements are done, there are a variety of ways how the present results can be interpreted. First, we assume that the data are correct, because all three determinations made so far (unpublished data by Orth and Attrep and by Claeys, and the present work) arrive at about the same results. Second, the low Ir values in all samples indicate either that the meteoritic dust was not incorporated in the Breccia, or it was diluted by the large mass of the Breccia, or that the projectile was of Ir-poor composition, either because it was an achondrite or a cometary nucleus. This work confirms how difficult it can be to detect an impact signature even in rocks of relatively uniform composition.

Acknowledgements

This work was supported by the Austrian Fonds zur Förderung der wissenschaftlichen Forschung, Projekt Y-58-GEO (to CK), and by general funds from the Department of Geology and Geological Engineering, Colorado School of Mines. We are grateful to the reviewers, Philippe Claeys and Enrique Diaz-Martinez, for helpful comments on an earlier version of this paper. We are also grateful to Philippe Claeys for sharing his unpublished Ir data of Alamo Breccia samples.

References

Blair TC, MacPherson JG (1999) Grain-size and textural classification of coarse sedimentary particles. Journal of Sedimentary Geology 69: 6-19

Grieve RAF, Langenhorst F, Stöffler D (1996) Shock metamorphism in nature and experiment: II. Significance in geoscience. Meteoritics and Planetary Science 31: 6-35

Huber H, Koeberl C, McDonald I, Reimold WU (2001) Geochemistry and petrology of Witwatersrand and Dwyka diamictites from South Africa: Search for an extraterrestrial component. Geochimica et Cosmochimica Acta 65: 2007-2016

Koeberl C (1998) Identification of meteoritical components in impactites. In: Grady MM, Hutchison R, McCall GJH, Rothery DA (eds) Meteorites: Flux with Time and Impact Effects. Geological Society of London, Special Publication 140: 133-152

Koeberl C, Huber H (2000) Optimization of the multiparameter γ-γ coincidence spectrometry for the determination of iridium in geological materials. Journal of Radioanalytical and Nuclear Chemistry 244: 655-660

Koeberl C, Reimold WU, McDonald I, Rosing M (2000) Search for petrographical and geochemical evidence for the late heavy bombardment on Earth in Early Archean rocks from Isua, Greenland. In: Gilmour I, Koeberl C (eds) Impacts and the Early Earth. Lecture Notes in Earth Sciences 91, Springer Verlag, Heidelberg, pp 73-97

Kuehner H-C (1997) The Late Devonian Alamo Impact Breccia, southeastern Nevada. Colorado School of Mines. Unpublished PhD thesis, 321 pp

Leroux H, Warme JE, Doukhan J-C (1995) Shocked quartz in the Alamo breccia, southern Nevada: Evidence for a Devonian impact event. Geology 23: 1003-1006

Montanari, A, Koeberl C (2000) Impact Stratigraphy: The Italian record. Lecture Notes in Earth Sciences, Vol. 93, Springer Verlag, Heidelberg, 364 pp

Morrow JR, Sandberg CA, Warme JE, Kuehner H-C (1998) Regional and possible global effects of sub-critical Devonian Alamo impact event, southern Nevada, USA. Journal of the British Interplanetary Society 51: 451-460

Reimold WU, Koeberl C, Bishop J (1994) Roter Kamm impact crater, Namibia: Geochemistry of basement rocks and breccias. Geochimica et Cosmochimica Acta 58: 2689-2710

Sandberg CA, Morrow JR, Ziegler W (2002) Late-Devonian sea-level changes, catastrophic events, and mass extinctions In: Koeberl C, MacLeod KG (eds) Catastrophic Events and Mass Extinctions: Impacts and Beyond. Geological Society of America, Special Publication 356: 473-487

Stöffler D, Langenhorst F (1994) Shock metamorphism of quartz in nature and experiment: I. Basic observations and theory. Meteoritics 29: 155-181

Warme JE, Sandberg CA (1995) The catastrophic Alamo breccia of southern Nevada: Record of a Late Devonian extraterrestrial impact. Courier Forschungsinstitut Senckenberg 188: 31-57

Warme JE, Sandberg CA (1996) Alamo megabreccia: record of a Late Devonian impact in southern Nevada. GSA Today 6 (1):1-7

Warme JE, Kuehner H-C (1998) Anatomy of an anomaly: The Devonian catastrophic Alamo impact breccia of southern Nevada. International Geology Review 40:189-216

Warme JE, Morgan ML, Kuehner H-C (2002) Impact-generated carbonate accretionary lapilli in the Late Devonian Alamo Breccia. In: Koeberl C, MacLeod KG (eds) Catastrophic Events and Mass Extinctions: Impacts and Beyond. Geological Society of America, Special Paper 356: 489-504

The Osmussaar Breccia in Northwestern Estonia – Evidence of a ~475 Ma Earthquake or an Impact?

Kalle Suuroja[1], Kalle Kirsimäe[2,3], Leho Ainsaar[3], Marko Kohv[3], William C. Mahaney[4], and Sten Suuroja[1,5]

[1]Geological Survey of Estonia, Kadaka tee 82, 12618 Tallinn, Estonia.
[2]Institute of Geography, University of Tartu, Vanemuise 46, 51014 Tartu, Estonia. (arps@ut.ee)
[3]Institute of Geology, University of Tartu, Vanemuise 46, 51014 Tartu, Estonia. (arps@ut.ee)
[4]Geomorphology and Pedology Lab, York University, 4700 Keele St., North York, Ontario, Canada, M3J IP3.
[5]Institute of Mining, Technical University of Tallinn, Kopli 82, 12623 Tallinn, Estonia.

Abstract. The Osmussaar Breccia occurs in beds of the ~475 Ma basal Middle Ordovician (Arenig and Llanvirn series) siliclastic-carbonate rocks of northwestern Estonia. The Breccia consists of fragmented and slightly displaced (sandy) limestones, which are penetrated by veins and bodies of strongly cemented, breccia-like, lime-rich sandstone injections. The rocks above (horizontally-bedded, hard limestone) and below (weakly cemented silt and sandstone) are undisturbed and do not contain the sediment intrusions. Osmussaar Breccia is found over an area of more than 5000 km^2 and is distributed in a west-east oriented elliptical half-circle centred approximately at Osmussaar Island (59°18´ N; 23°28´ E). The thickness of the brecciated unit ranges from 1-1.5 m on Osmussaar to a few (tens of) cm at ~70 km east of the island. Arenitic sandstone of the sediment injections contains quartz grains with planar deformation features (PDF). Several hypotheses concerning the origin of the Osmussaar event have been proposed: catastrophic earthquake, regional tectonic movements, tectonic movements occurring simultaneously with coastal processes, and an impact event. The latter hypothesis was suggested in connection with the discovery of the nearby-situated Neugrund impact structure. However, the sediment intrusions are stratigraphically ~60 Ma younger than the impact structure. Osmussaar Breccia does not correspond to any known impact structure of this age in Baltoscandia. Also, results of a seabed geophysical survey in the Baltic Sea for the search of a possible undiscovered feature did not identify any large structure in the area of the Osmussaar Breccia. Consequently, we suggest that a devastating ~475 Ma earthquake with an epicentre close to Osmussaar split the sea floor. It initiated underwater mud-flows eroding the primary Neugrund crater ejecta and/or crater rim walls, thus reworking the impact materials into the sedimentary injections, which is suggested by rounded morphology of the shocked quartz grains found in breccia matrix.

1
Introduction

Apart from meteorite impact craters, which provide direct evidence of an impact, distal ejecta and associated geochemical anomalies of platinum-group metals are able to provide information for recognition of impact events in the stratigraphic record (Grieve 1997). Because distal impact ejecta are distributed over areas orders of magnitude larger than the impact structure itself, the probability of finding impact signatures in dispersed segments of the terrestrial stratigraphic record is much higher. However, such layers are rather thin, and dispersion, modification due to weathering, redeposition, diagenetic and/or metamorphic processes make chemical and physical identification of the fallout signatures, and correlation to distinct impact events, difficult. Moreover, most unusual beds in the stratigraphic record are not related to impact events, but are the result of other geological processes, such as volcanic eruptions, turbidite flows, secondary sediment deformation due to compaction, etc. Schmitz et al. (1994) investigated more than 200 000 quartz grains from 57 individual bentonite and bentonite-resembling clayey layers from a 120-m thick (representing ~2 million yr) Early-Silurian carbonate sequence on Gotland Island, Sweden, for shocked quartz and did not find any. However, these authors, as well as Izett et al. (1993), suggested that stratigraphic intervals containing rip-up clasts or showing other signs of sea-floor turbulence would be more suitable for a search of impact events rather than clay beds.

In this contribution, we report a study of a breccia layer of early Middle Ordovician age in northwestern Estonia. This breccia, named Osmussaar Breccia, was originally interpreted as the result of earthquake-triggered seafloor destruction (Öpik 1927). In contrast, after the discovery of the Neugrund impact crater in the area coinciding with the breccia distribution, the occurrences of sand-filled cracks and brecciation, were re-interpreted as a commence of an impact event at ~475 Ma (Suuroja and Saadre 1995). However, more recent work revealed that the Neugrund Crater is infilled with older, Cambrian sediments and is of Early Cambrian age (~540 Ma) (Suuroja and Suuroja 2000). Consequently, the origin of the Osmussaar Breccia remains unclear.

2
The Osmussaar Breccia

2.1
Geology

The Osmussaar Breccia occurs in beds of the basal Middle Ordovician (Arenig) siliciclastic-carbonate rocks of northwestern Estonia (Meidla 1997; Webby 1998). The breccia is found in outcrops (on Osmussaar Island, Väike- and Suur Pakri islands, and Pakri Peninsula) and drill core sections over an area of more than 5000 km^2 (Fig. 1). It is distributed in a west-east oriented swath centred approximately on Osmussaar Island (Fig. 1), where the brecciated unit is also found to be thickest (1-1.5 m). Osmussaar (Odensholm) Island is situated on the southern side of the entrance to the Gulf of Finland (59°18´ N; 23° 28´ E). It is the highest part of a flooded peninsula of the Baltic Clint emerging over the sea level on the northwestern top of the submarine Osmussaar Bank. In the northern part of the island, at a ~5 km long cliff, the Ordovician (Arenig and Llanvirn series) limestones crop out (Öpik 1927; Meidla 1997). The breccia, which occurs in this section at the foot of the cliff, consists of fragmented and slightly displaced blocks of sandy limestones of the Billingen, Volkhov and Kunda regional stages (Figs. 2, 3 and 4) (Suuroja et al. 1999). This, up to 1.5 m thick, brecciated layer is unconformably cut by horizontally bedded and undisturbed Llanvirnian (Aseri, Lasnamägi, Uhaku regional stages) bioclastic limestones (Figs. 2 and 4) (Suuroja et al. 1999). The stratigraphic age of the brecciation is bio- and lithostratigraphically well determined to Kunda regional stage (Nõlvak 1997; Meidla 1997) and it is biostratigraphically confined to lower part of the *Cyathochitina regnelli* chitinozoan Zone of the late Arenig-earliest Llandvirn age (Nõlvak and Grahn 1993), which suggest an event age of ~475 Ma (Fig. 3) (Gradstein and Ogg 1996).

The fractured and crushed limestone blocks of the Osmussaar Breccia are penetrated by up to 2-3 m thick veins of strongly carbonate cemented sandstone. The thickness of the brecciated unit with sand intrusions ranges from 1 to 1.5 m in Osmussaar to a few tens of cm ~70 km east of the island. The intensity of the fracturing decreases laterally in the same direction. At some distance from Osmussaar (on the Suur- and Väike-Pakri islands and Pakri Peninsula) (Fig. 1) the brecciation is scarce and the sandstone intrusions are thinner (1-5 cm). On Osmussaar Island the veins and bodies of sand filled intrusions make up approximately 10% of the layer, areas on the Suur-Pakri Island the content of these is less than 1% (about 1 vein per every 10 meters). The carbonate lithoclasts in brecciated layer range in size from tens of centimetres to tens of metres (Fig. 4). Fine-grained material filling dike- and vein-like spaces between blocks is composed of lime-rich mud and sand-sized siliciclastic material, which contains small fragments of limestones of varying size from a few to tens of centimetres. On Osmussaar Island, the sharp-edged limestone fragments in the sand intrusions make up to 50% (on average 15%) of the intrusions volume.

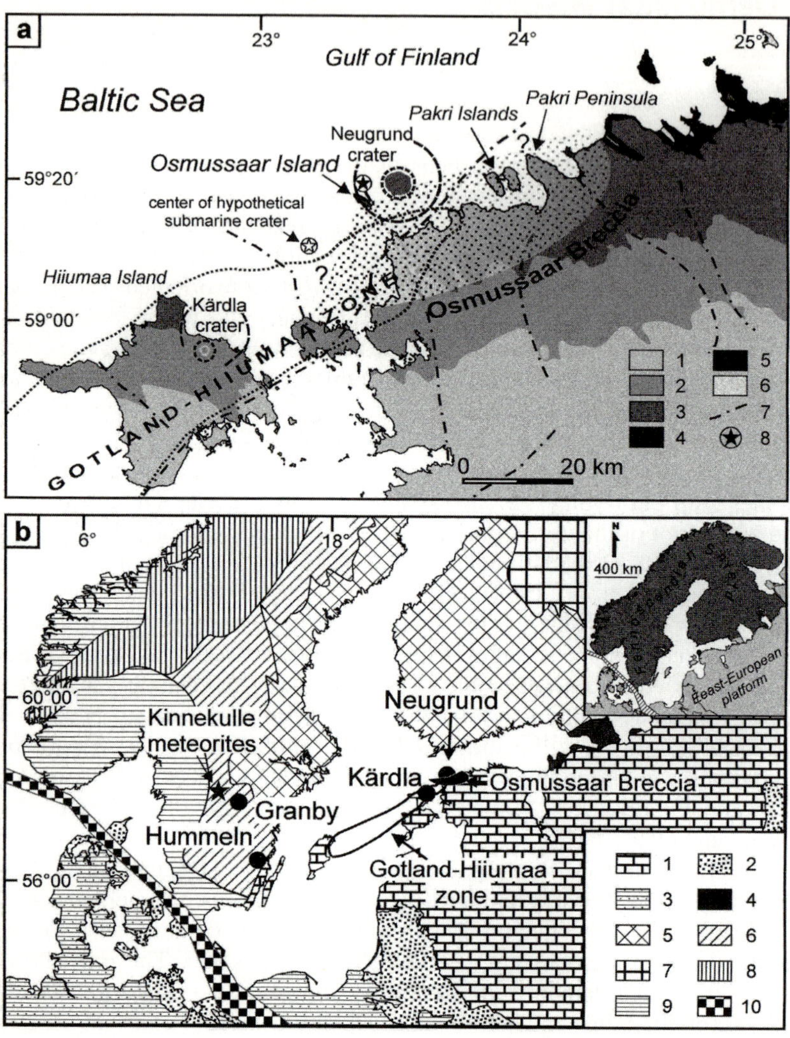

Fig. 1. Distribution of Osmussaar Breccia. (a) Simplified geological map of northwestern Estonia: 1 - Lower Silurian, 2 - Upper Ordovician, 3 - Middle Ordovician, 4 - Lower Ordovician, 5 - Cambrian, 6 - Vendian, 7 - major faults, 8 - epicentre of the Osmussaar earthquake of 1976; (b) general location of the impact craters and fossil meteorites discussed in the text, and main structural elements of the study area: 1 - Palaeozoic sediments, 2 - Mesozoic sediments, 3 - Cenozoic sediments, 4 - Upper Precambrian (Vendian) sediments, 5 - Archaean Domain, 6 - Svecofennian Domain, 7 - Transscandinavian Belt, 8 - Sveconorwegian Belt, 9 - Caledonides, 10 - Tornquist Line.

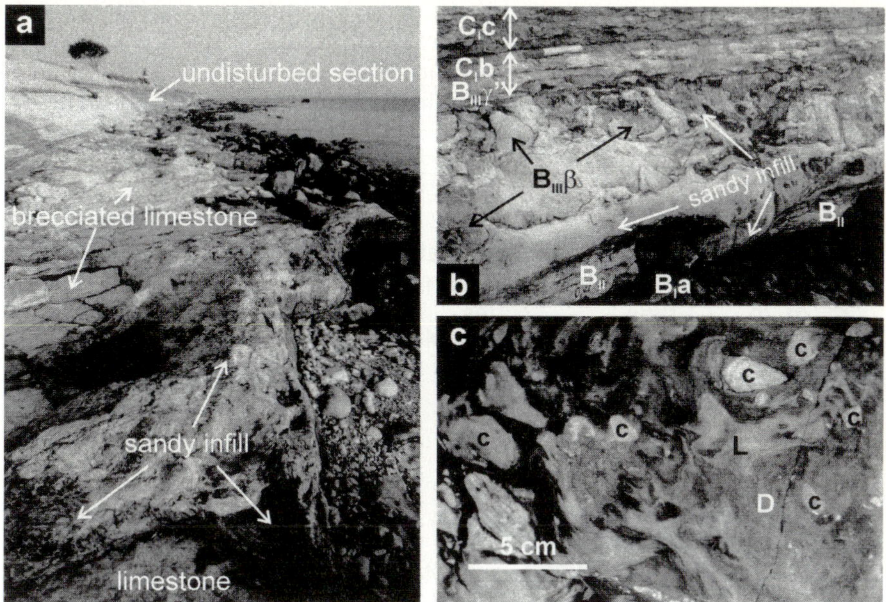

Fig. 2. Photographs of the Osmussaar Breccia. (a) Northeastern cliff of Osmussaar Island and eroded surface of the layer with breccia. To the left, the horizontally bedded and undisturbed layer of the post-Osmussaar event Middle Ordovician (Llanvirn) limestones; (b) brecciated beds in the section at the northeastern cliff of the Osmussaar Island. Undisturbed section: BIII ''- upper Kunda, CIb - Aseri and CIc - Lasnamäe regional stages. Brecciated section: BIII - middle Kunda, BII - Volkhov and BIa - Billingen regional stages; (c) detail of the infilling sediment with the signs of flow. Note the dark (D) and light (L) varieties of the sandstone. The clasts (C) are limestone of the Billigen, Volkhov and Kunda regional stages.

The rocks immediately above the brecciated layer (a flat bedded 4–6 m thick sequence of Llanvirnian limestones) are undisturbed (Figs. 2a, b, 3 and 4). Also, no significant disturbances are found in sediments right below the brecciated layer, in the weakly cemented massive glauconitic sandstone of the Hunneberg Stage or in the section below.

The pre-brecciation bedrock section consists of ~160 m sedimentary rocks overlying the crystalline basement of Paleo-Proterozoic metamorphic rocks. The sedimentary rocks in this area are mostly (>95% of the section) composed of Vendian, Lower Cambrian and Lower Ordovician, weakly cemented terrigenous rocks (sand-and siltstone, claystone). The present porosity of these rocks is ~20-30% (Jõeleht et al. 2002), which suggests that this section was (accounting only for mechanical compaction) about 30% thicker (up to ~200 m) at the time of the Osmussaar Event.

The upper part of this terrigenous complex is composed of Lower Ordovician, dark green, glauconitic sandstone overlain by a thin layer (1-1.5 m) of limestone and sandy limestone. This limestone bed consists of a 50–60 cm thick bed of glauconitic limestone of the Toila Formation corresponding to the Billingen and Volkhov regional stages, and of 20–60 cm thick structurally heterogeneous bed of the bioturbated slightly kerogeneous sandy limestone of the Pakri Formation corresponding to the Valaste substage of the Kunda regional stage (Figs. 3 and 4) (Nõlvak 1997; Meidla 1997).

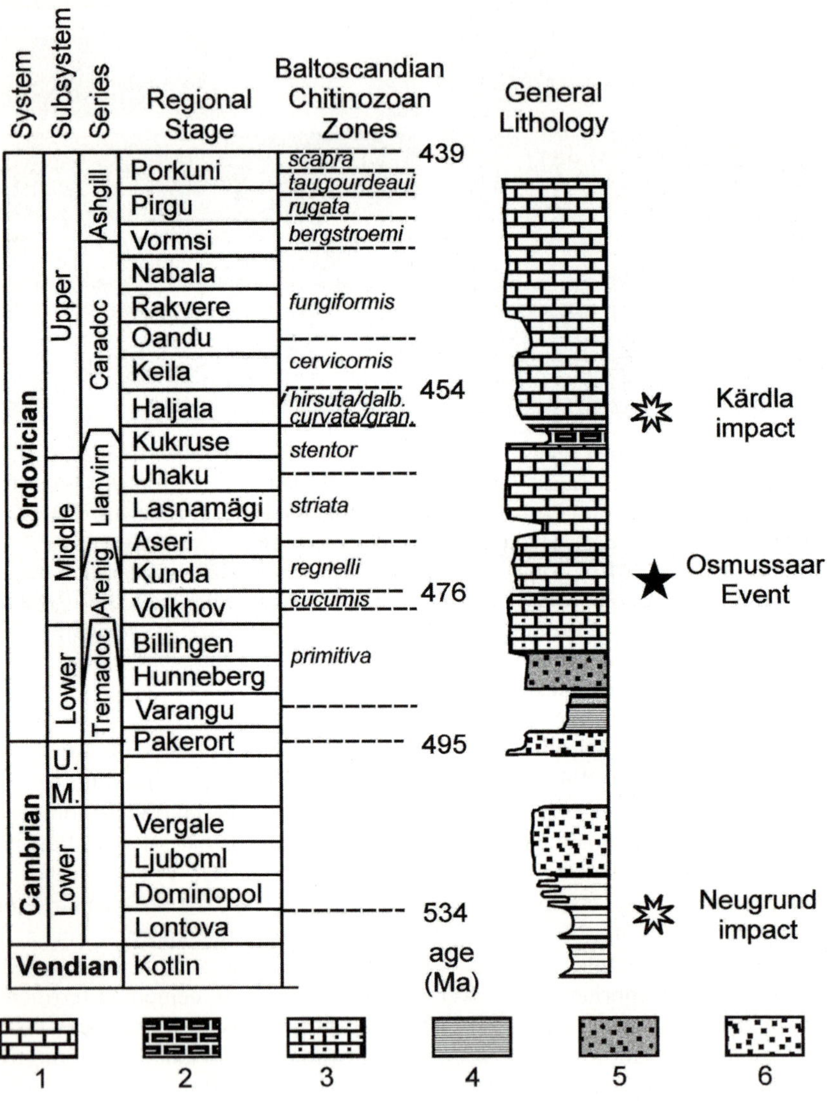

Fig. 3. A schematic stratigraphy of the sedimentary sequence in northwestern Estonia. Chitionozoan zones according to Nõlvak and Grahn (1993). 1- limestone, 2 - organic-rich limestone, 3 - glauconitic limestone, 4 - claystone, 5 - glauconitic sandstone, 6 - sandstone.

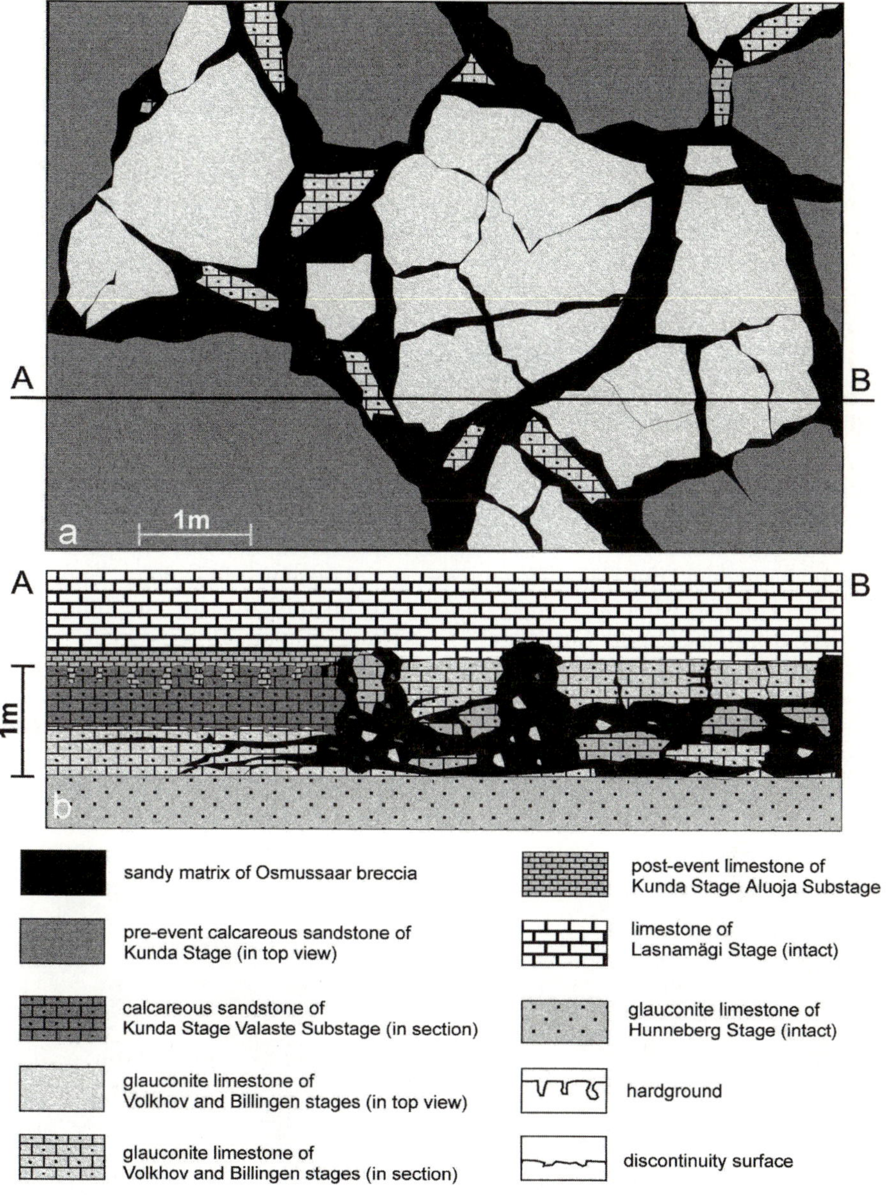

Fig. 4. Schematic drawing of brecciated layer in Osmussaar Island. (a) The top view; (b) cross-section.

The matrix of the sediment consists of a mixture of two types of lime-rich sandstone - light grey and dark brownish, both showing flow structures (Fig. 2c). The dark brownish type prevails in the thin (1-20 cm) veins injected into the blocks and situated horizontally between displaced beds of limestone. The light-grey type is the prevailing type in mixtures in the thicker veins and bodies. The lime-rich sandstone matrix in Osmussaar Island makes up 50-100 vol% (average 85 vol%) of sediment. The sediment in the other places (Väike- and Suur-Pakri islands, Pakri Peninsula) consists only of light lime-rich sandstone. Composition and structure of the sandy sediment in the drill core sections in the mainland area of northwestern Estonia are difficult to assess due to low core recovery from this part of the section. However, the dark variety seems to dominate in the breccia matrix in the mainland area.

2.2
Composition of the Sediment

The compositions of the sediment and host rocks were studied by means of X-ray diffractometry (XRD), optical and scanning (SEM) microscopy and standard sedimentological analysis. XRD patterns from powdered mounts of the sandstones were analysed on a Dron-3M diffractometer at the Institute of Geology, University of Tartu, Estonia. Mineral compositions were quantified using the full-profile Rietveld technique based Siroquant$^{(TM)}$ software (Taylor 1991). Grains in the coarse silt and fine sand fractions were examined with a Jeol SEM-EDS system at the Geomorphology and Pedology Lab at York University, Canada. The grain size of the terrigenous component was determined by sieve and pipette analyses after dissolving the calcite matrix of the rock with 3% HCl.

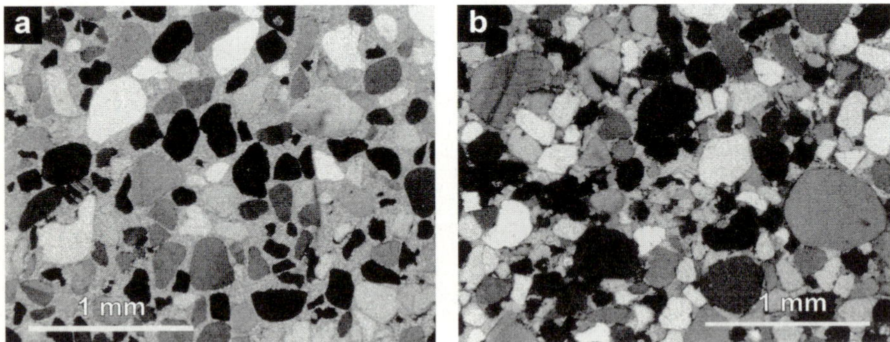

Fig. 5. Photomicrographs of the thin sections of carbonate cemented quartz sandstone matrix filling the sedimentary intrusions. (a) Dark sandstone, sample N-50; (b) Light sandstone, sample N-55. Both from Osmussaar Island. Crossed polars.

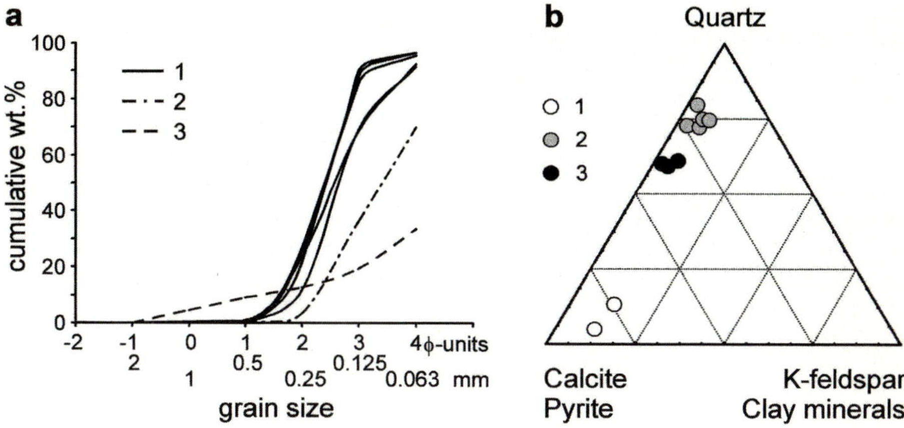

Fig. 6. Grain size distribution and composition of the breccia matrix and host rock: (a) grain size cumulative frequency curves of the insoluble residue: 1 – breccia matrix sediments, 2 – sand-rich limestone of the Pakri Formation, 3 - limestone of the lower Kunda stage; (b) mineral composition by whole rock X-ray diffraction: 1 - host rock, 2 - light grey type of the sandstone, 3 - dark brownish type of the sandstone.

The terrigenous component is composed mainly of subangular to well-rounded quartz sand (Mahaney 2002). The structure of siliciclastic is mainly matrix-supported and the matrix is composed of calcite with small amounts of clay (Fig. 5a,b). The fine-grained micritic calcite is completely recrystallized. Skeletal fragments have not been observed in the matrix of the sediment, but they are abundant in limestone rock fragments embedded in sandstone. The insoluble residue is very fine- to medium-grained sand. The sandy material is fairly well sorted and, on average, more than 90% of the grains are in the range from 63 μm to 500 μm, where the content of fine sand (125-250 μm) dominates (>50% of the sediment) (Fig. 6a).

The results of whole rock XRD analyses show that the quartz content in the dark brownish type of lime-rich sandstone is nearly constant at ~60% and in light gray type varies from 70 to 79% (Fig. 6b). Calcite contents in dark and light sandstone vary from 25 to 32% and 15 to 21%, respectively. Content of K-feldspar and clay minerals (illite, illite-smectite and traces of kaolinite) is subordinate and does not exceed 5%. Also, traces of -tridymite and cristobalite were found. The dark sandstone, compared to the light variety, contains considerably more pyrite (5.3-7.2% and 1.6-2.7%, respectively). According to optical microscopic examination, the dark sandstone contains a small amount of organic material (occasionally up to 3 % by volume), which is obviously the reason for the colour difference between the two sandstone facies.

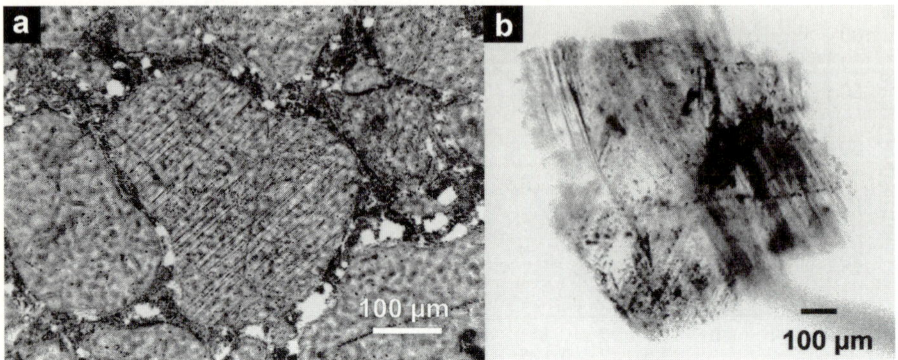

Fig. 7. Photomicrographs of quartz grains from the breccia matrix in Osmussaar Island showing planar deformation features (PDF); (a) – quartz grain with one set of PDFs (sample N-55, Osmussaar Island, inverted phase-contrast image of a thin section, parallel polarizers); (b) – quartz grain with two sets of PDFs (sample N-56, Osmussaar Island, parallel polarizers, in immersion liquid).

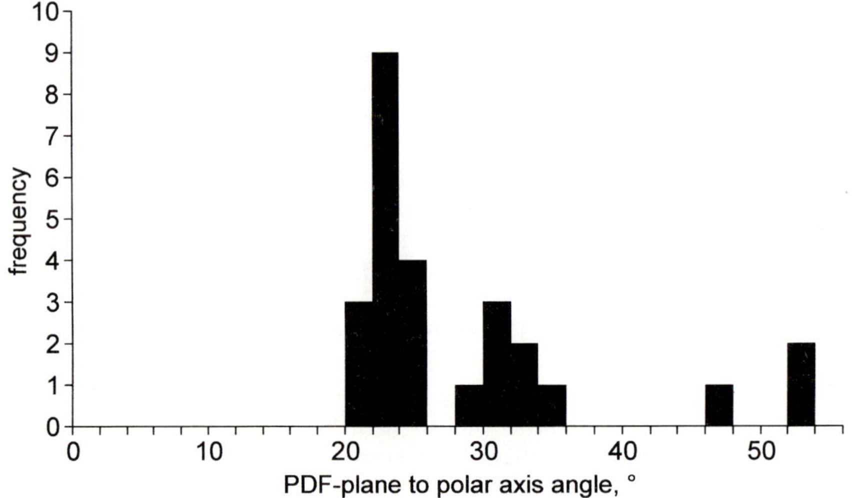

Fig. 8. Crystallographic orientation of PDFs in 22 quartz grains (26 sets) from breccia matrix from Osmussaar Breccia.

Thin section and grain mount analysis of the sandstone reveals quartz grains and a single K-feldspar grain with planar fractures (PF) and, possibly, planar deformation features (PDF). Most of quartz grains found have one set of PDF's

(Fig. 7a), but also two and rarely up to three sets of frequently decorated PDF's were found (Fig. 7b). PDF's were found in lime-rich sandstone fill of the sediment in all area of Breccia distribution from Osmussaar Island to Pakri peninsula (Fig. 1). The frequency of grains with shock signatures is <0.3% from 1000 to 1500 grains counted in each thin section, but the number increases up to 2.5% in one exceptional thin section of Osmussaar Island sediment intrusion. The orientation of crystallographic planes was measured on an optical microscope equipped with a universal-stage in 22 quartz grains. Most of the systems show crystallographic plane-to-pole axis angles typical to PDF's at ~23°(±3°), ~31°(±3°) and probably also ~52° (±5°), which correspond to plane indices $\{10\bar{3}1\}$, $\{10\bar{2}1\}$ and $\{10\bar{1}1\}$, respectively (Fig. 8). PDF's were not observed in the host rock to the sediment.

3
Discussion and Conclusions

The origin of the Osmussaar Breccia has been discussed since the middle of the 19th century (Eichwald 1840). Several hypotheses have been proposed: an earthquake (Öpik 1927; Suuroja et al. 1999), slumping/compaction structures (Orviku 1960), and desiccation and/or earthquake cracks formed during several episodes and modified by karst (Puura and Tuuling 1988). Such sheet-like clastic injections (sills) and dikes cutting host rock form either by forceful hydraulic injection of the fluidized material into fractures along gradients of decreasing pressure (true clastic injections), by filling of cavities and cracks from above under influence of gravity, or by upward intrusion due to dewatering and load casting (Potter and Pettijohn 1977; Talbot and von Brunn 1989). Hydraulic intrusion can also be initiated by tectonic earthquakes or by impacting meteorites (Talbot and von Brunn 1989; Sturkell and Ormö 1997).

Sedimentary dikes formed by upward intrusion of sandy deposits due to dewatering or load casting are common features in carbonate rocks in the Baltoscandian sedimentary rock record (e.g., Dubar and Levin 1971; Larsson 1975). However, upward intrusion of Osmussaar Breccia sediments is excluded as the sandstone below the brecciated unit is a dark-green, glauconitic sediment, but the sandy sediment filling the space between carbonate blocks in brecciated unit does not contain glauconite grains, other than in fragments of glauconitic limestone of the Toila Formation embedded in sandy matrix of the sediment. The most intense brecciation of the Osmussaar Breccia occurs close to the 8-km diameter submarine Neugrund impact crater ~10 km east of Osmussaar Island. Therefore, after the discovery of Neugrund crater in 1995 the brecciation was interpreted as a result of this event (Suuroja and Saadre 1995). However, the Neugrund structure was later recognized as being of Early Cambrian age (~540 Ma; Suuroja and Suuroja 2000). Moreover, submarine exploration at the Neugrund site has revealed that the limestones of the Arenig Serie (Billingen, Volkhov and Kunda regional stages) occur inside of the crater, but are undisturbed and unusually thick (~6 m) compared to the same sediments on Osmussaar Island, where they occur in thicknesses of ~2 m. This is probably due to the additional sediment accommodation space created in the crater depression by the higher compaction rate of the crater filling impact clastic breccias compared to the normal sequence outside of the crater. Well-rounded morphology of the detrital grains in sediment suggest that sedimentological processes have been significantly reworked the material, and it might have been transported for long distances. Thus, an impact origin of Osmussaar Breccia relatied to the Neugrund event can be excluded.

However, the occurrence of possibly shocked quartz grains with PFs and PDFs suggests that a relationship of the Osmussaar Breccia to an impact event is still possible. The Osmussaar Event is biostratigraphically confined to the *Cyathochitina regnelli* chitinozoan Zone of the late Arenig-Llandvirn age (Fig. 3) (Nõlvak and Grahn 1993). This coincides well with the age of the Granby crater in south-central Sweden (Fig. 1), which is placed in the lower part of the same biostratigraphic zone (Grahn et al. 1996). Also, the Hummeln crater in southwestern Sweden has been dated biostratigraphically to the same period (Lindström et al. 1999). However, the Granby and Hummeln impact craters are only ~4 km and 1.2 km in diameter, respectively, and ~500 km away from Osmussaar. It is not probable that the Granby or Hummeln impacts alone would have caused the brecciation in northwestern Estonia. It is interesting enough that fossil meteorites found in Lower Ordovician beds at Kinnekulle, Sweden, lie within the *Amorphognatus variabilis – Microzarkodina flabellum* conodont subzone (Schmitz et al. 1996), which corresponds to the same *Cyathochitina regnelli* chitinozoan Zone to which the Osmussaar Breccia is dated. This could indicate a higher meteoritic flux during this period and, thus, suggest a higher probability of large impact events during this time and in this area. On the other hand, biostratigraphic zonation does not allow precise correlation (±5 Ma) and, thus, a causal connection between these events is not proven. Still, assuming an impact origin for the Osmussaar Breccia, we may speculate that a large ~475 Ma impact structure could exist in the Baltic Sea, west of Osmussaar. We selected a possible candidate from the map of aeromagnetic anomalies of the coastal areas of northwestern Estonia (see Fig. 5, p. 285 in Suuroja et al. 2002), which reveals a large, ~10-km diameter, roughly circular feature southwest of Osmussaar, about half way on a line connecting Kärdla and Neugrund craters (Fig. 1). Seismoacoustic profiling of the seabed in this area was undertaken in summer 2001. However, our preliminary results do not support the presence of an impact structure in this area. Consequently, until a suitable meteorite crater proves an impact origin of the Osmussaar Breccia, it is reasonable to assume that the shocked quartz grains in Osmussaar Breccia originate from Neugrund impactites. We suggest that a ~475 Ma earthquake, with an epicentre close to Osmussaar Island, split the seafloor and initiated tsunami-induced underwater mudflows in the area of the Neugrund crater. This would have caused erosion of the Neugrund impact ejecta sediments and/or crater rim wall, and reworking of impact materials. Kunda time (upper part of the Arenig) marks a prominent sea-level fall in the Baltic Paleobasin (Männil 1966), which also suggests that the erosion of the uplifted rim-wall areas of the Neugrund crater might have started earlier, and earthquake initiated currents have redistributed material seawards around the structure. The material filling the sedimentary injections differs from deposits in the surrounding area at this time (Pakri Formation) and was obviously injected from suspensions carried by forceful mudflows. The structure of the sea floor itself, composed of a thin (up to 2 m) layer of the lithified limestones on top of the thick complex (~200 m) of weakly cemented and water-saturated low-strength terrigenous sediments favored the brecciation as a result of the passage of seismic surface waves, which caused crushing and splintering of the brittle uppermost limestones on the top of ductile unconsolidated clays, silts, and sands.

The occurrence of such a devastating single earthquake or series of earthquakes within a short time period is interpreted to be related to regional tectonic movements within the Gotland–Hiiumaa structural zone (Fig. 1). The Gotland structure, interpreted as an Early Palaeozoic uplifted zone, extends over hundreds of kilometres from northwestern Estonia via Hiiumaa Island towards Gotland Island, Sweden (Männil 1966; Jaanusson 1973, 1995). Geological mapping (Suuroja et al. 1991) shows that Lower Ordovician and lowermost Middle Ordovician sedimentary beds with a thickness of up to 10 m have been eroded away within the southeastern part of this structural zone. The time of erosion approximately coincides with the formation of the sediment at Middle Ordovician Kunda time (~475 Ma), suggesting a possible, local tectonic uplift in that period. An argument supporting a seismic origin of the Osmussaar Breccia is the Osmussaar Earthquake in 1976 with an epicentre located 5 to 7 km northeast of the island and a magnitude of 4.7. It is the strongest earthquake ever recorded in historical times in generally unseismic Estonia, and was interpreted as the result of block movements along fault zones crossing this area (Miidel and Vaher 1997).

In conclusion, we suggest that the Osmussaar Breccia in northwestern Estonia is possibly connected to a strong Middle Ordovician (~475 Ma) earthquake or series of earthquakes, which caused seabed fracturing, brecciation, and sand infilling. Material for the sedimentary infill was most probably reworked from the 7-km diameter Early Cambrian (~540 Ma) Neugrund Crater located in the area of the breccia distribution. This would explain the well-rounded morphology of the shocked quartz grains found in sediments.

Acknowledgements

This paper is part of the marine geological mapping of northwestern Estonia with financial support from the Geological Survey of Estonia (K. Suuroja and S. Suuroja). The authors are grateful to Tom Floden and Monica Björkeus (University of Stockholm), and to Vello Mäss and his colleagues (Estonian Maritime Museum) for help and assistance in carrying out marine geological investigations. Igor Tuuling (University of Tartu) and Jens Ormö (International Research School of Planetary Sciences, Pescara, Italy) are thanked for helpful discussions during the preparation of this contribution. We thank reviewers Wolf Uwe Reimold and Birger Schmitz for the most helpful comments and editor

Christian Koeberl for handling the manuscript. The work was supported in part by Estonian Science Foundation grants ESF-4541, ESF-5192, and by Research Theme TBGGG-1826 (K. Kirsimäe).

References

Dubar G, Levin A (1971) Injectional clastic dykes in the northern Baltic region. Proceedings of the Estonian Academy of Sciences. Chemistry. Geology 20: 239–242 (in Russian)

Eichwald E (1840) Kurze Anzeige einer geognostischen Untersuchung Estlands und einiger Inseln der Ostsee. Die Urwelt Russlands, Heft 1. St.-Petersburg: 1-24

Gradstein FM, Ogg J (1996) A Phanerozoic time scale. Episodes 19: 3-5

Grahn Y, Nõlvak J, Paris F (1996) Precise chitinozoan dating of Ordovician impact events in Baltoscandia. Journal of Micropaleontology 15: 21-35

Grieve RAF (1997) Extraterrestrial impact events: the record in the rocks and the stratigraphic column. Palaeogeography, Palaeoclimatology, Palaeoecology 132: 5-23

Izett GA, Cobban WA, Obradovich JD, Kunk MJ (1993) The Manson structure: 40Ar/39Ar age and its distal ejecta in the Pierre Shale in southeastern South Dakota. Science 262: 729-732

Jaanusson V (1973) Aspects of carbonate sedimentation in the Ordovician of Baltoscandia. Lethaia 6: 11-34

Jaanusson V (1995) Confacies differentiation and upper Middle Ordovician correlation in the Baltoscandian Basin. Proceedings of the Estonian Academy of Sciences. Geology 44: 73-86

Jõeleht A, Kirsimäe K, Shogenova A, Đliaupa S, Kukkonen IT, Rastenienë V, Zabele, A. (2002) Thermal conductivity of Cambrian siliciclastic rocks from the Baltic basin. Proceedings of the Estonian Academy of Sciences. Geology 51: (in press)

Larsson K (1975) Clastic dykes from the Burgsvik Beds of Gotland. Geologiska Föreningens i Stockholm Förhandlingar (GFF) 97: 125-134

Lindström M, Floden T, Grahn Y, Hagenfeldt S, Ormö J, Sturkell EFF, Törnberg R (1999) The Lower Palaeozoic of the probable impact crater of Hummeln, Sweden. Geologiska Föreningens i Stockholm Förhandlingar (GFF) 121: 243-252

Mahaney WC (2002) Atlas of Sand Surface Microtextures and Applications, Oxford, Oxford University Press (in press)

Männil R (1966) Evolution of the Baltic Basin during the Ordovician. Valgus, Tallinn, 248 p (in Russian)

Meidla T (1997) Hunneberg Stage. Billingen Stage. Volkhov Stage. Kunda Stage. In: Raukas A, Teedumäe A (eds) Geology and mineral resources of Estonia. Estonian Academy Publishers, Tallinn, pp 58-66.

Miidel A, Vaher R (1997) Neotectonics and crustal movements. In: Raukas A, Teedumäe A (eds) Geology and mineral resources of Estonia. Estonian Academy Publishers, Tallinn, pp 177-183

Nõlvak J (1997) Ordovicia. Introduction. In: Raukas A, Teedumäe A (eds) Geology and mineral resources of Estonia. Estonian Academy Publishers, Tallinn, pp 52-55

Nõlvak J, Grahn Y (1993) Ordovician chitinozoan from Baltoscandia. Reviews of Palaeobotany and Palynology 79: 245-269

Öpik A (1927) Die Inseln Odensholm und Rogö. Ein Beitrag zur Geologie von NW-Estland. Acta et Commentationes Universitatis Tartuensis A XII: 1-69

Orviku K (1960) Lithostratigraphy of the Toila and Kunda Formations in Estonia. Studies of the Institute of Geology of the Estonian Academy of Sciences V: 45-85 (in Russian)

Potter PE, Pettijohn FJ (1978) Paleocurrents and Basin Analysis. Academic Press, New York, 320 p

Puura V, Tuuling I (1988) Geology of the Early Ordovician clastic dikes of Osmussaar. Proceedings of the Estonian Academy of Sciences. Geology 37: 1-9 (in Russian)

Schmitz B, Jeppson L, Ekvall J (1994) A search for shocked quartz grains and impact ejecta in early Silurian sediments on Gotland, Sweden. Geological Magazine 131: 361-367

Schmitz B, Lindström M, Asaro F, Tassinari M (1996) Geochemistry of meteorite-rich marine limestone strata and fossil meteorites from the lower Ordovician at Kinnekulle, Sweden. Earth and Planetary Science Letters 145: 31-48

Sturkell EFF, Ormö J (1997) Impact-related clastic injections in the marine Ordovician Lockne impact structure, Central Sweden. Sedimentology 44: 793-804

Suuroja K, Saadre T (1995) The gneiss-breccia erratic boulders from north-western Estonia as witnesses to an unknown impact structure. Bulletin of the Geological Survey of Estonia 5/1: 26-28

Suuroja K, Suuroja S (2000) Neugrund Structure – the newly discovered submarine Early Cambrian impact crater. In: Gilmour I, Koeberl C (eds) Impacts and the Early Earth, Lecture Notes in Earth Sciences v. 91, Springer, Berlin-Heidelberg, pp 417-445

Suuroja K, Saadre T, Kask J (1999) Geology of Osmussaar Island. Estonian Maritima 4: 39-63

Suuroja K, Koppelmaa H, Niin M, Kivisilla J (1991) Geological mapping of the crystalline basement in Island of Hiiumaa. Geological Survey of Estonia, Tallinn (in Russian) 291 pp

Suuroja S, All T, Plado J, Suuroja K (2002) Geology and magnetic signatures of the Neugrund impact structure, Estonia. In: Plado J, Pesonen L (eds) Impact into Precambrian Shields, Impact Studies v. 2, Springer, Berlin-Heidelberg, pp 277-294

Talbot CJ, von Brunn V (1989) Melanges, intrusive and extrusive sediments and hydraulic arcs. Geology 17: 446-448

Taylor JC (1991) Computer programs for the standardless quantitative analysis of minerals using the full powder diffraction profile. Powder Diffraction 6: 2-9

Webby BD (1998) Steps toward a global standard for Ordovician stratigraphy. Newsletter of Stratigraphy 36: 1-33

Printing: Mercedes-Druck, Berlin
Binding: Stein+Lehmann, Berlin